ARCHIWORLD

Introduction

Architecture competition is for selecting the most suitable design for the objective project through competing with applicants. It is, in addition, to choose an architect who understands and deals with client's purpose of the construction. Real construction begins as the suitable design and architecture are selected after diverse competitions.

Architecture competition is a method of architecture which is conducted generally in the world. Also, it has a great influence on development of recent architecture as well as current. We publish no.19 and 20 of <Architecture Competition Annual> with various data on architecture competition. The motto of the book is to promote the value of the data as it forms the basis of comparing and analyzing awarded design work and all applied design works with originality. It also provides a public measure to judge the level of contemporary architectures.

<Architecture Competition Annual> selected 90 works from design competitions held through the year of 2022, which put into two volumes with a use-based classification: Public·Office facility, Housing, Neighborhood living facility, Education·Library·Daycare Center, Culture·Exhibition facility and Other facilities. We hope to have a great view of architecture through <Architecture Competition Annual>. We are deeply grateful for architects and architecture companies that offer the information and data to us to publish those annual books.

Publisher Jeong Kwang-young

머 리 말

설계경기는 응모안들의 경합을 통하여 대상 프로젝트의 성격에 가장 부합되는 설계안을 뽑는 일이다. 아울러 건축에 관한 클라이언트의 목적을 잘 이해하고 해결할 수 있는 건축가를 선정하는 일이기도 하다. 여러 가지 형태의 설계경기 경합을 벌여 그 중 가장 적합한 설계안과 그리고 건축가를 선정함으로써 본격적인 건축행위가 이루어진다.

설계경기는 전 세계적으로 행해지는 보편타당한 건축행위의 한 방법으로, 근대는 물론 현대건축의 발전에 절대적인 영향을 미쳤다. 이런 설계경기의 자료들을 모아 〈설계경기연감〉 19, 20권을 발간하게 되었다. 이는 훌륭한 당선안과 건축의 창의성이 돋보이는 다양한 응모안들을 한 자리에 모아 비교, 분석할 수 있는 토대를 마련함으로써 자료적인 가치를 높여 나가는 작업이다. 또한, 당대 건축의 수준을 가늠할 수 있는 척도를 제공하는 공시성을 갖는다.

〈설계경기연감〉은 2022년 한 해 동안 열린 설계경기 중 우수한 프로젝트 90작품을 엄선해 선정했으며, 공공·업무시설, 주거시설, 근린생활시설, 교육시설·도서관·어린이집, 문화·전시시설, 기타시설의 용도별 분류를 통해 2권으로 나누어 편집했다. 본 〈설계경기연감〉을 통하여 건축의 혜안을 얻을 수 있기를 바라며, 끝으로 본 연감이 출판되기까지 협조하여 주신 건축가와 자료를 제공해 주신 건축사사무소에 이 지면을 빌어 깊은 감사를 드린다.

발행인 정광영

Contents

Education·Library·Daycare Center
교육시설·도서관·어린이집

10	Ulsan Future Education Center 울산미래교육관
16	Jinju Innovation Complex Culture Library 진주혁신복합문화도서관
22	Pyeongchon Library 평촌도서관
28	Research Building Environment Improvement Project Ⅳ_Winner 연구동 환경개선사업 Ⅳ_1등작
36	Research Building Environment Improvement Project Ⅳ_2nd Prize 연구동 환경개선사업 Ⅳ_2등작
44	Jeollabuk-do Representative Library 전라북도 대표도서관
50	Seosan Central Library 서산 중앙도서관
56	Sosabeol 2 Middle School 소사벌 2중학교
62	Bucheon Okgil Middle & High School Integrated Management School 부천 옥길 중·고 통합 운영학교
68	Janghang Elementary School 장항초등학교
74	Agricultural Research & Extension Services in Gyeongsangbuk-do 경상북도농업기술원
82	Joint Research Centre 공동연구센터
88	MCI Campus MCI 캠퍼스
92	Singil District 12 Social Welfare Facility & Kindergarten_Winner 신길12구역 사회복지시설 및 유치원_1등작
96	Singil District 12 Social Welfare Facility & Kindergarten_2nd Prize 신길12구역 사회복지시설 및 유치원_2등작
102	Wanju-gun Agricultural Technology Center 완주군 농업기술센터

ACA
Architecture Competition Annual XX

Culture·Exhibition
문화·전시시설

- 108   Jinghe New City Culture & Art Centre
  징허 신도시 문화& 예술 센터
- 114   Construction Project of Third KINTEX Exhibition Hall
  킨텍스 제3전시장
- 120   National Digital Heritage Center
  국립디지털문화유산센터
- 126   National Jeongdong Theater_Winner
  국립 정동극장_1등작
- 132   National Jeongdong Theater_2nd Prize
  국립 정동극장_2등작
- 138   House for Film & Media in Stuttgart
  슈투트가르트 영화 & 미디어 하우스
- 142   Montreal Holocaust Museum
  몬트리올 홀로코스트 뮤지엄
- 146   Turku Music Centre
  투르쿠 음악 센터
- 154   Wuxi Art Museum
  우시 미술관
- 158   Philharmonic Hall
  필하모닉 홀
- 162   Jinju Public Science Museum
  진주공립 전문과학관
- 166   Saskatoon New Central Library
  새스커툰 뉴 센트럴 라이브러리
- 172   Terezín Ghetto Museum
  테레진 게토 박물관
- 178   National Intangible Heritage Center Miryang Branch
  국립무형유산원 밀양 분원
- 186   Namdo Righteous Army History Museum
  남도의병역사박물관

Other
기타

- 196   Boramae Hospital Safe Respirator Center
  보라매병원 안심호흡기전문센터
- 202   Keflavik Airport
  케플라비크 공항
- 206   Ferry Terminal Turku
  투르쿠 페리 터미널
- 212   Chongqing Cuntan International Cruise Center
  충칭 쿤탄 국제 크루즈 센터
- 218   Gyeongsangbuk-do Training Center
  경상북도 수련원
- 222   Sinan-dong Sports Complex
  신안동 복합스포츠타운
- 228   MZ Sports Plaza
  MZ 스포츠플라자
- 234   Sports Hall in Zatec
  자테츠 스포츠홀
- 240   Barclay Tower
  바클레이 타워
- 248   Ørestad Church
  외레스타드 교회
- 252   Fragments of Nostalgia
  노스탤지어의 조각
- 258   Magnifica Fabbrica
  웅장한 공장
- 266   Yantai Seafront Garden
  옌타이 해안 정원

Contents

Public·Office
공공·업무시설

10	Haeundae-gu New Office Building	해운대구 신청사
16	Namhae County Office Building	남해군 신청사
24	Changwon Agricultural Technology Center_Winner	창원시 농업기술센터_1등작
30	Changwon Agricultural Technology Center_2nd Prize	창원시 농업기술센터_2등작
36	NION	니온
42	Biotech Research & Incubation Center	바이오테크 리서치 앤 인큐베이션 센터
48	Chungju Knowledge Industry Center_Winner	충주 지식산업센터_1등작
54	Chungju Knowledge Industry Center_2nd Prize	충주 지식산업센터_2등작
60	Osan Post Office	오산우체국
68	BRAINERGY HUB	브레이너지 허브
74	Goyang New City Hall	고양시 신청사
80	Samjuk-myeon Community Service Center_Winner	삼죽면 행정복지센터_1등작
86	Samjuk-myeon Community Service Center_2nd Prize	삼죽면 행정복지센터_2등작
92	Bundang-gu Health Center	분당구 보건소
96	Nanjing Nexus	난징 넥서스
100	Suwon Nambu Police Station	수원남부경찰서
104	Yuseong-gu Veterans Hall	유성구 보훈회관
110	Xiaomi Headquarters	샤오미 본사
114	Viettel Group Headquarters	비에텔그룹 본사
118	SHCCIG (Hainan) International Industrial Headquarters	SHCCIG (하이난) 국제 산업 본부
122	Nexum	넥슘
126	Daerim 3-dong Public Complex Building	대림3동 공공복합청사
132	Pujiang Civic Center	푸장 시민회관

Housing
주거시설

138	Goyang Changneung S-3BL Apartment House	고양창릉 S-3BL 공동주택
144	Ansan Jangsang A-2BL Apartment House	안산장상 A-2BL 공동주택
150	Goyang-Changneung A-1BL Apartment House	고양창릉 A-1BL 공동주택
156	Belsenpark	벨센파크
162	Namyangju Wangsuk District A-19BL Apartment House	남양주 왕숙지구 A-19BL 공동주택
168	Oasis Towers	오아시스 타워
174	Wasserhauser	바사하우저
180	Palas Residential	팔라스 주거시설
186	Residential Quarter Stadtpark Gersthofen	게르스트호펜 공원 주거 지구
192	Sejong-si Global Leader Training Center	세종시 글로벌 리더 연수센터
198	Almada Housing	알마다 주거시설
204	Bristol Housing	브리스톨 주거시설

Neighborhood Living
근린생활시설

210	Support Center for Childcare_Winner	육아종합지원센터_1등작
214	Support Center for Childcare_2nd Prize	육아종합지원센터_2등작
220	Complex Community Center & Urban Regeneration Eoulim Center	복합커뮤니티센터 및 도시재생어울림센터
224	Dangjin Sports Complex & Bandabi Sports Training Center	당진종합체육관 및 반다비국민체육센터
232	Gurye Wellness Complex Center	구례 웰니스 복합센터
236	Jeju Citizens' Hall Life SOC Complex Facility	제주시민회관 생활SOC복합화시설
242	Mullae-dong Base Facility_Winner	문래동 거점시설_1등작
248	Mullae-dong Base Facility_3rd Prize	문래동 거점시설_3등작
252	Andeok-myeon Culture Sports Complex Center_Winner	안덕면 문화체육복합센터_1등작
258	Andeok-myeon Culture Sports Complex Center_2nd Prize	안덕면 문화체육복합센터_2등작
266	Chungnam Disabled Family Healing Center	충남 장애인가족 힐링센터

10	Ulsan Future Education Center 울산미래교육관
16	Jinju Innovation Complex Culture Library 진주혁신복합문화도서관
22	Pyeongchon Library 평촌도서관
28	Research Building Environment Improvement Project Ⅳ_Winner 연구동 환경개선사업 Ⅳ_1등작
36	Research Building Environment Improvement Project Ⅳ_2nd Prize 연구동 환경개선사업 Ⅳ_2등작
44	Jeollabuk-do Representative Library 전라북도 대표도서관
50	Seosan Central Library 서산 중앙도서관
56	Sosabeol 2 Middle School 소사벌 2중학교
62	Bucheon Okgil Middle & High School Integrated Management School 부천 옥길 중·고 통합 운영학교
68	Janghang Elementary School 장항초등학교
74	Agricultural Research & Extension Services in Gyeongsangbuk-do 경상북도농업기술원
82	Joint Research Centre 공동연구센터
88	MCI Campus MCI 캠퍼스
92	Singil District 12 Social Welfare Facility & Kindergarten_Winner 신길12구역 사회복지시설 및 유치원_1등작
96	Singil District 12 Social Welfare Facility & Kindergarten_2nd Prize 신길12구역 사회복지시설 및 유치원_2등작
102	Wanju-gun Agricultural Technology Center 완주군 농업기술센터

Ulsan Future Education Center
울산미래교육관

ING ARCHITECTURE / Hyunseok Jo + MPT ENGINEERING ARCHITECTURE / Wonhyo Kim
㈜아이엔지건축사사무소 / 조현석 + ㈜엠피티엔지니어링 건축사사무소 / 김원효

With a sustainable landscape where nature and architecture interact, a new educational and cultural complex Ulsan Future Education Center is a cultural facility located on the border of the old city center and the East Sea. It was necessary to consider connection with the existing auditorium, land use planning considering the characteristics of the site, separation of pedestrian and vehicle routes, and various frontalities in response to the surrounding environment. Therefore, we tried to realize this with the design concept a new landscape that adds the experience of a sustainable way of life to the infinite dialogue between nature and architecture.

First, it is a paradigm multi-place of future-oriented educational space. It is arranged to respond to nature and the city center and is created to be a place for sustainable education and local communities. Second, it is the harmony between sustainable environment and education. It proposes organic interior and exterior spaces where both students and local community visitors, who are the main users of Ulsan Future Education Center, create natural interactions and a sustainable community. Third, it is a boundless interior space that unfolds without boundaries. It intends to create a continuous circulation system of various educational spaces by establishing a circulation from the core learning library of the facility remodeled to the existing auditorium to the experience space, project, and adventure space.

Ulsan Future Education Center, a future-oriented and sustainable facility, will welcome students, teachers, and general users with a flexible mass arrangement that responds to both the city and nature, and express the vision of sustainable future education through continuous streamlined mass.

자연과 건축이 대화하는 지속가능한 풍경, 새로운 복합교육문화공간 인피너스 울산미래교육관은 구도심과 동해 바다의 경계에 자리한 문화시설로 기존 강당과의 연계, 대지의 특성을 고려한 토지이용계획, 제한된 조건의 보차분리 및 주변 환경에 대응할 수 있는 다양한 정면성 등의 고려가 필요했다. 이에 자연과 건축의 무한한 대화 속에 지속가능한 삶의 방식의 경험을 더하는 새로운 풍경을 뜻하는 인피너스라는 설계개념을 통해 이를 구현하고자 하였다.

첫 번째, 미래지향적 교육공간의 패러다임 멀티 플레이스이다. 자연과 도심에 대응하는 배치형태와 지속가능한 교육, 지역 커뮤니티의 장이 되는 멀티 플레이스를 조성하였다. 두 번째, 지속가능한 환경과 교육의 조화이다. 울산미래교육관의 주 이용자인 학생들과 지역사회방문자 모두 이용하며 상호 자연스런 교류와 지속가능한 커뮤니티가 형성되는 유기적인 내·외부공간을 제안하였다. 세 번째, 경계없이 펼쳐지는 내부공간 바운드리스이다. 기존의 강당을 리모델링한 시설의 핵심공간인 러닝도서관부터 체험터, 프로젝트, 모험터까지 연계되는 동선을 구축하여 다양한 교육공간이 연속되는 순환체계를 조성하고자 하였다.

미래지향적이고 지속가능한 시설인 울산미래교육관은 도심과 자연 모두 대응하는 유연한 형태의 매스배치로 학생과 교사, 일반 이용자 모두를 담아내고, 끊임없이 이어지는 유선형의 매스를 통해 지속가능한 미래 교육의 비전을 표현하는 시설로 거듭날 것이다.

Location : 1119-6, Muryong-ro, Buk-gu, Ulsan, Korea I **Function** : Education & Research facility I **Site area** : 11,292㎡ I **Bldg. area** : 2,862㎡ I **Total floor area** : 9,114㎡ I **Stories** : B1, 3FL I **Structure** : Reinforced concrete I **Finish** : Low-E pair glass, App' stone, Terracotta, Metal louver

위치 : 울산시 북구 무룡로 1119-6 I **용도** : 교육연구시설 I **규모** : 지하1층, 지상3층 I **구조** : 철근콘크리트 I **마감** : 로이복층유리, 지정석재마감, 테라코타, 금속루버 I **설계팀** : 윤승민, 조하정, 홍도희, 박하연

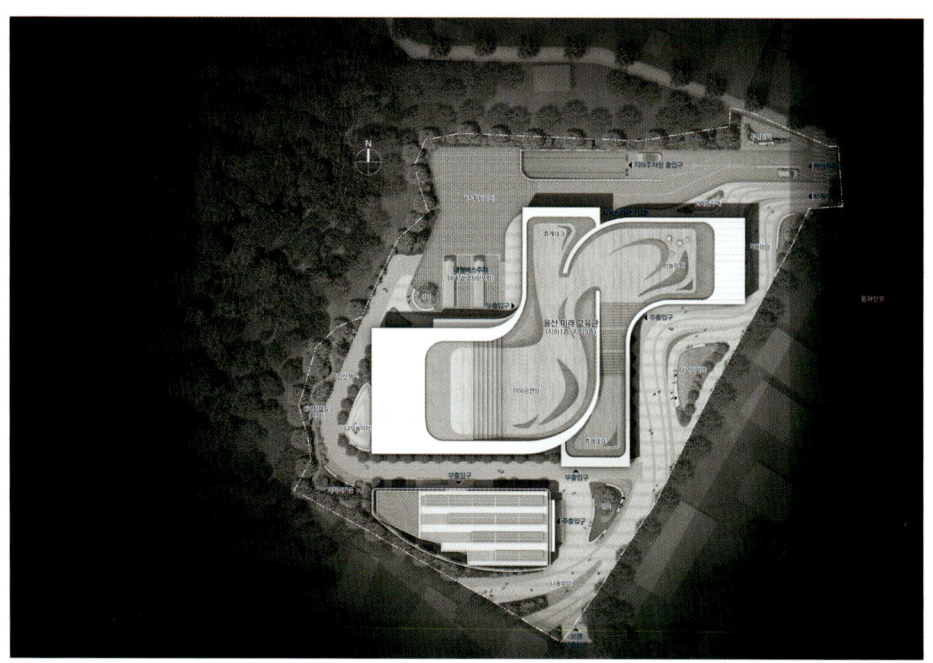

Site plan

■ Site Analysis

- Acting as a cultural facility in the old city center

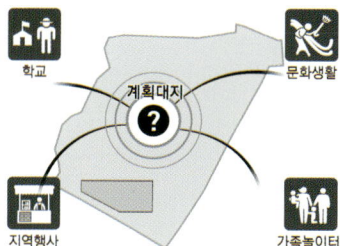

- Consideration of connection with existing auditorium (learning library)

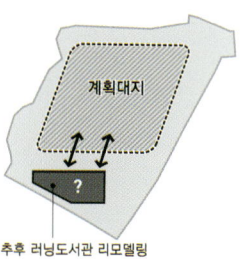

- Land use plan considering the characteristics of the site

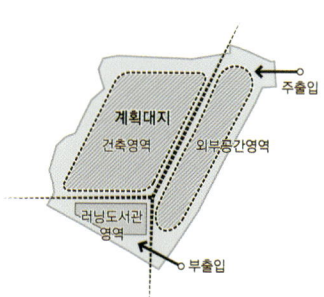

- Car separation plan under limited conditions

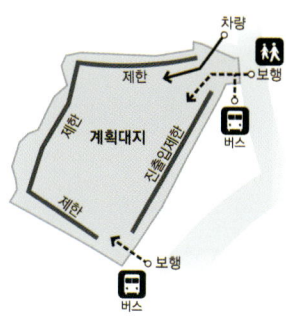

- Consideration of multi-faceted frontality in response to the surrounding environment

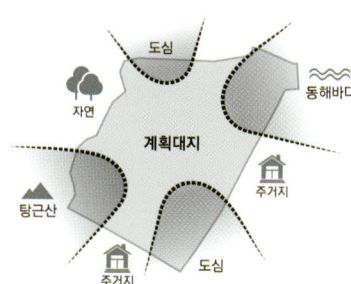

Winner _ 당선작

▌Masterplan System

- Need for awareness
- Facility zoning
- Linkage of functions
- Inflow of external space

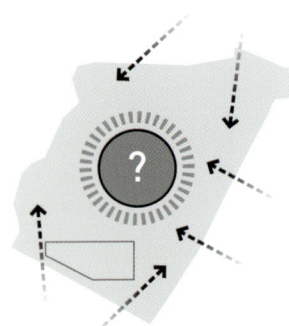
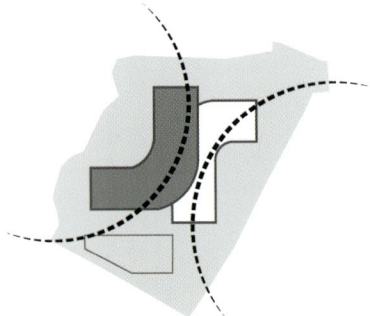
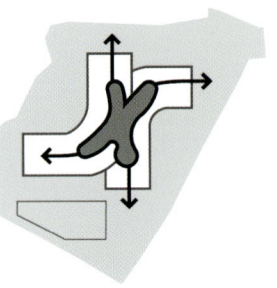
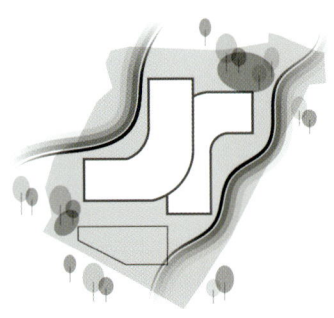

- 다방면에서의 인지성요구
- 자연과 도심축을 기반으로 시설별 시설조닝
- 중앙홀을 통한 체험공간 연계
- 대지를 둘러싼 파노라마 풍경연출

Ulsan Future Education Center >>

Winner _ 당선작

East elevation

West elevation

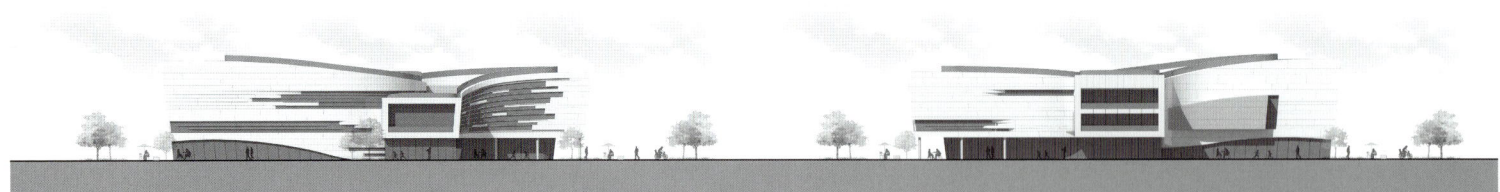

South elevation

North elevation

1. Parking
2. Machine room
3. Electrical room
4. Lobby
5. Play space
6. School clinic
7. Slow food room
8. Idea room
9. Cafe
10. Water tank
11. Storage
12. Experience center
13. Story center
14. Special exhibitions room
15. Woodwork preparation room
16. Maker & Media zone
17. Adventure center
18. Adventure zone
19. Maintenance office

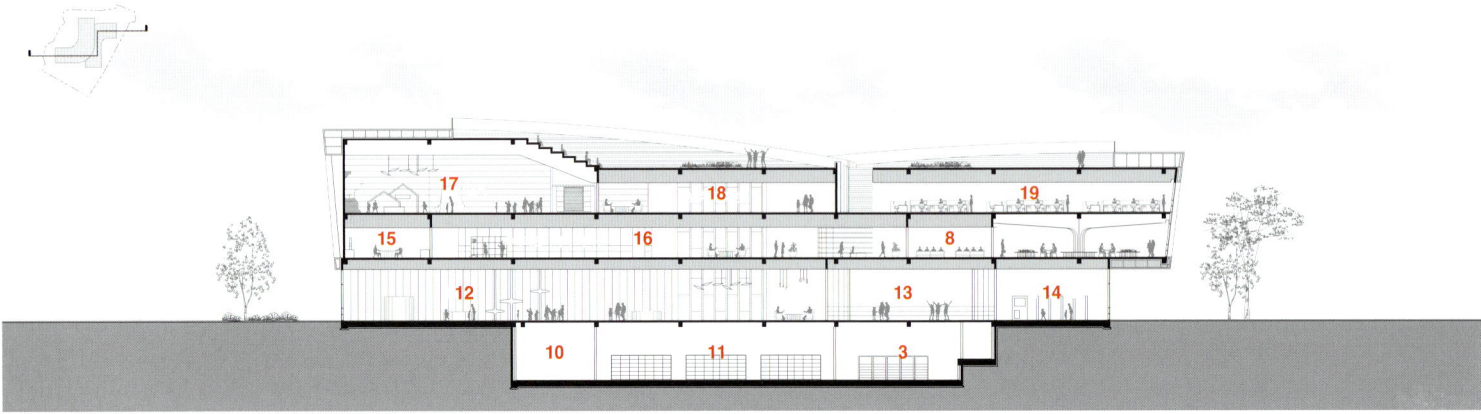

Cross section

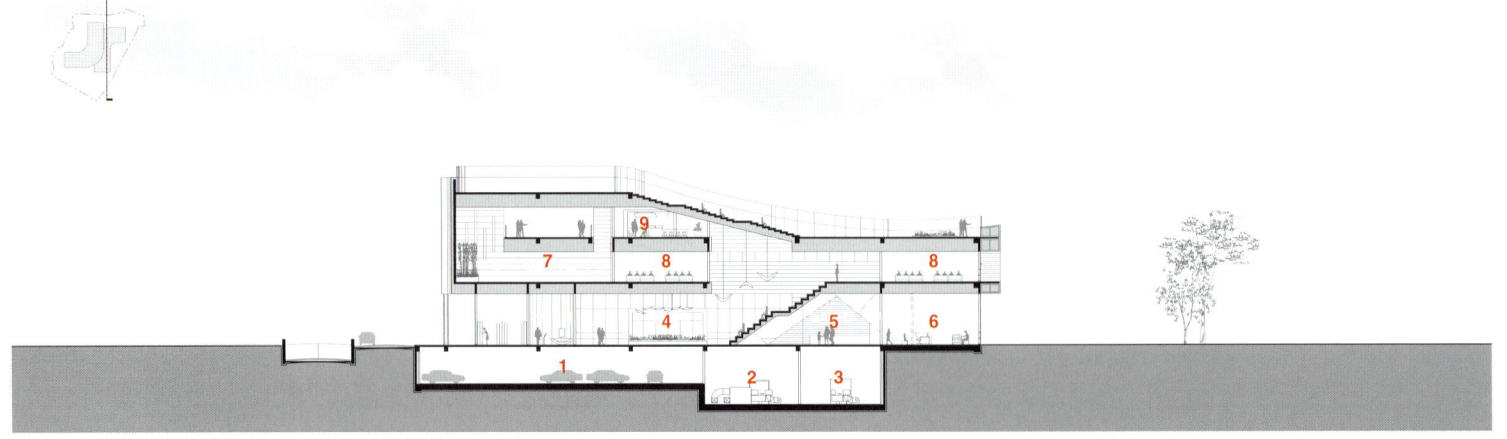

Longitudinal section

Ulsan Future Education Center >>

1. Running library
2. Experience center
3. Idea room
4. School clinic
5. Lobby
6. Customer support division
7. Special exhibitions room
8. Workshop
9. Woodwork room
10. Digital creative room
11. Media room
12. Slow food room
13. Adventure center
14. Maintenance office

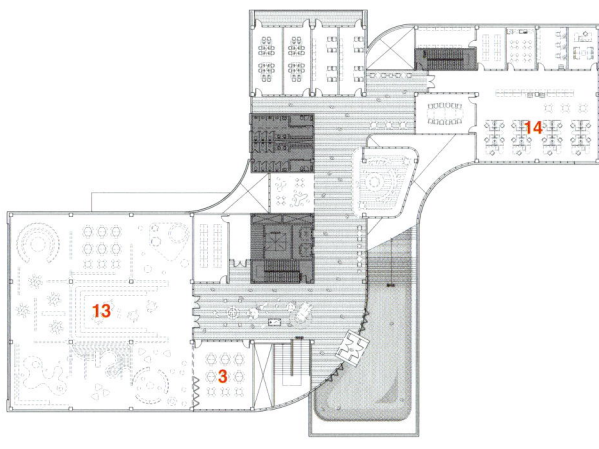

3rd floor plan

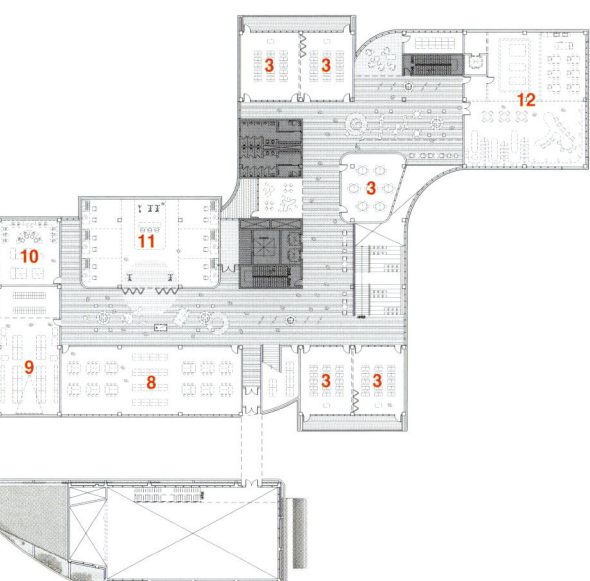

2nd floor plan

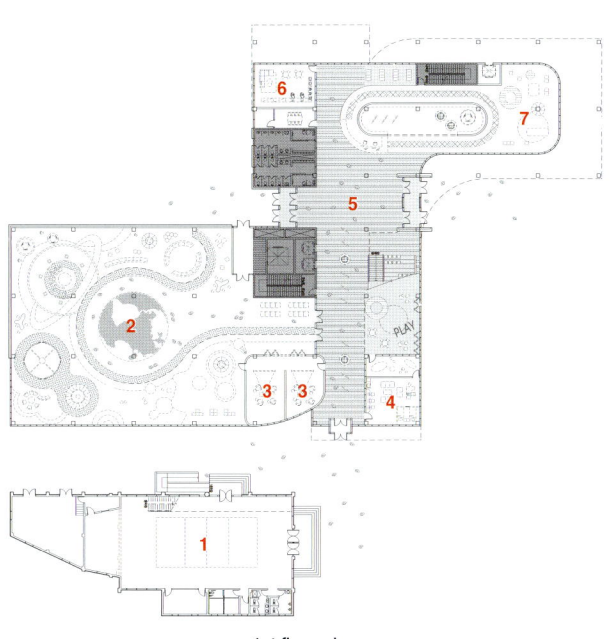

1st floor plan

Winner _ 당선작

Jinju Innovation Complex Culture Library
진주혁신복합문화도서관

HAENGLIM ARCHITECTURE & ENGINEERING / Yongho Lee + Dong-A University / Guemhong Seo
㈜행림종합건축사사무소 / 이용호 + 동아대학교 / 서금홍

Become a root of historical city growth
The Jinju Innovation Complex Culture Library symbolizes the root of Jinju's history and tradition, inducing a two-way communication arrangement that expands to the city while embracing nature, and a playground containing various cultural spaces that communicate with nature serve as nourishment for the cultural city.

A cultural platform that connects performances and daily life
Resting, lectures, and VR experiences were induced in connection with the performance hall, and a reference room related to Jinju was placed adjacent to it to inspire history and culture naturally. The spacious deck will be used as a central space for cultural daily life.

A cultural playground in the forest
A children's library was planned on the first floor, a cultural space on the second floor, and a smart library for free creative activities on the third and fourth floors. A cultural playground was implemented for reading books and enjoying creations with nature.

역사도시 성장의 근원이 되다
진주혁신복합문화도서관은 진주의 역사와 전통의 숨결을 간직한 뿌리를 상징화해 자연을 포용하며 도시로 확장하는 쌍방향 소통의 배치형태를 유도하고, 자연과 소통하는 다양한 문화공간을 담은 놀이터를 계획하여 문화도시로의 자양분이 되는 역할을 한다.

공연과 일상을 연계한 문화플랫폼
공연장과 연계하여 휴식, 강연, 공연 VR체험을 유도하고, 인접하여 진주관련 자료실을 배치해 자연스러운 역사문화 고취가 일어나도록 하였다. 넓은 데크는 문화적 일상을 담는 중심공간으로 활용될 것이다.

숲속에서 즐기는 문화놀이터
1층에는 어린이도서관을, 2층에는 문화공간을, 3층과 4층에는 자유로운 창작활동을 위한 스마트도서관을 계획하여 자연과 함께 책을 읽고 창작을 즐기는 문화놀이터를 구현하였다.

Location : 184, Chungmugong-dong, Jinju-si, Gyeongsangnam-do, Korea I **Function** : Education & Research facility I **Site area** : 7,875m I **Bldg. area** : 3,741m I **Total floor area** : 12,365m I **Stories** : B1, 4FL I **Structure** : Reinforced concrete, Light steel frame I **Finish** : Ceramic panel, Aluminium louver, Low-E pair glass

위치 : 경상남도 진주시 충무공동 184 I **용도** : 교육연구시설 I **규모** : 지하1층, 지상4층 I **구조** : 철근콘크리트, 경량철골 I **마감** : 세라믹패널, 알루미늄루버, 로이복층유리 I **설계팀** : 이상혁, 박찬원, 최명환, 조세희, 이한정, 김민영, 김재린, 조수익, 윤이사야, 성창현

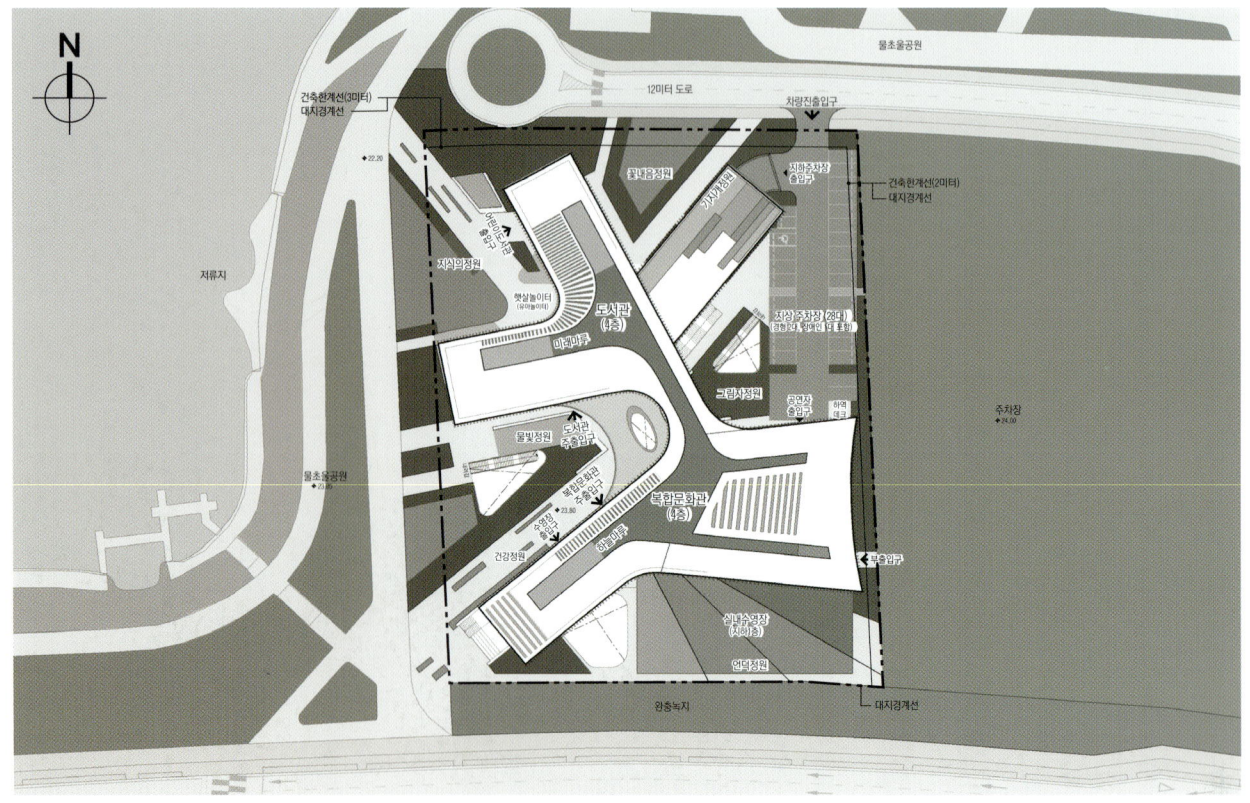

Site plan

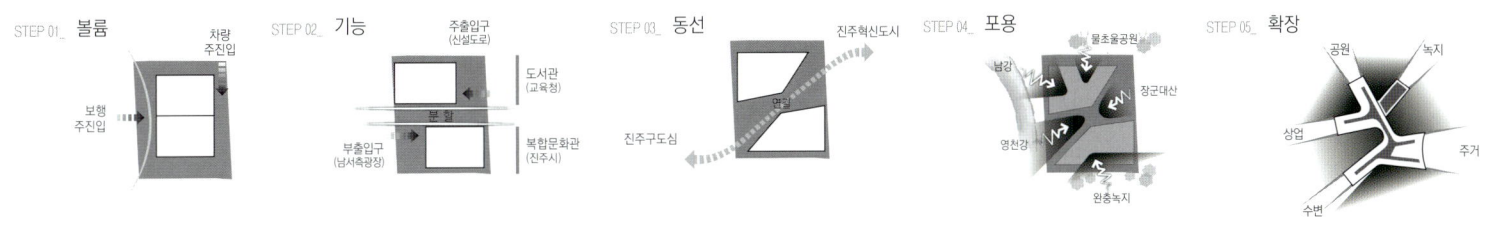

▌Arrangement Plan

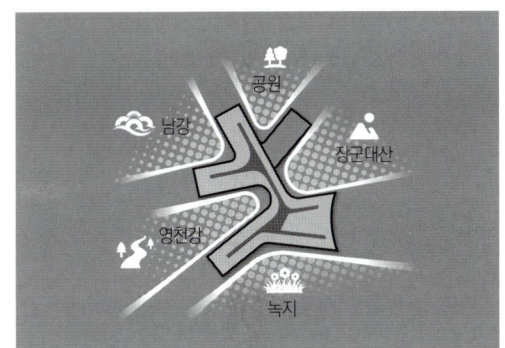

· Penetration and expansion scenarios

- 자연_주변 환경을 포용하는 자연친화적 배치

- 사람_보행흐름을 받아들이는 유기적 배치

- 도시_도시풍경을 담아내는 소통하는 배치

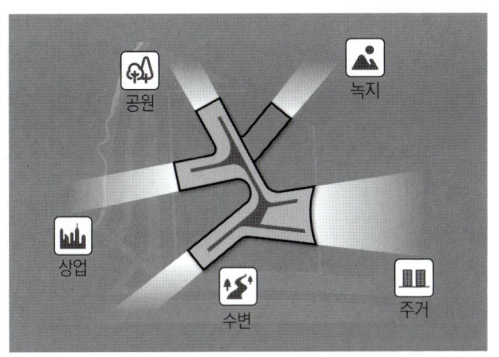

Winner _ 당선작

▌Elevation Plan

- Landscape panorama that connects nature and urban landscapes

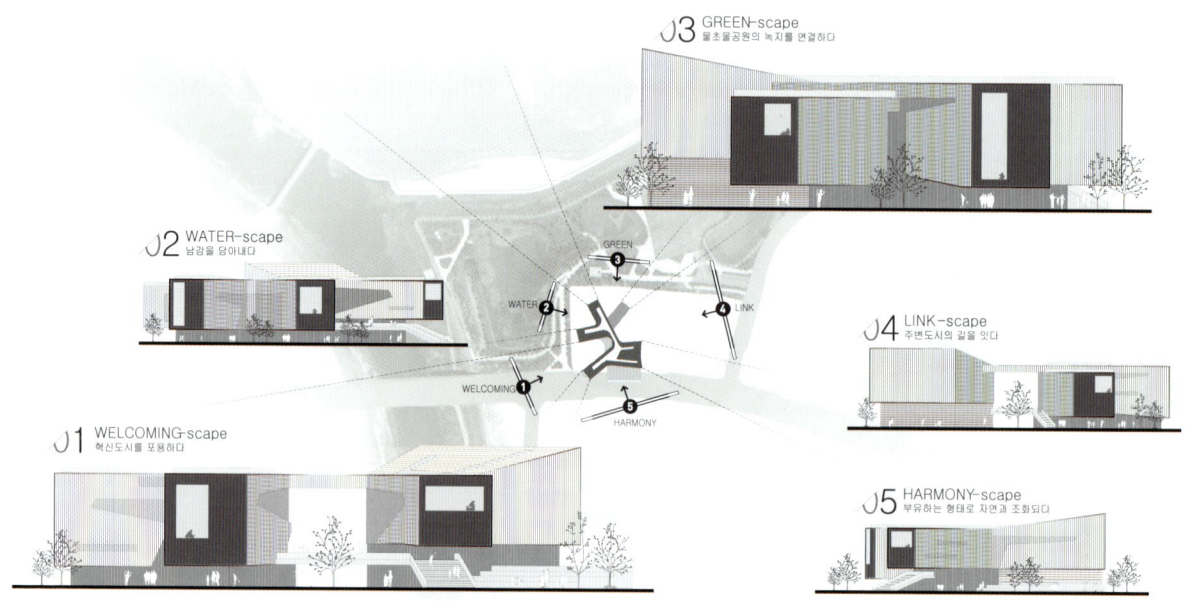

— 주변 자연을 끌어들임과 동시에 도시로 다시 확장하는 형태는 내부에서는 도시로의 조망을 즐기고 외부에서는 진주시민들이 바라보는 한 폭의 자연풍경으로 연출되었다.

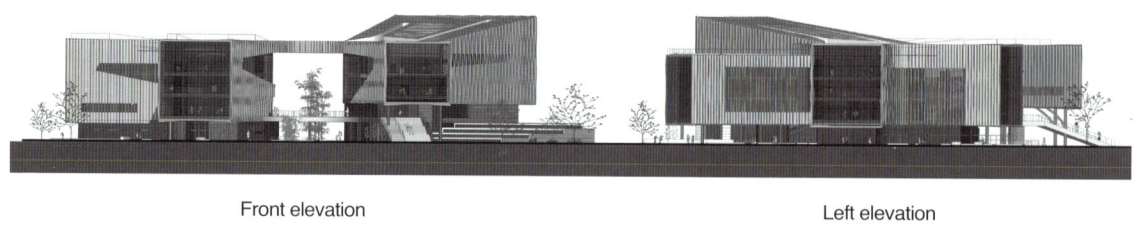

Front elevation　　　　　　　　　　　　Left elevation

Rear elevation　　　　　　　　　　　　Right elevation

Model

Section Plan

- Site to breathe

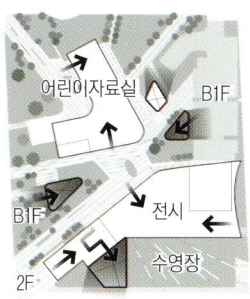

– 전방향 보행 흐름을 연결하는 길과 선큰

+

- Sector-type cultural complex

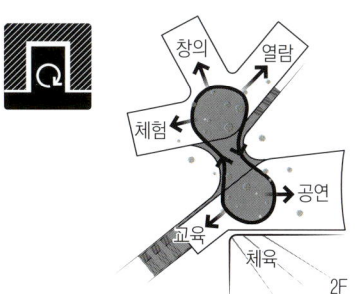

– 프로그램별 분리 및 연계에 유리한 공간구조

+

- Extended spatial experience

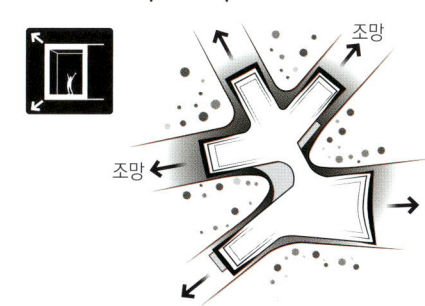

– 내외부 연계로 공간 경험을 확장하는 형태

1. Creative park
2. Collaboration park
3. Information commons
4. Characterization reference room
5. On-Jinju
6. Children's department
7. Lobby
8. Machine room
9. Parking
10. Lecture room
11. Yard
12. Exhibition facility
13. Shower / Changing room
14. Changing room
15. Swimming pool
16. Hall

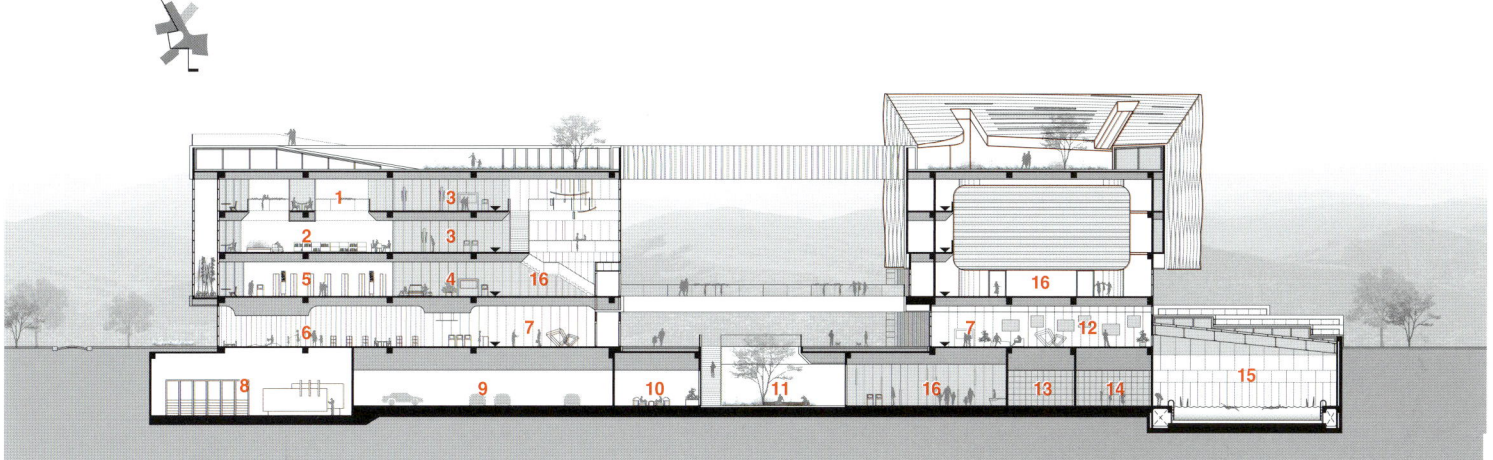

Longitudinal section

Winner _ 당선작

Floor Plan

- **Life-friendly space with independent accessibility**

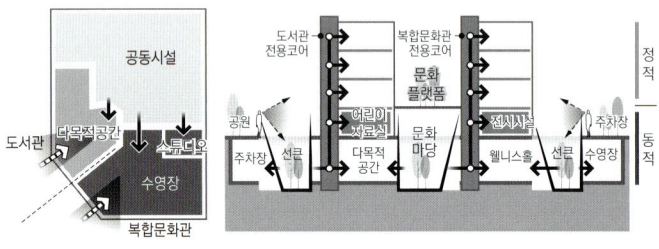

– 운영주체가 각각 다른 다목적공간과 수영장을 분리한 후 로비를 통해 연계하고, 지상 선큰 및 지하 주차장에서 직접 진입하는 동선을 계획해 주·야간, 학기·방학 등 멈추지 않는 지역주민의 생활을 담는다.

- **Open children's library and exhibition facilities**

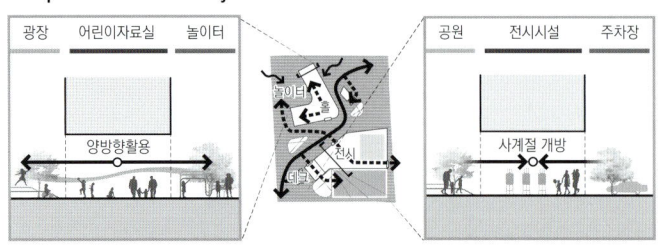

– 어린이도서관을 1층에 배치하여 놀이와 교육을 연계하고, 전시시설을 전방향 보행동선과 연결함으로써 문화로비로의 기능을 수행할 수 있도록 개방성을 극대화하였다.

- **Cultural life through a cultural platform**

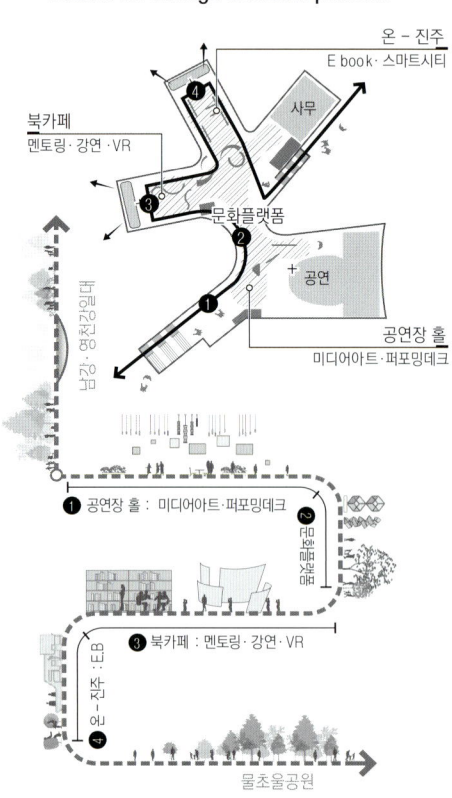

Jinju Innovation Complex Culture Library >>

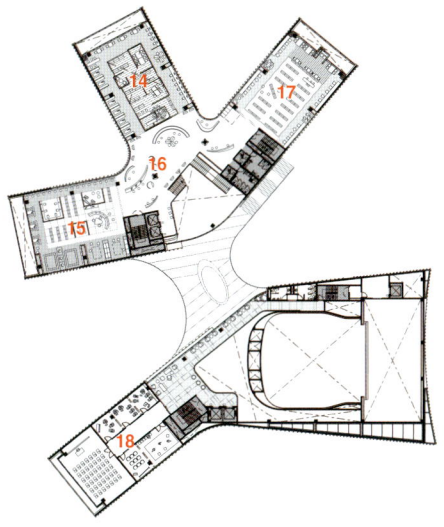
3rd floor plan

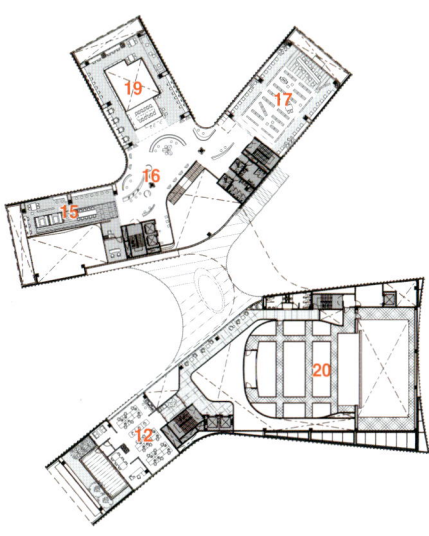
4th floor plan

1. Children's department
2. Lobby
3. Mom's station (cafeteria)
4. Parking
5. Exhibition facility
6. Neighborhood living facility (cafeteria)
7. Storage
8. On-Jinju
9. Book cafe
10. Characterization reference room
11. Hall
12. Office
13. Stage
14. Collaboration park
15. Maker space
16. Information commons
17. Running commons
18. Community center
19. Creative park
20. Cat walk

1st floor plan

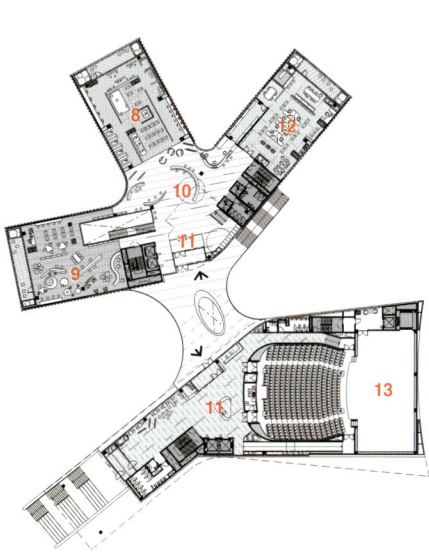
2nd floor plan

Pyeongchon Library
평촌도서관

HAK Architecture / Jaehak Kim, Daehwan Jang + LAON ARCHITECTURE&ENGINEERING / Kidong Kim
㈜건축사사무소 학건축 / 김재학, 장대환 + ㈜라온엔지니어링 건축사사무소 / 김기동

Add something special to everyday life of Pyeongchon
Pyeongchon Library pursues an open library that accepts the diverse lifestyles of modern people by renewing as a regional central pavilion among 10 existing municipal libraries in Anyang. As a facility that goes beyond providing information and delivers complex functions such as exhibitions, culture, lecture, and leisure activities according to the paradigm change of public libraries, we propose a living-friendly space with culture in our daily lives.

RITUAL LIBRARY that adds something special to the daily life of citizens
The first floor of the library is created as a central space for citizens to gather, rest, talk, share ideas, and engage in education and creative activities.

LIVING LIBRARY like a living room in Pyeongchon
The second floor of the library is created as an inspirational space for all generations to share time, learn about the past with books, experience the present, and prepare for the future.

RE-LIBRARY you want to revisit
We provide an event space in our daily life where you can engage in various outdoor activities by making the most of the adjacent Pyeongchon Park and the open view environment, and express the symbolism as a regional central pavilion by exposing the local representative preservation library, which will contain the story of Anyang, to the outside.

평촌일상에 특별함을 더하다
평촌도서관은 안양의 기존 10개의 시립도서관 중 지역중앙관으로 거듭남으로써, 현대인의 다양한 라이프스타일을 수용하는 열린 도서관을 추구한다. 공공도서관의 패러다임 변화에 따라 정보 제공에 중점을 두던 과거의 역할을 넘어 전시·문화·강좌·여가활동 등 복합적인 기능을 제공하는 시설로서, 일상에 문화가 있는 생활 밀착형 공간을 제안한다.

시민들의 일상에 특별함을 넣어주는 RITUAL LIBRARY
도서관의 1층은 시민들이 모여서 쉬고, 이야기를 나누고, 생각을 공유하며, 교육과 창작활동을 하는 시민의 중심 공간으로 조성

평촌의 거실과 같은 LIVING LIBRARY
도서관의 2층은 전 세대가 시간을 공유하고, 책을 통해 과거를 배우며, 현재를 경험하고, 미래를 준비하는 영감의 공간으로 조성

다시 오고 싶은, 다시 보고 싶은 RE-LIBRARY
인접한 평촌 공원과 열린 조망 환경을 적극적으로 활용하여 다양한 야외활동이 가능한 일상 속 이벤트 공간을 제공하고, 안양의 이야기를 담아나갈 지역대표 보존서고를 외부로 드러내어 지역의 중앙 관으로써의 상징성을 표현

Location : 213, Gwanpyeong-ro, Dongan-gu, Anyang-si, Gyeonggi-do I **Function** : Education & Research facility I **Site area** : 9,900m² I **Bldg. area** : 3,905m² I **Total floor area** : 10,013m² I **Stories** : B1, 3FL I **Structure** : Reinforced concrete I **Finish** : Slim ceramic panel, Off-white brick

위치 : 경기도 안양시 동안구 관평로 213 I **용도** : 교육연구시설 I **규모** : 지하1층, 지상3층 I **구조** : 철근콘크리트 I **마감** : 박판세라믹패널, 미색벽돌 I **설계팀** : 최천, 진규태, 양희진, 장진이, 정진호, 이예원, 안민주, 신재은, 우나린

Site plan

■ Concept

- **Pyeongchon daily life 'Harbor'**

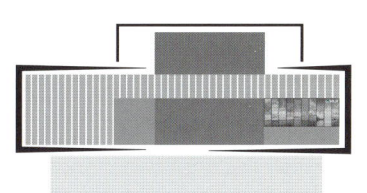

– 평촌과거, 현재, 미래를 품다

- **Pyeongchon daily life 'Meet'**

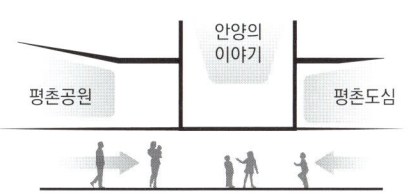

– 평촌의 사람, 공원, 도시가 만나다

- **Pyeongchon daily life 'Enjoy'**

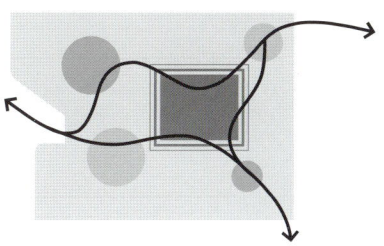

– 평촌의 공간을 누리다

Winner _ 당선작

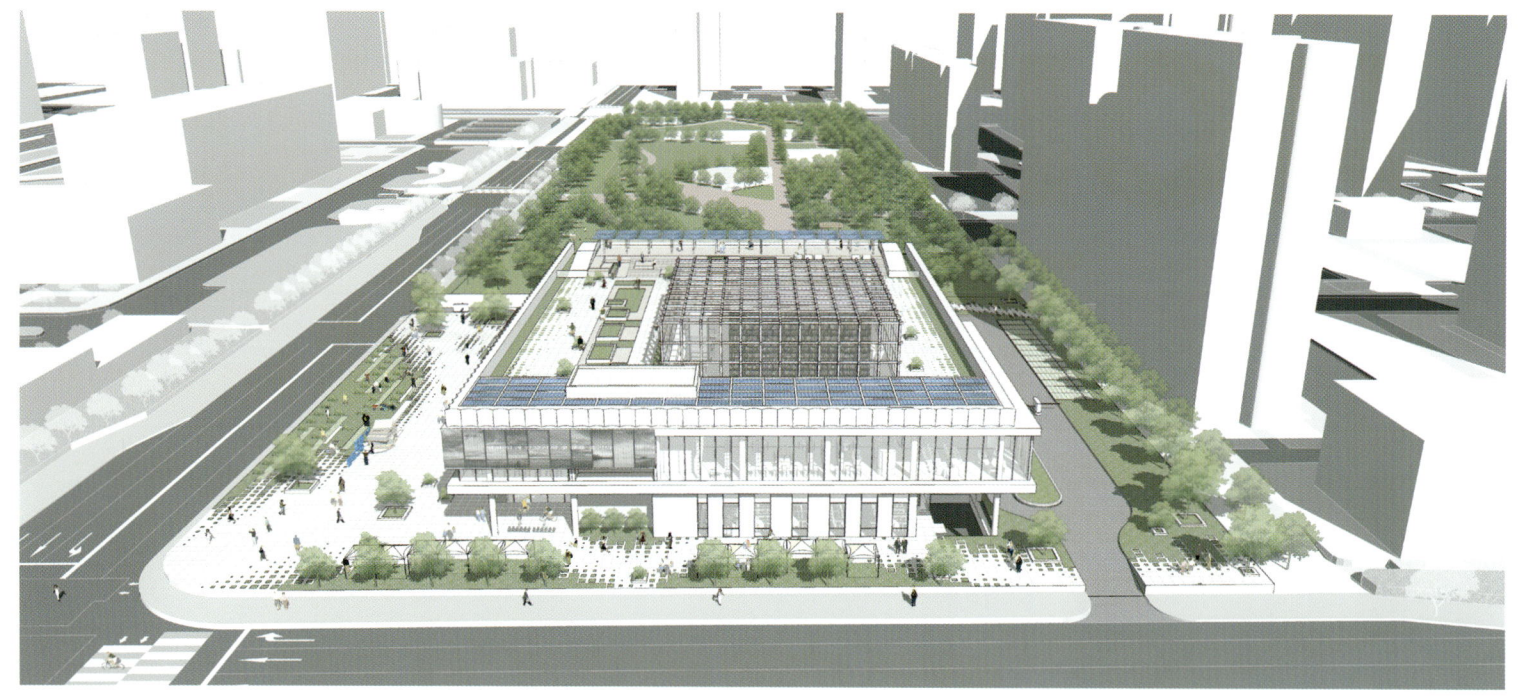

East elevation

South elevation

West elevation

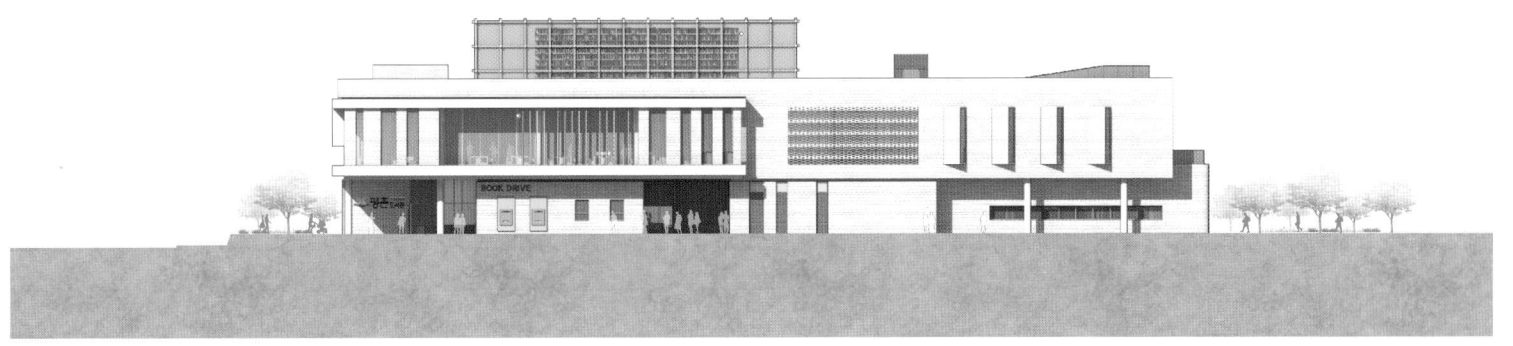

North elevation

Winner _ 당선작

1. Parking
2. Book cafe
3. Multipurpose space
4. Information commons
5. General data reading room
6. Rooftop garden
7. Machine room
8. Studio
9. Market place
10. Lactation room

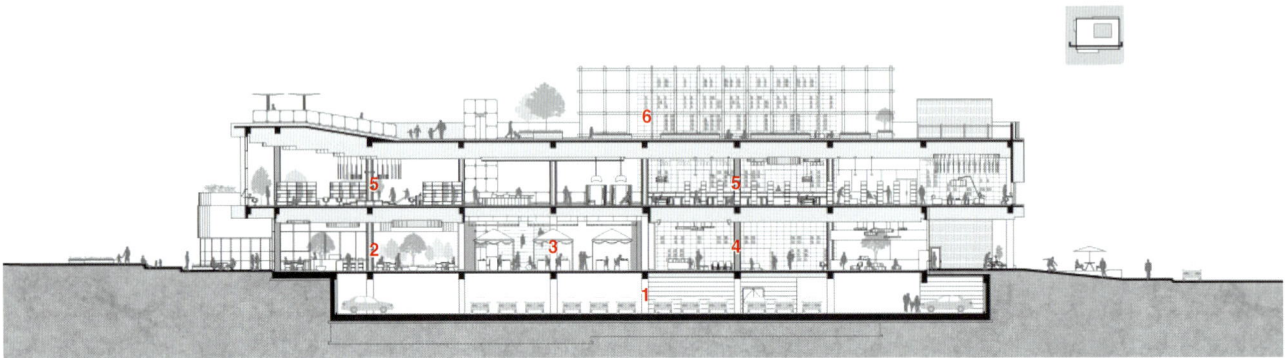

Cross section

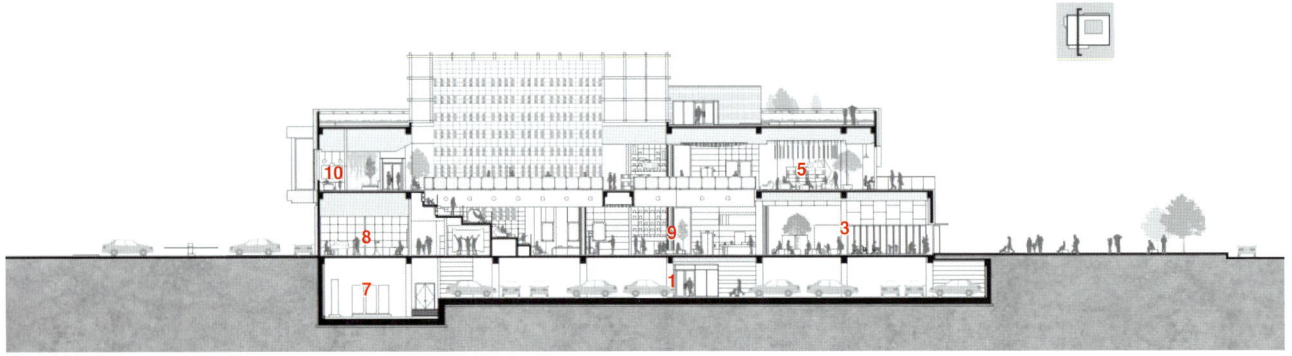

Longitudinal section

Pyeongchon Library >>

1. Book cafe
2. Multipurpose space
3. One-stop service
4. Class space
5. Information commons
6. Creative ground
7. Volunteer room
8. Take in data / Organization room
9. General data reading room
10. Cooperative storage facility
11. Kid zone
12. Baby zone
13. Office
14. Rooftop garden
15. Book floor
16. Event yard

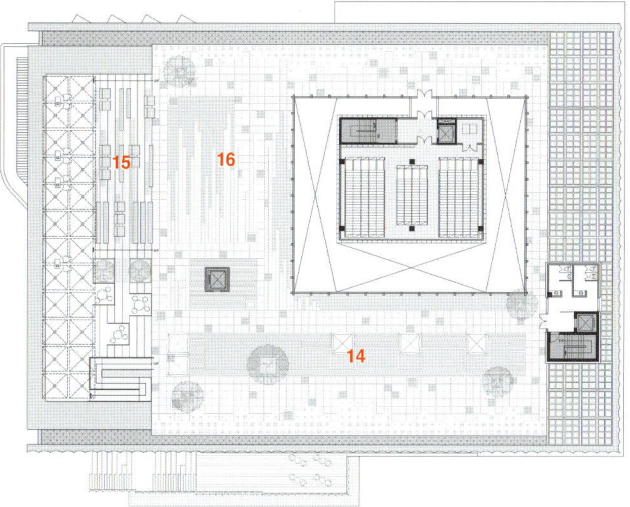

3rd floor plan

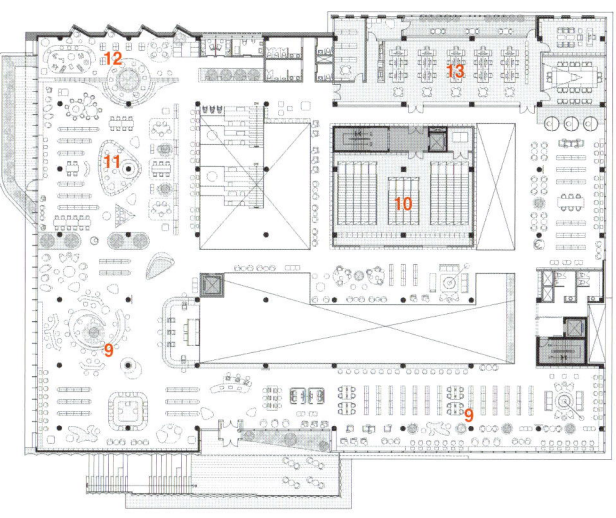

2nd floor plan

1st floor plan

Research Building Environment Improvement Project IV
연구동 환경개선사업 IV

dA Group Urban Design & Architecture / Hyunho Kim + Design Group Five / Sungmin Cho
㈜디에이그룹엔지니어링종합건축사사무소 / 김현호 + ㈜건축사사무소 오 / 조성민

With the main theme of 'a community of changing knowledge', the goal of the overall design is to maximize the power of the 'collective brain' of various researchers and to voluntarily and mutually share knowledge through daily and nondaily exchanges.

The open space is planned in the front to reflect the locational characteristics of the research building located in the central space of the Korea Institute of Science and Technology (KIST) campus. By linking and integrating this with other spaces on the existing campus, a new cultural center is provided for the future of KIST.

The new research institute should be a place for researchers to realize their ideas. Various creative spaces are arranged using the 'eureka moment', a concept of brain science that refers to the process in which new stimuli through extraordinary experiences are connected to creative ideas. By analyzing individual, team, and group behaviors, the space is divided into the use of immersion and relaxation, and the character and size of each space are set. Creative spaces arranged throughout the institute will maximize the creativity of researchers by inducing accidental encounters and various activities.

The goal of the design of the new institute is to have a structure to predict and respond to changes in users and work areas. The optimal research environment is provided by reflecting the requirements of researchers through post residence evaluation, and flexible research and experimental space is realized by intensively arranging the facility space outside to respond to changes in the future research environment.

'변화하는 지식의 공동체'라는 중심 주제로, 다양한 연구원들의 '집합적 두뇌'의 힘이 극대화되며 일상적·비일상적 교류에 의한 지식의 자발적 상호 공여가 이루어지는 연구소를 전체 설계 방향으로 삼았다.

한국과학기술원(KIST) 캠퍼스의 중심공간에 위치한 연구동의 입지적 특성을 반영하여 전면에 개방된 공간을 계획했다. 이를 기존 캠퍼스의 다른 공간과 연계 및 융합하여, 한국과학기술연구원의 미래를 준비하는 새로운 문화 중심공간을 마련하였다.

새로운 연구소는 연구원들의 사고 발현의 장이 되어야 한다. 비일상적 경험을 통한 새로운 자극이 창의적 아이디어로 연결되는 과정을 일컫는 뇌 과학의 개념, '유레카 모멘트'를 활용하여 다양한 창의 공간들을 마련하였다. 개인, 팀, 그룹별 행위 분석에 의해 몰입과 이완의 공간으로 분류하고, 공간의 성격과 크기를 설정했다. 연구소 곳곳에 배치된 창의 공간들은 우연적인 만남과 다양한 활동을 유발하여 연구원들의 창의력을 극대화 시킬 것이다.

사용자 및 업무영역의 변화를 예측하고 대응할 수 있는 구조를 갖추는 것을 새로운 연구실 설계의 목표로 삼았다. 사후 거주평가를 통한 연구원들의 요구 조건을 반영하여 최적의 연구 환경을 제공하였으며, 설비공간을 외부로 집중하여 미래 연구 환경 변화에 대응 가능한 유연한 연구·실험공간을 구현했다.

Location : 5, Hwarang-ro 14-gil, Seongbuk-gu, Seoul, Korea I **Function** : Education & Research facility I **Site area** : 271,714㎡ I **Bldg. area** : 2,822㎡ I **Total floor area** : 21,983㎡ I **Stories** : B3, 6FL I **Structure** : Reinforced concrete I **Finish** : Low-E pair glass, Aluminium louver, Slim ceramic panel

위치 : 서울시 성북구 화랑로 14길 5, L2 연구동 I **용도** : 교육연구시설 I **규모** : 지하3층, 지상6층 I **구조** : 철근콘크리트 I **마감** : 로이복층유리, 알루미늄루버, 박판세라믹패널 I **발주처** : 한국과학기술연구원(KIST) I
설계팀 : ㈜디에이그룹엔지니어링종합건축사사무소 / 이태민, 김상, 이근명, 이근택, 최재혁, 이혁, 정연욱, 강대규, 정동혁, 송예한, 김채린 + ㈜건축사사무소 오 / 조용준, 박종현

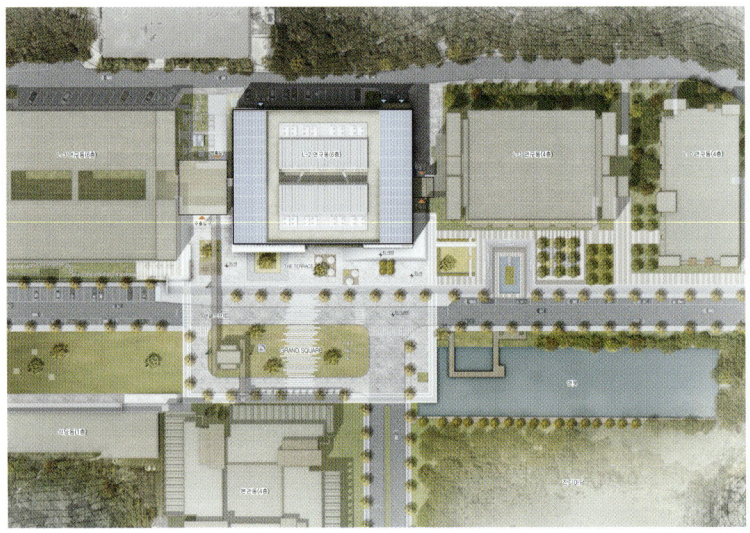

Site plan

Direction Plan

- Campus node

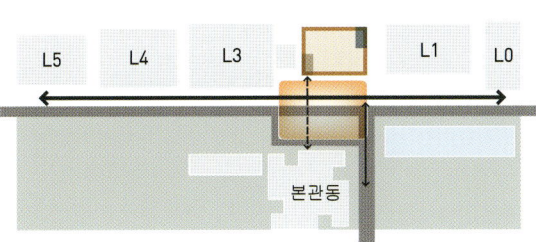

- Accidental encounter

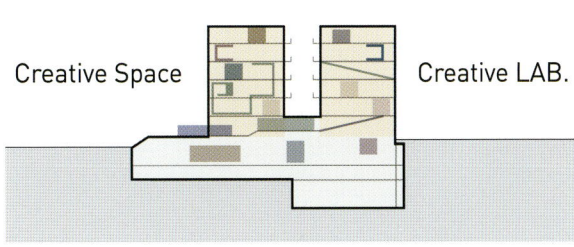

- Future lab.

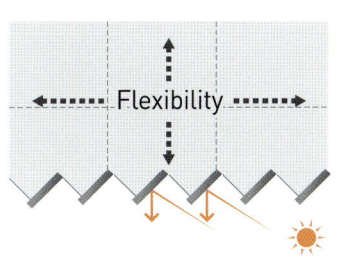

Design Process

- Volume

- Function

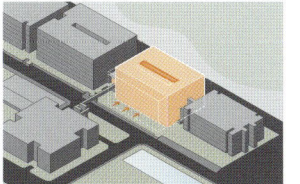

- Icon

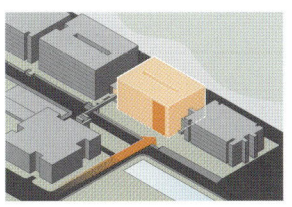

- Square

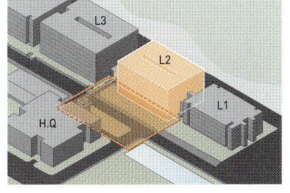

- Interaction

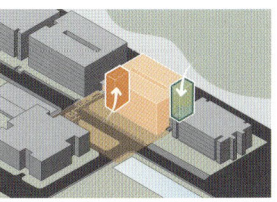

Winner _당선작

Zoning

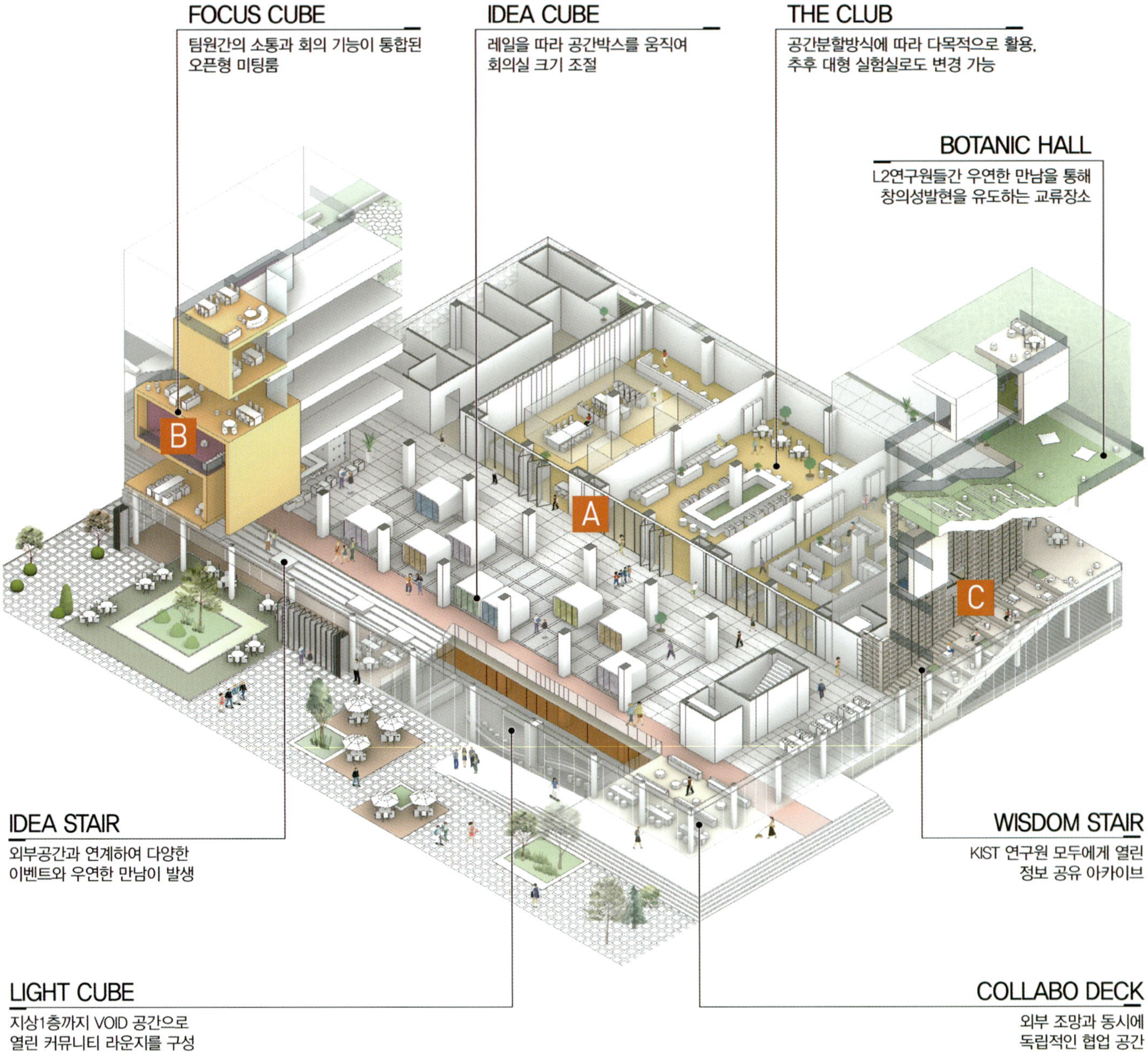

FOCUS CUBE
팀원간의 소통과 회의 기능이 통합된 오픈형 미팅룸

IDEA CUBE
레일을 따라 공간박스를 움직여 회의실 크기 조절

THE CLUB
공간분할방식에 따라 다목적으로 활용, 추후 대형 실험실로도 변경 가능

BOTANIC HALL
L2연구원들간 우연한 만남을 통해 창의성발현을 유도하는 교류장소

IDEA STAIR
외부공간과 연계하여 다양한 이벤트와 우연한 만남이 발생

WISDOM STAIR
KIST 연구원 모두에게 열린 정보 공유 아카이브

LIGHT CUBE
지상1층까지 VOID 공간으로 열린 커뮤니티 라운지를 구성

COLLABO DECK
외부 조망과 동시에 독립적인 협업 공간

Research Building Environment Improvement Project IV >>

1. Electric room
2. Parking
3. Robot laboratory
4. Light cube
5. Micronano fab
6. Animal laboratory
7. The station
8. Research, Experiment
9. Service gallery
10. LMO / Reagent storage room
11. Equipment analysis room

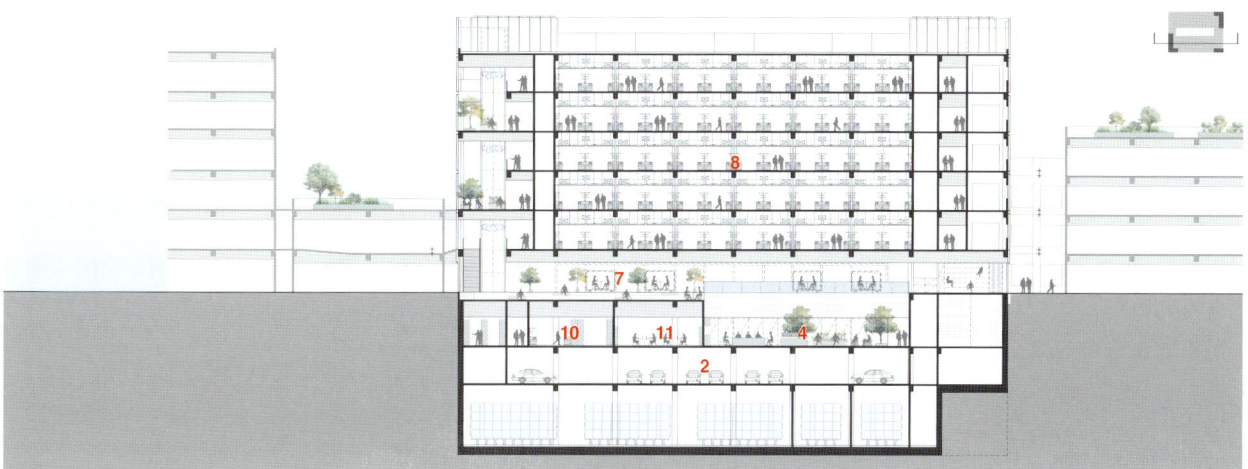

Cross section

Longitudinal section

Winner _ 당선작

Circulation Plan

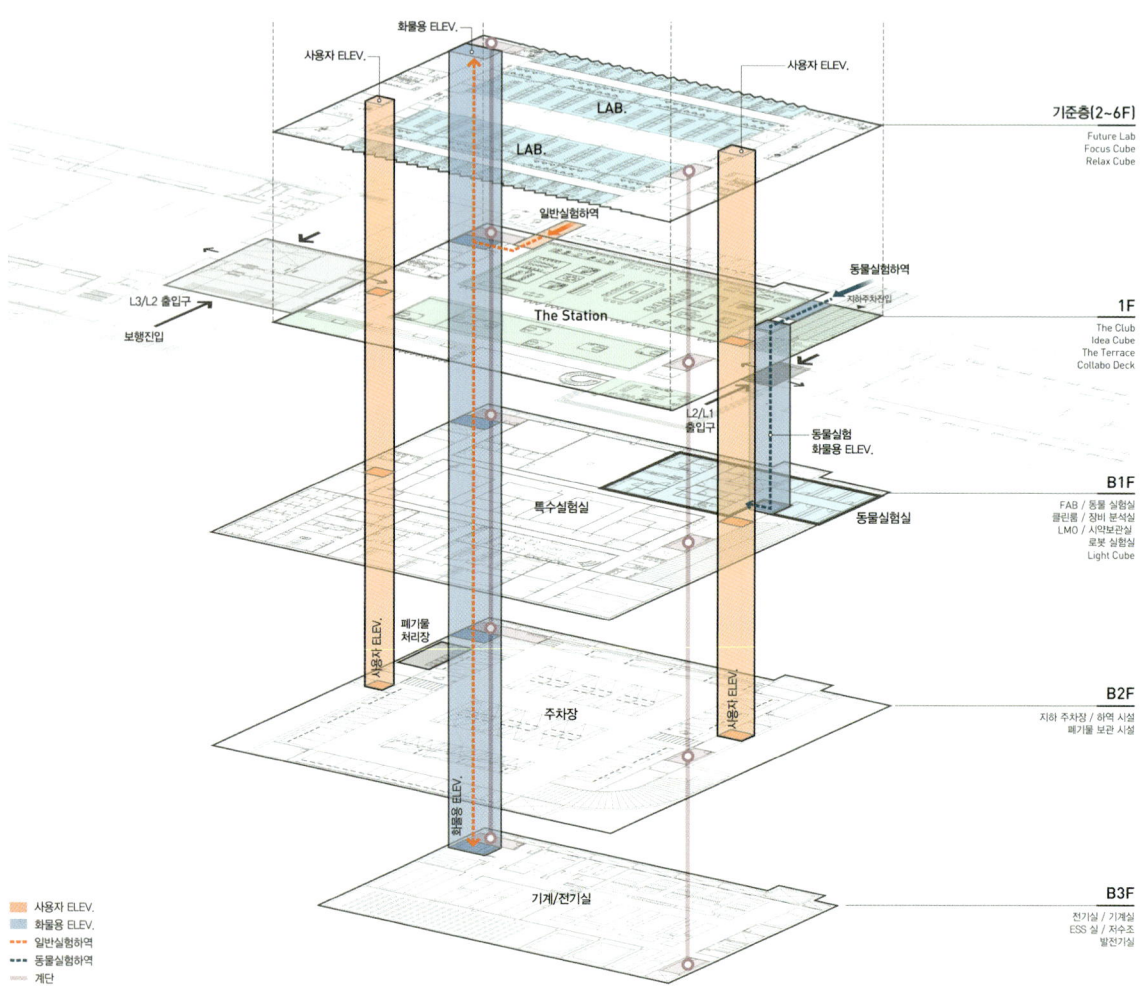

Research Building Environment Improvement Project Ⅳ >>

Convergence Lab

- Continuous flow

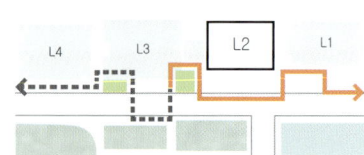

- 기조성된 연구동의 가로와 정원을 연속적으로 배치

- Borrowing landscape

- 내 외부 경계를 허물고 풍경을 활용한 다채로운 정원 배치

- Extension

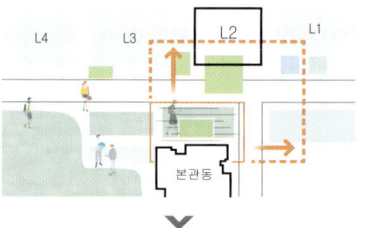

- 다양한 기능의 정원들과 연계하여 중심공간의 영역을 확장

- Grand square

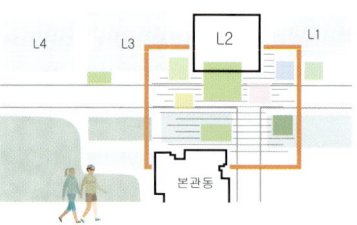

- 연구동과 본관동을 통합시켜 새로운 문화중심 공간으로 제안

Research Building Environment Improvement Project IV >>

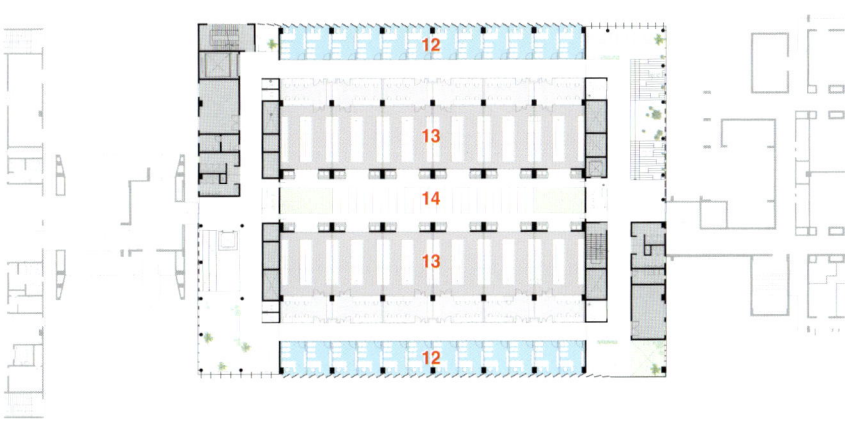

1. LMO / Reagent storage room
2. Equipment analysis room
3. Robot laboratory
4. Light cube
5. Micronano fab
6. Clean room
7. Animal laboratory
8. Idea stair
9. Collaboration deck
10. Idea cube
11. The club
12. Dry lab
13. Wet lab
14. Service gallery

2nd floor plan

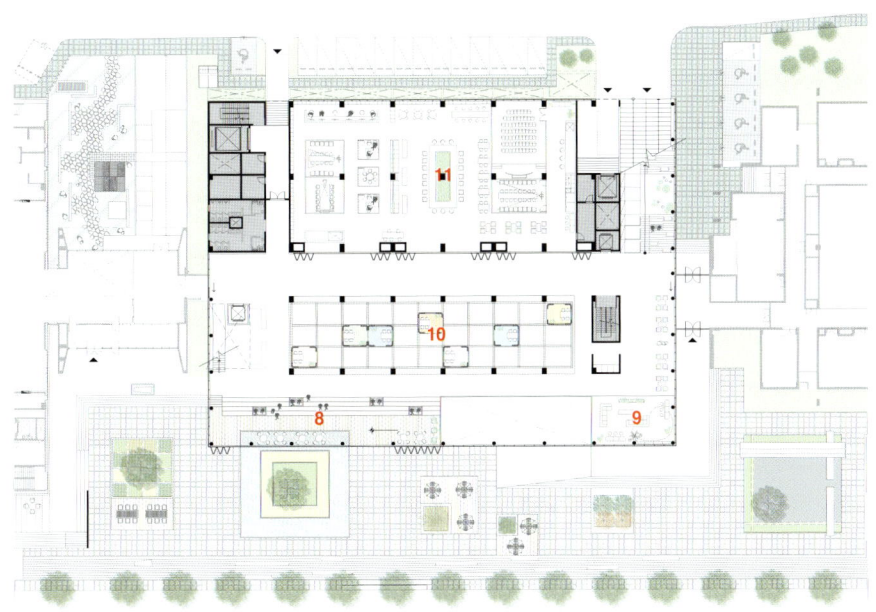

1st floor plan

B1 floor plan

2nd Prize _ 2등작

Research Building Environment Improvement Project Ⅳ
연구동 환경개선사업 Ⅳ

Sun and Partners / Yongmin Lee
㈜종합건축사사무소 선기획 / 이용민

The KIST research complex has buildings centered on research and administration in the north-south direction. The proposal is to put the flow of nature from Cheonjang Mountain in the research complex into the central L2 research building so that the existing research buildings, the L1 and L0 research buildings to be developed later, the external space and public/shared spaces leading to the main building coexist. It is intended to create a necessary community space within the existing research complex. In the process of the master plan, it inherits the architectural and historical values that are functionally intended and aesthetically created, while capturing the image of a future-oriented, cutting-edge research institute.

All laboratories in the research-only module are arranged facing the outside air to create a pleasant and safe research environment. The L2 research building is a multifunctional place in the research complex that is linked with the elements in the complex, and has a mediating spatiality for variable spatial composition according to changes. The three-dimensional and eco-friendly elevation is planned by adding a vertical louver and a screen wall layer for solar control of the laboratory to the layer that reflected the vertical and horizontal architectural language used in the existing research buildings. It inherits the spirit of elevation of the existing research building and expresses the spirit of the L2 research building for the future.

KIST 연구 단지는 남북으로 이어지는 연구축과 행정축을 중심으로 연구동들이 배치되어있다. 이러한 연구 단지 내에 천장산에서부터 이어지는 자연의 흐름을 단지 내 구심점이 되는 L2연구동에 담아 기존의 연구동들과 추후 개발될 L1, L0연구동, 본관동으로 이어지는 외부공간 및 공공/공유공간이 조화롭게 공존하고, 기존 연구단지 내 부족한 커뮤니티 공간을 제안한다. 이러한 마스터플랜의 과정에서 기능적으로 의도되고 미적으로 조성된 건축적, 역사적 가치를 계승하는 동시에 미래를 지향하는 최첨단 연구기관의 이미지를 담아내었다.

연구 전용모듈 내 모든 연구, 실험실은 외기에 면하도록 배치하여 쾌적하고 안전한 연구 환경을 조성한다. L2연구동은 단지 내 요소들과 링크(LINK)되어 연구단지 내 다기능적 장소로서, 변화에 따라 가변적 공간구성이 가능한 매개적 장소성을 갖는다.

입면은 기존의 연구동들에 사용된 수직, 수평적 건축언어를 수용한 레이어에 연구실의 일사조절을 위한 수직루버와 스크린월 레이어를 더하여 입체적이고 친환경적인 입면을 계획하였으며, 기존 연구동의 입면정신을 계승하고 미래를 지향하는 L2연구동의 정신을 표현하였다.

Location : 5, Hwarang-ro 14-gil, Seongbuk-gu, Seoul, Korea | **Function** : Education & Research facility | **Site area** : 271,714㎡ | **Bldg. area** : 38,616㎡ | **Total floor area** : 154,027㎡ | **Stories** : B3, 6FL | **Structure** : Reinforced concrete | **Finish** : Low-E pair glass, Aluminium louver, Clay brick, High density wood panel

위치 : 서울시 성북구 화랑로 14길 5, L2 연구동 | **용도** : 교육연구시설 | **규모** : 지하3층, 지상6층 | **구조** : 철근콘크리트 | **마감** : 로이복층유리, 알루미늄루버, 점토벽돌, 고밀도목재패널 | **설계팀** : 김승태, 김태현, 이창연, 김종섭, 허장미, 박신영, 박가영, 양용준, 홍석효, 김하연

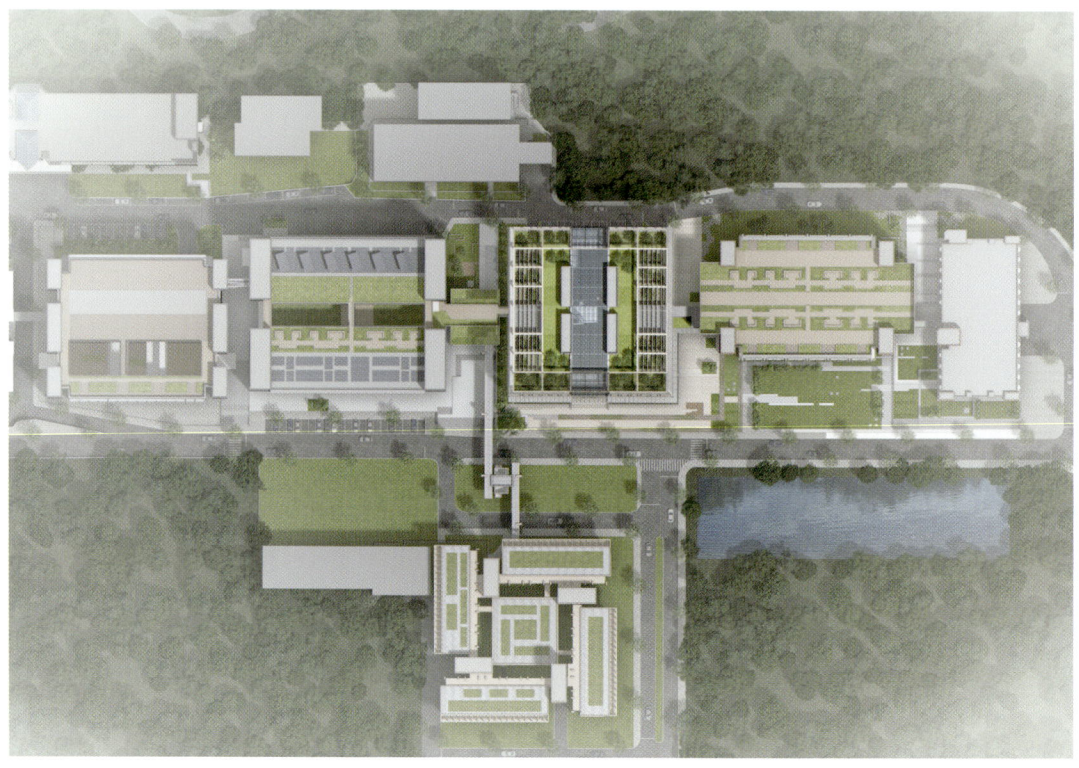

Site plan

Design Concept

- Planning based on the site context of KIST

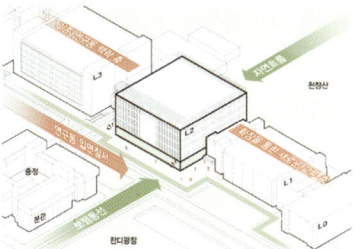

- Functional module plan considering research characteristics

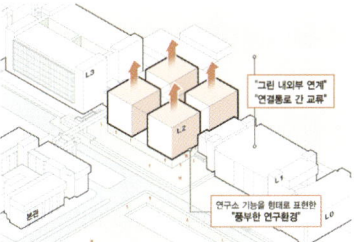

- Various vertical green community spaces

Zoning

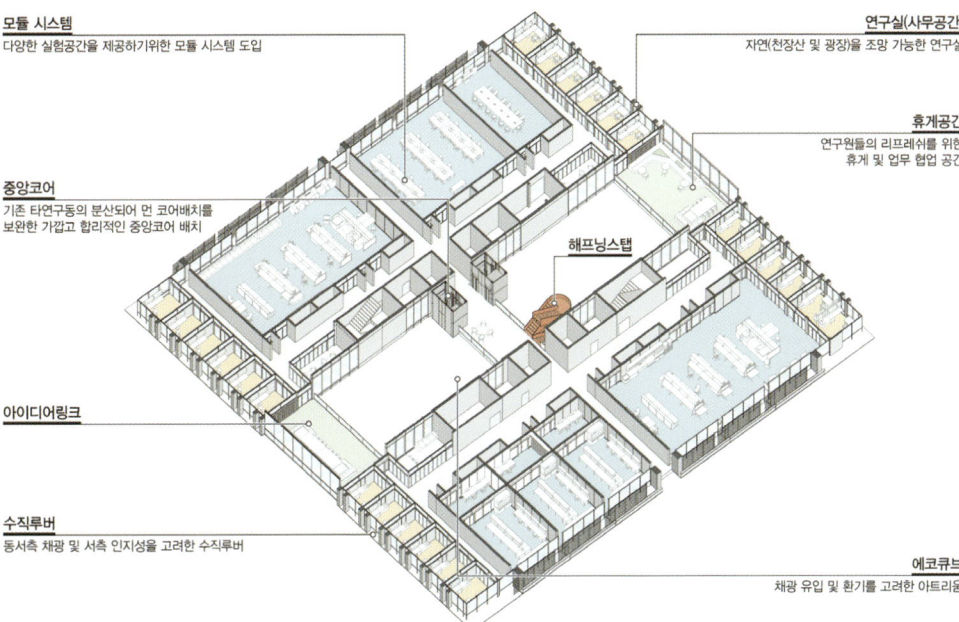

2nd Prize _ 2등작

Research Building Environment Improvement Project Ⅳ >>

39

2nd Prize _ 2등작

■ Elevation Plan

- Facade system

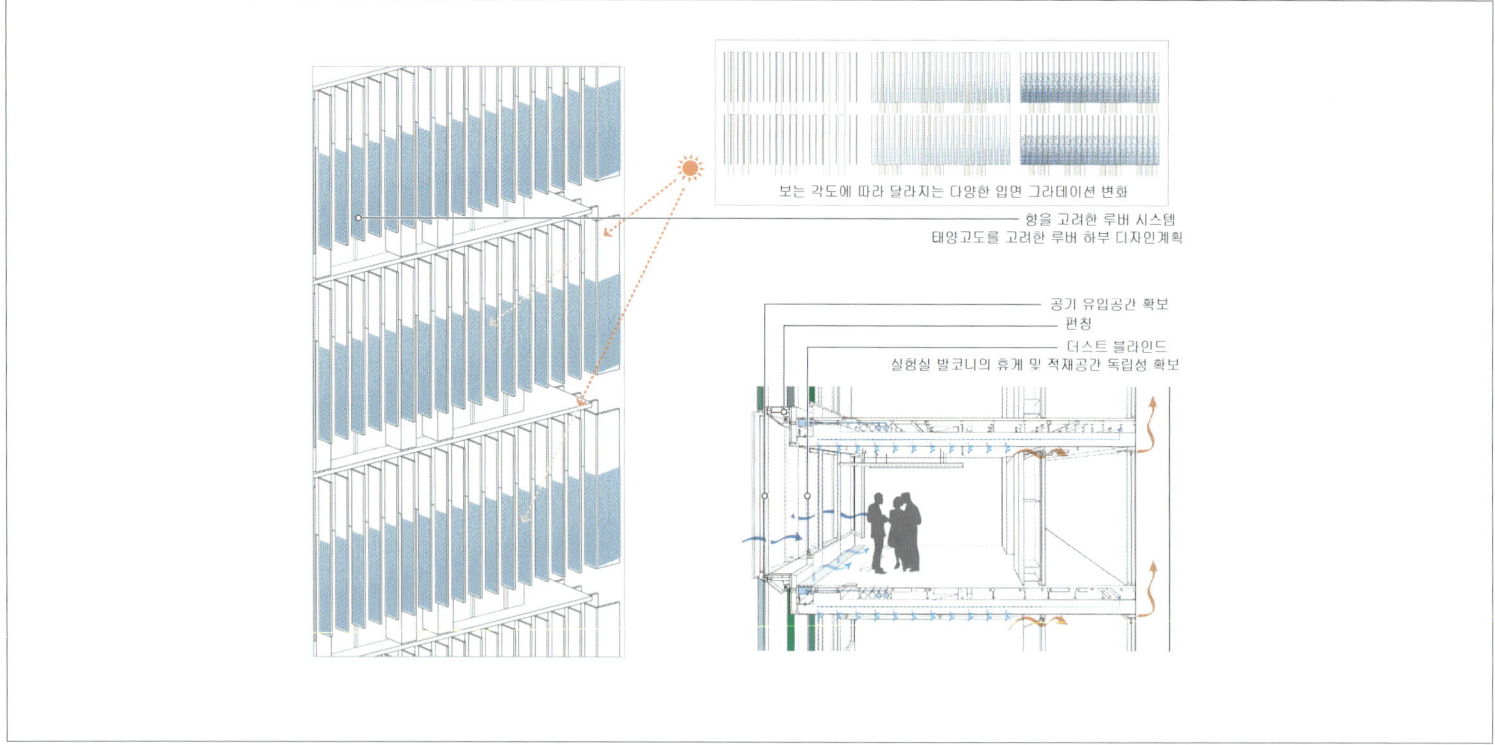

- 수직루버를 통한 파사드는 KIST 연구시설로서의 정체성을 드러내며 다양한 형태의 입면을 제공한다.
- 중첩된 여러 핀의 루버는 차양으로 일사량을 조절하여 내부 열부하를 낮추며, 외부조명, 내부에서의 오픈 뷰 확보 등의 기능을 한다.

Front elevation   Left elevation

Rear elevation   Right elevation

Research Building Environment Improvement Project Ⅳ >>

Section Perspective View

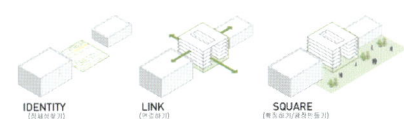

1. Machine room
2. Parking
3. Animal laboratory
4. Fitting room
5. Corridor
6. Reagent storage room
7. Clean room
8. Rest deck
9. Warehouse
10. Healing link
11. Meeting room
12. Laboratory
13. Analytical laboratory
14. Eco cube atrium
15. Forest link
16. Rest space
17. Lecture room
18. Laboratory + Analysis room
19. Open laboratory
20. Small laboratory
21. Integrated analytical laboratory
22. Electrical room
23. Dynamo room
24. UPS room
25. Emergency room
26. Treatment room
27. Air handling unit room
28. Office
29. Library
30. Development room
31. Bridge

Cross section   Longitudinal section

2nd Prize _ 2등작

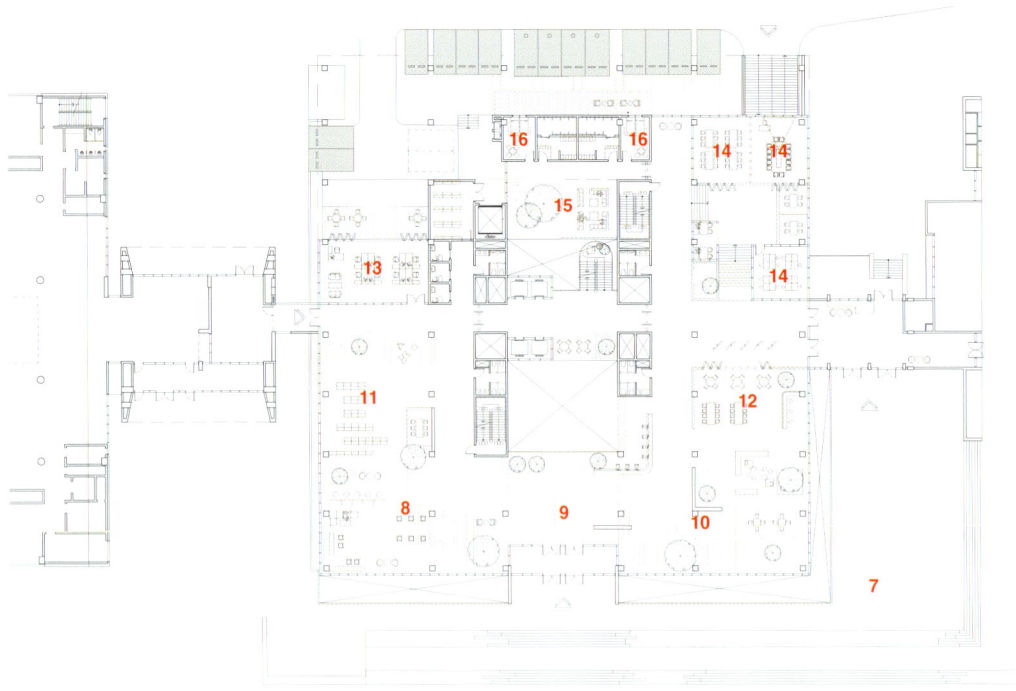

1st floor plan

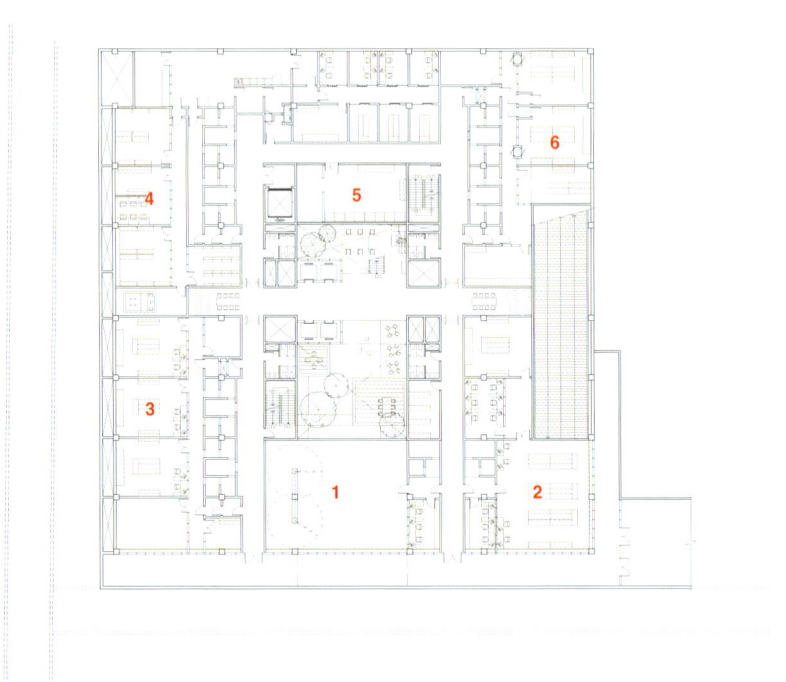

B1 floor plan

Research Building Environment Improvement Project IV >>

1. Robot laboratory
2. Equipment analysis room
3. Micronano fab
4. Animal laboratory
5. Reagent storage room
6. Clean room
7. Community link
8. Business link
9. Lobby
10. Green link
11. Library
12. Cafeteria
13. Office
14. Meeting room
15. Healing link
16. Sleeping room
17. Development room
18. Idea link
19. Lecture hall
20. Laboratory
21. Analytical laboratory
22. Program room
23. Club room
24. Forest link

Typical floor plan

2nd floor plan

Winner _ 당선작

Jeollabuk-do Representative Library
전라북도 대표도서관

Yi architects / Eunyoung Yi + A-GROUP ASSOCIATES CO.,LTD. / Minkwan Lee
Yi architects / 이은영 + ㈜종합건축사사무소에이그룹 / 이민관

The work of refining the basic form of architecture that has been passed down for thousands of years and clearly revealing its essential elements is incomparably more necessary than ever in our era of disorientation and drift. Also, as important as the discovery of the prototype of pure architecture is the 'genius loci'. It is intended to capture the values of the type and space of architecture in Jeonju with a special historical background of 600 years. In particular, Gyeonggijeon and Jeonjuhyanggyo have many functionalities that can be shared with the library, so in addition to discovering types, we apply the things that can be shared, such as the combination of khan, the correlation between architectural elements, and the proportion, and refine it with the spirit of our times. Since it is comfortably nested in Yeopsoon Park and faces the main road, the main body of the representative library is created as a monolith in a clear and pure form, worthy of revealing the vision and spiritual value of a new era. In response to the 3m difference in topography, the base part is formed as a roof garden that is naturally connected to the surrounding park, and at the end with a panoramic view of the lake, the Jeollabuk-do's representative book space with precious historical value is located like a pavilion by the water. The two main masses show a tangible reversal as a massive monolith with a corridor-like courtyard in the middle and a pavilion with a massive center.

수천 년간 전달되어 오던 건축의 기본형을 정제해 내고, 그 본질적 요소를 다시 극명하게 드러내 주는 작업은 방향을 잃고 표류하는 우리 시대에 그 어느 시대에도 비교할 수 없을 만큼 절실하다. 또한, 순수한 건축의 원형의 발견 못지않게 중요한 것이 장소의 혼 'genius loci'이다. 600년의 특수한 역사적 배경을 가진 전주에 새겨져 있는 건축의 유형과 공간 등에 들어있는 가치들을 담아내고자 했다. 특히 경기전과 전주향교는 기능적으로도 도서관과 함께할 수 있는 면이 많이 있으므로, 유형발견에서뿐 아니라 칸의 조합, 건축 요소 간의 상관관계, 비례 등에서도 공유할 수 있는 것들을 적용하며, 우리 시대의 시대정신으로 정제하는 과정을 거쳤다. 엽순공원에 편안하게 품어져 있으면서도 주도로에 면하여 새로운 시대의 비전과 정신적 가치를 드러내기에 합당하게 대표도서관의 본체를 명쾌하고 순수한 형태의 모노리스(Monolith)로 빚어내었다. 약 3m의 지형 차에 맞게 기단부를 주변의 공원과 자연스레 연계되는 옥상정원으로 형성하였고, 호수로 시야가 트이는 끝 편에 소중한 역사적 가치를 품는 전북학도서 공간이 물가의 정자와 같이 자리 잡게 하였다. 두 개의 주 매스는 중앙에 회랑형 중정을 갖는 매시브한 모노리스와 매시브한 구심을 갖는 정자로써 유형적인 반전을 보여준다.

Location : 1110, 1090, Jang-dong, Deokjin-gu, Jeonju-si, Jeollabuk-do, Korea I **Function** : Education & Cultural facility I **Site area** : 29,400m I **Bidg. area** : 5,776m I **Total floor area** : 12,599m I **Stories** : B1, 4FL I **Structure** : Reinforced concrete I **Finish** : Precast concrete, Terrazzo floor, Pandomowall

위치 : 전라북도 전주시 덕진구 장동 1090, 1110번지 I 용도 : 교육문화시설 I 규모 : 지하1층, 지상4층 I 구조 : 철근콘크리트 I 마감 : 프리캐스트 콘크리트, 테라쪼, 판도모월 I 발주처 : 전라북도 I 설계팀 : Yi architects / 장영수, 오준형 + ㈜종합건축사사무소에이그룹 / 권현철

Site plan

■ Circulation Plan

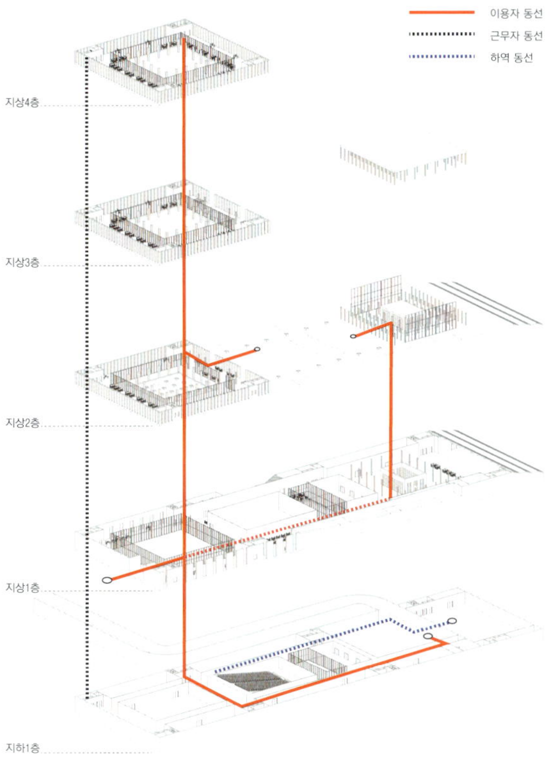

■ Landscape Concept

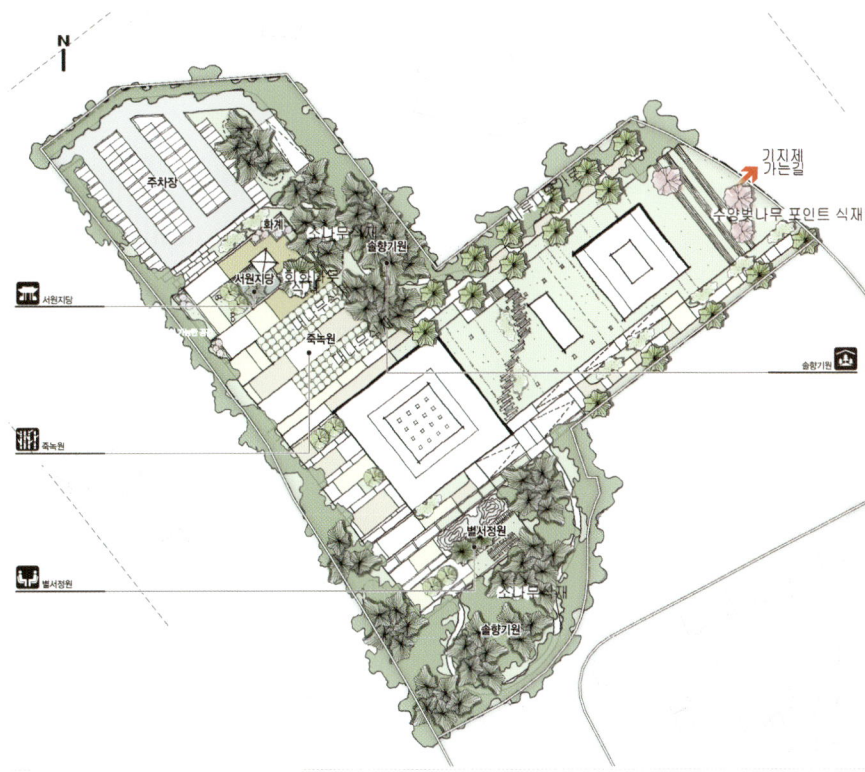

Winner _ 당선작

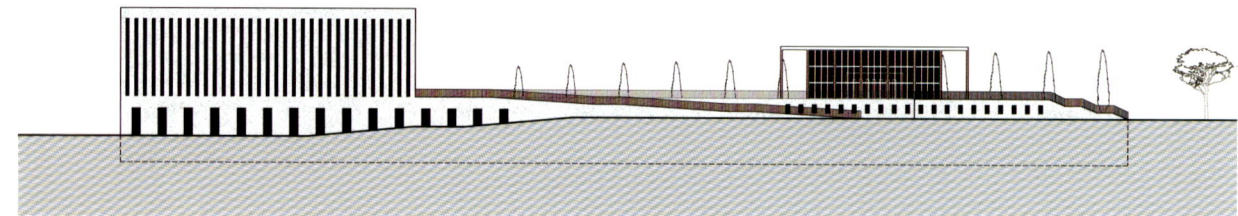

East elevation

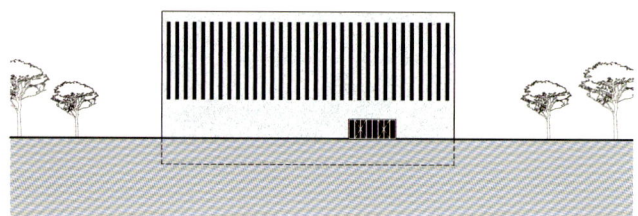

South elevation

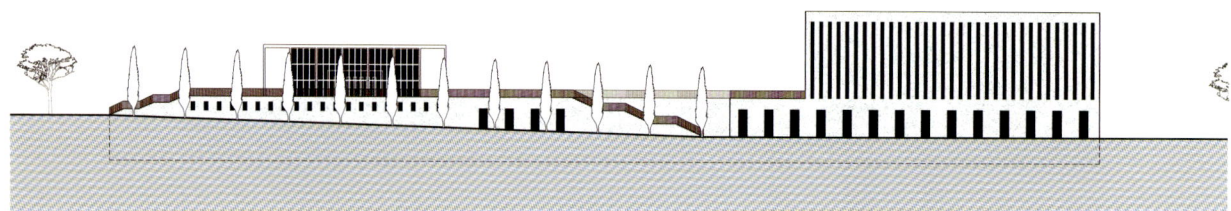

West elevation

Winner _ 당선작

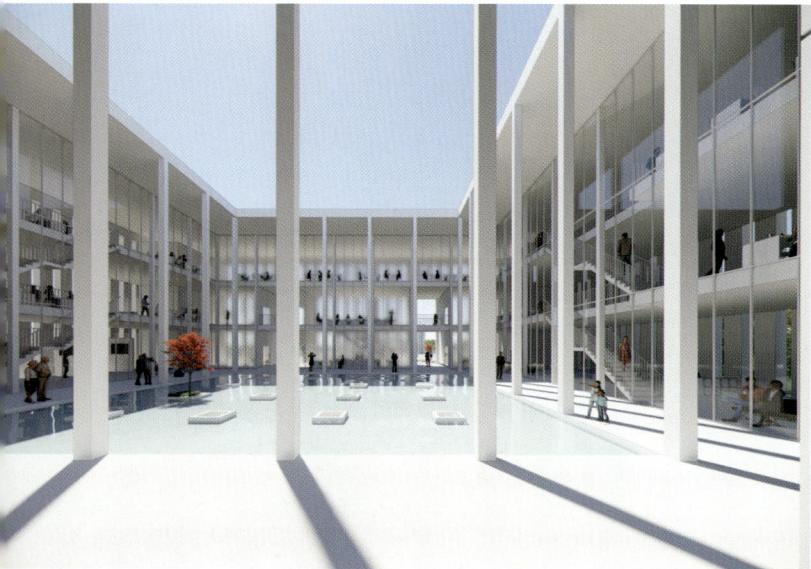

1. Warehouse
2. Machine room / Electrical room
3. Senior / Disabled reference
4. Innovation lab
5. VR experience center
6. Childhood / Story space
7. Children's department
8. General data room
9. Club room
10. Parking
11. Cooperative storage facility
12. Entrance space
13. Multipurpose room / AV room
14. Classroom
15. Jeonbuk history archive
16. Jeonbuk studies archive
17. Lounge / Restaurant

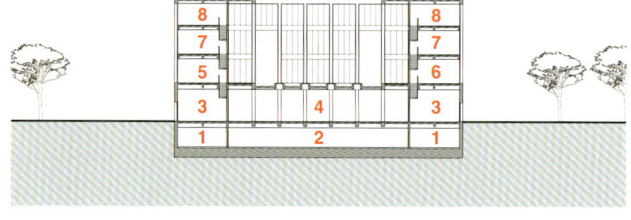

Section I

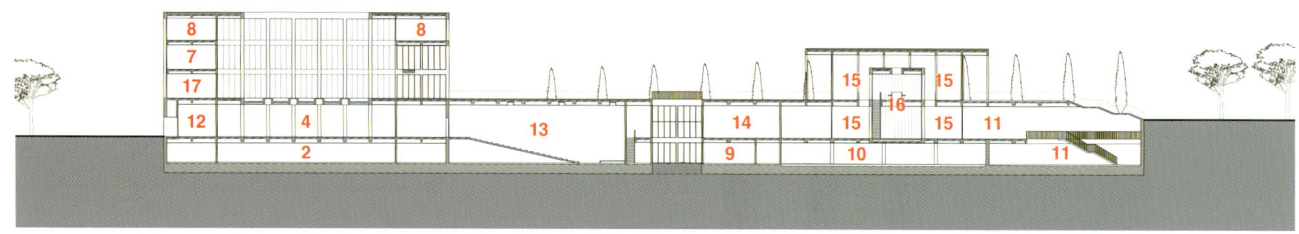

Section II

Jeollabuk-do Representative Library >>

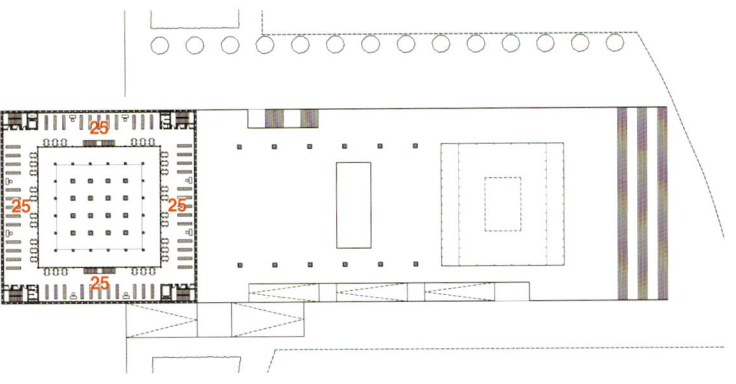

4th floor plan

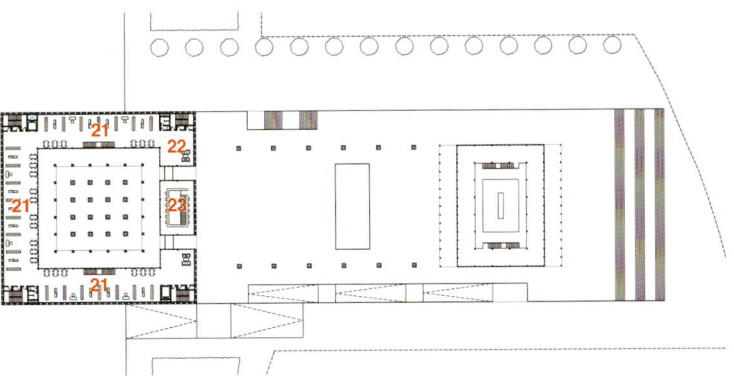

3rd floor plan

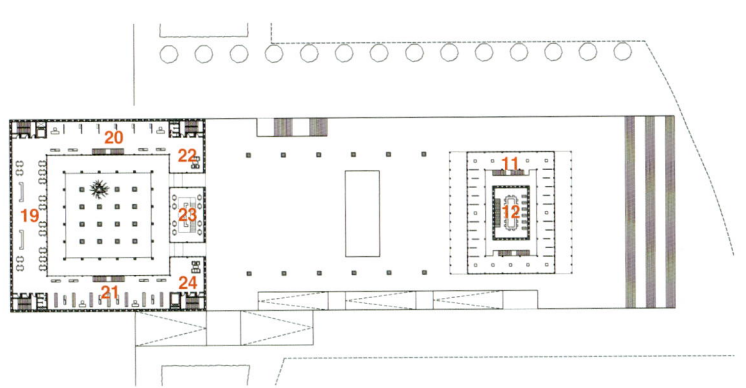

2nd floor plan

1st floor plan

1. Information desk
2. Book rental
3. Book returning
4. Senior / Disabled reference
5. Innovation lab
6. Volunteer room / Lounge
7. Multipurpose room / AV room
8. Free learning room
9. Classroom
10. Office
11. Jeonbuk history archive
12. Jeonbuk studies archive
13. Director's office
14. Meeting room
15. Document storage room
16. Server room
17. Cooperative storage facility
18. Operation room
19. Lounge / Restaurant
20. VR experience center
21. Children's department
22. Rest space
23. Book cafe
24. Story space
25. General data room
    / Continual publication room

Seosan Central Library

서산 중앙도서관

ARCHITECTURE Espace / Seonghwan Park
㈜건축사사무소 에스파스 / 박성환

Find a road in the library

The starting point of the design is the 'road'. A 'road' does not appear because it already exists, but appears spontaneously as one person passes by and another person passes by. In other words, while the artificial 'road' is a concept of border and control, the natural 'road' is not only a link that connects a purpose that has not yet been defined, but also a neutral space that does not represent the value of existence itself.

Based on this concept, a 'library' is a question of oneself about a purpose that does not exist from a simple desire and inquiry for knowledge, and it creates various sharing and acquisition through conversation with others.

Seosan Central Library that draws a three-dimensional village road

The reading culture provides an opportunity for encounters and is like a space for mental healing to experience the emptied time. This starts from the interpretation of the new overlapping courtyard combined with the nature that we encounter while walking by overlapping various spaces.

According to the law of nature, which should be empty in order to be filled, an empty space is filled with time. Therefore, by cultivating the open space adjacent to the target site as a forest embracing nature, and establishing a continuous and close relationship between the building and the user's body movements, the composition of functional space follows one continuous 'road' from outside to inside to outside, and induces the coexistence of various activities of staying and passing. Therefore, we suggest a new way of reading and communicating.

Create a link between local flow and nature with thorough site analysis

Boundary has the dictionary meaning of division, but the flow and behavior of various people can be made between two adjacent areas through a boundary. The site is located at the boundary between the central park, which connects nature and the city, and various low-rise and high-rise residential complexes and neighborhood living facilities.

Seosan Central Park has four roads that are the daily flow of people: 'Pine Forest Road' for healing, 'Community Road' for exhibition and relaxation, 'Health Road' for exercise and training, and 'View Road' overlooking Central Park Lake.

Through these roads, the nodes of various pedestrian flows and the surrounding living streets become the site. Therefore, by presenting a three-dimensional walking platform that actively brings the various daily flows and surrounding living streets of the park to the site, it provides a cultural landmark as a communication space that connects culture, education, and leisure networks.

Location : 1255-1, Yecheon-dong, Seosan-si, Chungcheongnam-do, Korea | **Function** : Education & Research facility | **Site area** : 5,000m² | **Bldg. area** : 2,959m² | **Total floor area** : 7,336m²
Stories : B1, 5FL | **Structure** : Reinforced concrete, Steel frame | **Finish** : Red brick, Low-E pair glass, U-glass, Old brick

위치 : 충청남도 서산시 예천동 1255-1 | 용도 : 교육연구시설 | 규모 : 지하1층, 지상5층 | 구조 : 철근콘크리트, 철골 | 마감 : 적벽돌, 로이복층유리, 유글라스, 고벽돌 | 설계팀 : 김은지, 임희영, 박성찬, 박현수

Site plan

도서관에서 길을 묻다

본 설계의 출발점은 '길'이다. '길'이란 이미 존재하여 나타나는 것이 아니라 한 사람이 지나고 또 다른 사람이 지나며 자연발생적으로 나타나는 것이다.
즉, 인위적 '길'이란 경계와 통제의 개념이라면 자연적 '길'이란 아직 규정되지 않은 목적을 연결하는 고리일 뿐, 그 자체로서 존재의 가치를 나타내는 것이 아닌 중성의 공간이다.
이와 같은 개념을 토대로 '도서관'이란 단순한 지식의 갈구와 탐구로부터 존재하지 않는 목적에 대한 스스로의 질문이자 타자와의 대화를 통한 다양한 나눔과 얻음일 것이다.

입체적 마을길을 그리는 서산 중앙도서관

독서의 문화는 조우하는 기회를 제공하고 이를 통한 비워진 시간을 경험하는 정신적 치유의 공간이다. 이는 다양한 공간의 중첩을 활용하여 거닐며 우연히 마주하는 자연과 결합된 새로운 감각의 중첩된 중정의 해석으로부터 출발한다.
채워지기 위해서는 비워져야 하는 자연의 법칙과 같이 비워진 공간은 시간을 담는 채워진 공간이다. 따라서 본 설계에서는 대상지와 인접한 공지를 자연을 품는 숲으로 경작하고 본 건물과 사용자의 신체 움직임에서 연속적이고 밀접한 관계성을 구축함으로써, 기능적 공간의 구성을 외부-내부-외부로 연속되는 하나의 '길'을 따라 머무르는 동시에 지나가는 다양한 행위가 공존하며 나타나도록 유도함으로써, 새로운 방식의 책읽기와 소통의 방법을 제시하고자 한다.

명확한 대지분석을 통한 지역의 흐름과 자연의 연결고리 형성

경계는 구분이라는 사전적 의미를 지니지만, 경계를 둠으로 인해 인접한 두 영역 사이에서 다양한 사람들의 흐름과 행위가 일어나게 하는 계기가 되기도 한다.
본 대지는 자연과 도시의 사이인 중앙공원과 다양한 저층, 고층 주거단지와 근린생활시설이 위치한 사이의 경계에 위치한다.
서산중앙공원은 주변 사람들의 일상의 흐름인 네 가지 길이 존재한다.
첫째, 사람들의 사람들이 힐링되는 '소나무 숲 길'. 둘째, 전시와 휴게가 가능한 '커뮤니티 길'. 셋째, 운동과 트레이닝이 가능한 '건강의 길'. 넷째, 중앙공원호수를 바라보는 '조망의 길'이다.
이러한 길을 통해 다양한 보행의 흐름과 주변 생활 가로와 만나는 결절점이 우리의 대지이다. 따라서 주변에 존재하는 공원의 다양한 일상의 흐름과 생활 가로를 부지에 적극적으로 끌어들이는 입체적 보행 플랫폼을 제시함으로써 문화, 교육, 여가 네트워크를 잇는 소통공간으로 문화적 랜드마크 공간을 제공한다.

Winner _ 당선작

Design Strategy

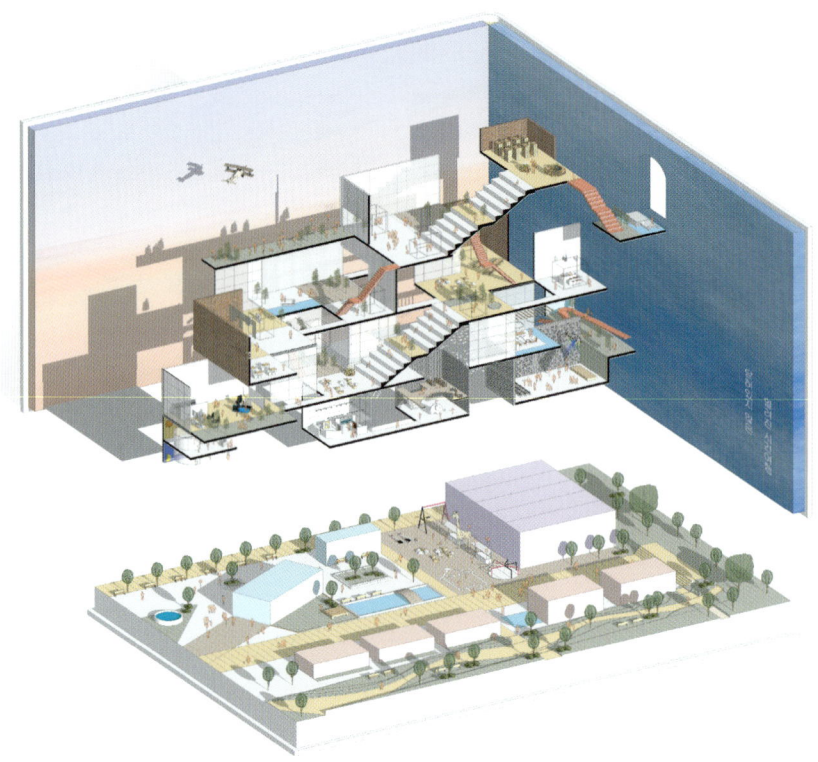

Design Description

- Multiple space between where various actions coexist

- Link to lake park through an open window

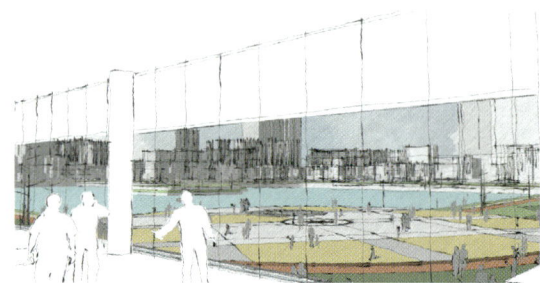

Seosan Central Library >>

East elevation

South elevation

North elevation

Winner _ 당선작

1. Machine room	8. Child reading room	15. Cultural classroom	22. Continual publication
2. Corridor	9. Communication lounge	16. Community forest	23. Observatory deck
3. Cooperative storage facility	10. Multipurpose room	17. Outdoor rest deck	24. Open reading room
4. Volunteer waiting room	11. Studio	18. Meeting room	25. Individual study space
5. Lobby	12. Teenager space	19. Adult reading room	26. Reading step
6. Parents waiting room	13. Outdoor garden	20. Hall	27. Comprehensive reading room
7. Wait lounge	14. Showroom / Seminar	21. Data tabulation / Stockroom	

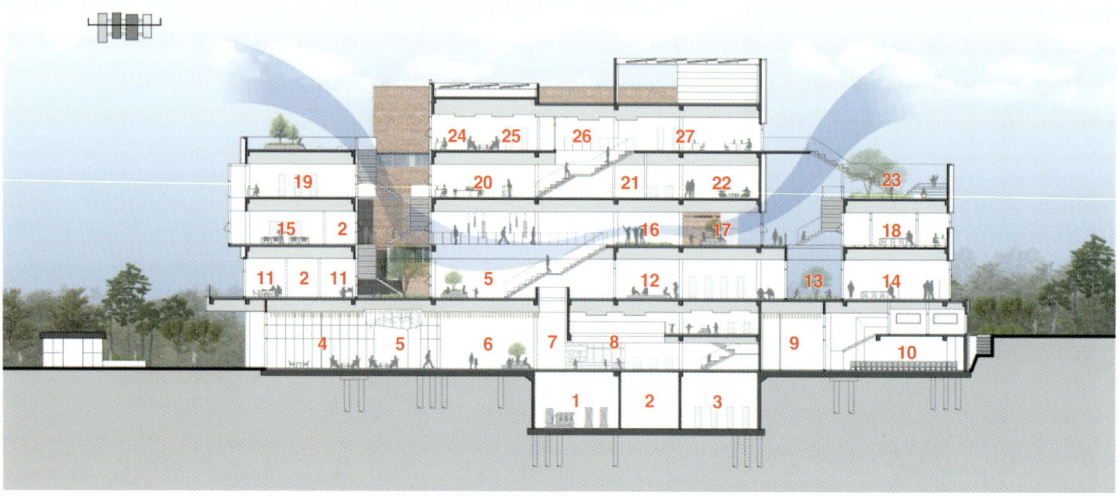

Section

Seosan Central Library >>

5th floor plan

1. Lobby
2. Kids cafe
3. Mom cafe
4. Indoor playground
5. Child playbook library
6. Imagination step
7. Office
8. Children's cultural class
9. Shared kitchen
10. Data tabulation
11. Multipurpose room
12. Family bathroom
13. Public toilet
14. Craft workroom
15. Counseling space
16. Multipurpose space
17. Outdoor reading deck
18. Teenager carrel
19. Teenager media reading room
20. Community cafe
21. Showroom / Seminar
22. Running commons
23. Lecture preparation room
24. Cultural classroom
25. Cleaning service office
26. Community forest
27. Club
28. Computer room
29. Meeting room
30. Data reading room
31. Digital reading space
32. Hall
33. Data tabulation / Stockroom
34. Rural area data space
35. Small forest library
36. Carrel
37. Comprehensive reading room
38. Study room
39. Book cafe
40. Continual publication

3rd floor plan

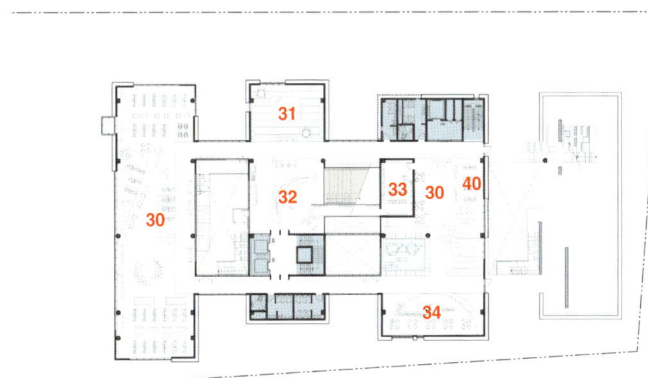

4th floor plan

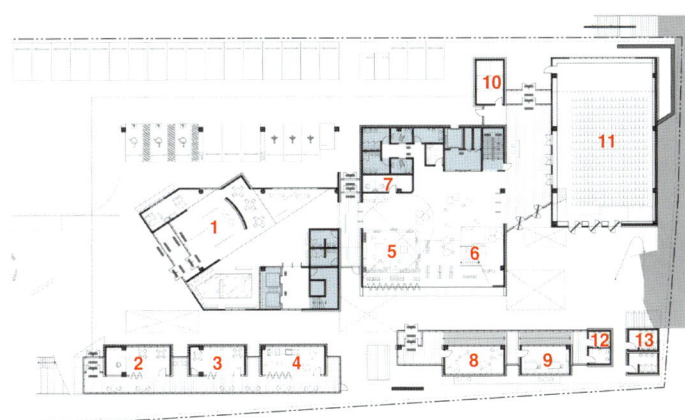

1st floor plan

2nd floor plan

Winner _ 당선작

Sosabeol 2 Middle School
소사벌 2중학교

HANDeul ARCHITECTS & PLANNERS / Young-gun Kim
㈜한들종합건축사사무소 / 김영근

In the district unit plan of the Sosabeol District of Pyeongtaek, the number of students is rapidly increasing every year, and nearby middle schools are accommodating students by installing temporary modular classrooms on the existing school grounds, and accordingly, an additional middle school is needed. The business site is currently surrounded by apartment houses and is the center of the Sosabeol District in Pyeongtaek, accessible from various directions. Through the newly built Sosabeol 2 Middle School, it is intended to provide a new educational and cultural center space where students and local residents can study and exchange freely.

A 'Cluster Square' creates a convenient learning environment for students through clustering and home-based arrangement of facilities with similar subject characteristics. A 'Community Square' creates a school open to students and local residents through the intensive arrangement of a square complex around a Community Street. Various external space plans such as community gardens and meeting yards considering openness and facility comfort creates 'Green Square', an eco-friendly educational space. It proposes a 'Dream Square' to become a new educational and cultural space where students and local residents form a village community by providing a better learning environment.

In a special plan for Gyeonggi Future School, a square complex is intensively arranged to facilitate various exchanges. With the low-rise arrangement of the facility open to the residents, a space plan is made for both students and local residents to hold various events such as exhibitions and events. In addition, the eco-friendly school forest provides a pleasant learning environment for students and a resting space for local residents by organically connecting with the nature of the neighborhood park on the southern side of the site.

평택 소사벌지구 지구단위계획구역 내에는 현재 매년 학생수가 급증해 인근 중학교에서는 기존 학교 운동장에 임시 모듈러 교실을 설치하여 학생들을 수용하고 있으며 이에 따라 중학교 추가 신설이 필요한 상황이다. 사업부지는 현재 주변이 공동주택으로 둘러싸여 있으며, 다양한 방향에서의 접근이 이루어지는 평택 소사벌지구의 중심지이다. 새롭게 신축되는 소사벌2중 신축을 통해 학생들과 지역주민이 자유롭게 공부하고 교류하는 새로운 교육문화 중심공간을 제공하고자 한다.

교과특성이 비슷한 시설을 클러스터화 및 홈베이스 중심배치를 통해 학생들의 편리한 학습환경을 조성한 'Cluster Square', 커뮤니티스트리트를 중심으로 복합형 광장시설 집중배치로 학생과 지역주민에게 열린 학교가 되는 'Community Square', 개방감 및 시설쾌적성을 고려한 커뮤니티정원, 만남의 마당 등의 다양한 외부공간계획으로 친환경 교육공간인 'Green Square'계획으로 보다 나은 학습환경을 제공하여 학생 및 지역주민이 마을공동체를 이루는 새로운 교육문화공간이 되는 'Dream Square'를 제안하였다.

경기미래학교 특화계획으로 복합형 광장시설을 집중 배치하여 다양한 교류가 일어나도록 계획하였고 주민개방시설의 저층배치로 학생 및 지역주민 모두가 전시 및 행사 등의 다양한 이벤트가 가능한 공간계획을 하였다. 또한 친환경 학교숲은 대상지 남측의 근린공원의 자연과 유기적으로 연계되어 학생들의 쾌적한 학습환경 및 지역주민들의 휴게공간을 제공하였다.

Location : 1010, Bijeon-dong, Pyeongtaek-si, Gyeonggi-do, Korea I **Function** : Education & Research facility I **Site area** : 11,859㎡ I **Bldg. area** : 3,788㎡ I **Total floor area** : 11,345㎡ I **Stories** : B1, 5FL I **Structure** : Steel frame, Reinforced concrete I **Finish** : Stone, Clay brick, Cement-based wide brick, Metal panel, Color steel plate

위치 : 경기도 평택시 비전동 1010 I 용도 : 교육연구시설 I 규모 : 지하1층, 지상5층 I 구조 : 철골, 철근콘크리트 I 마감 : 석재, 점토벽돌, 시멘트계와이드벽돌, 금속패널, 컬러강판 I 설계팀 : 하홍원, 임준혁, 김지수, 김진호, 김태우

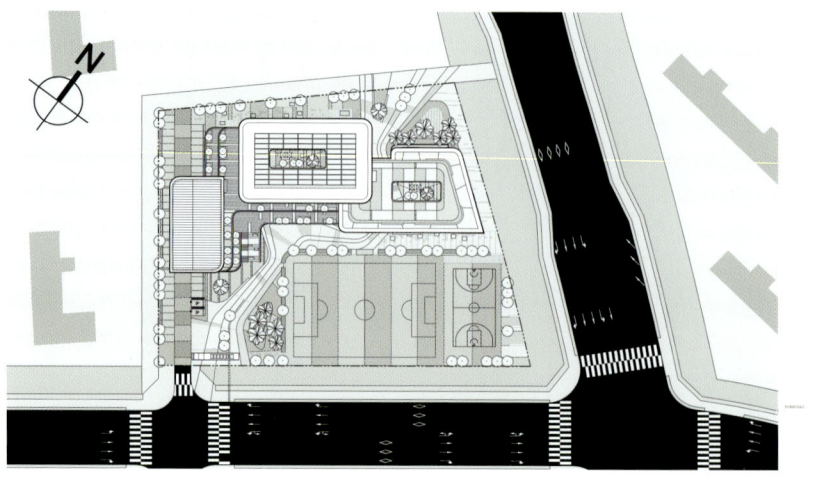

Site plan

■ Concept

- **Cluster square**

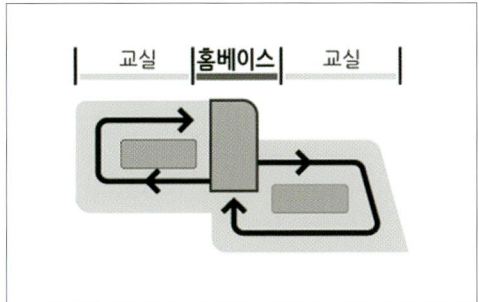

– 교과특성이 비슷한 시설을 클러스터화 및 홈베이스 중심배치를 통해 학생들의 편리한 학습환경 조성

- **Community square**

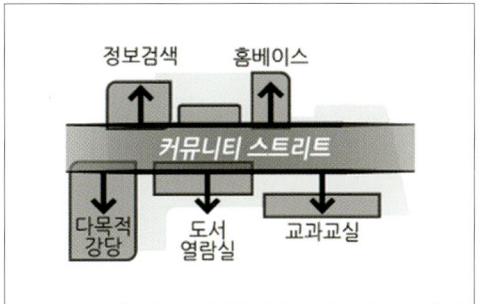

– 커뮤니티스트리트를 중심으로 한 복합형 광장시설 집중배치로 학생과 지역주민에게 열린 학교 조성

- **Green square**

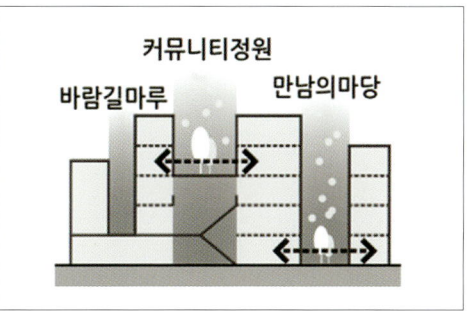

– 개방감 및 시설쾌적성을 고려한 커뮤니티정원, 만남의 마당 등의 다양한 외부공간 계획으로 친환경 교육공간 조성

■ Space Plan

- **Complex square facility**

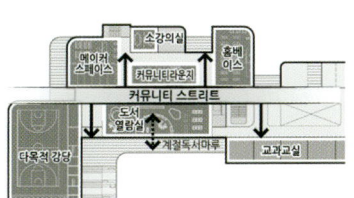

– 커뮤니티스트리트를 중심으로 실내광장시설을 집중 배치하여 다양한 교류가 일어나는 복합 커뮤니티공간 조성

- **Resident open facility**

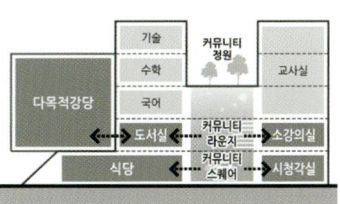

– 커뮤니티스퀘어를 중심으로 지역주민 개방시설 배치하여 학생, 지역주민의 전시 및 행사 등의 다양한 이벤트가 가능한 공간계획

- **School forest plan to extend the nature of neighborhood parks**

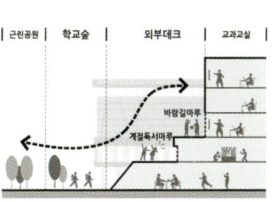

– 대상지 남측 근린공원의 자연을 연장하여 조성된 학교숲 계획으로 학생들의 쾌적한 학습환경 및 지역주민들의 휴게공간을 제공

- **School forest detailed space plan**

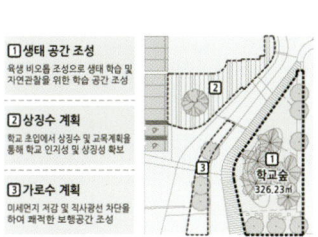

– 기능군 성격에 따라 독립적이면서도 상호 유기적으로 연계되는 학교숲 계획을 통해 편안한 힐링공간 제공

Winner _ 당선작

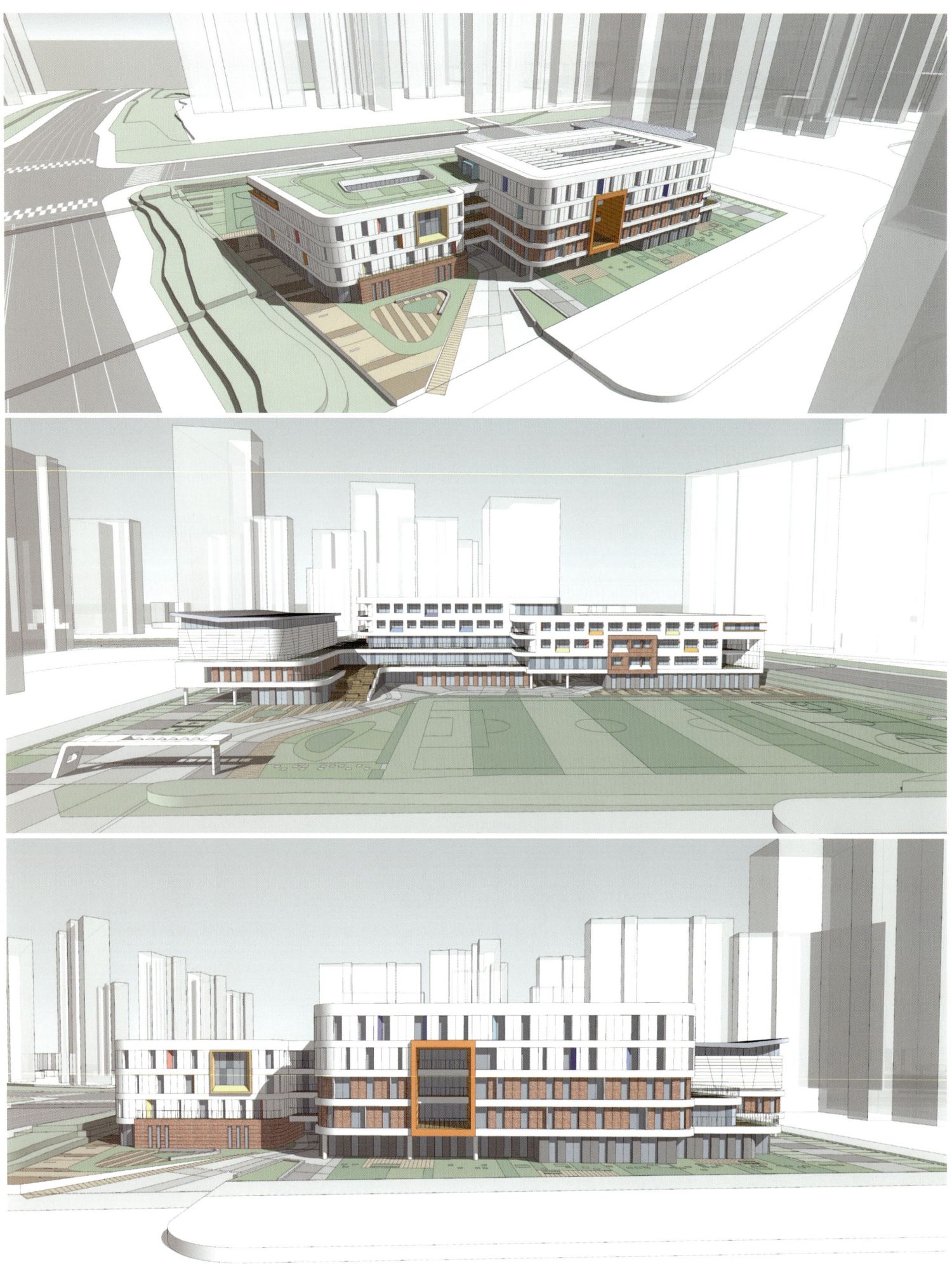

Sosabeol 2 Middle School

■ Floor Plan

- Special school affairs / care space considering safe school life

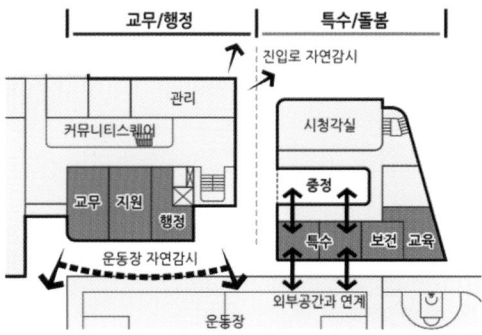

- 교무 / 행정영역 운동장 인접배치로 자연감시 및 사고 예방

- Open dining plan for multipurpose use

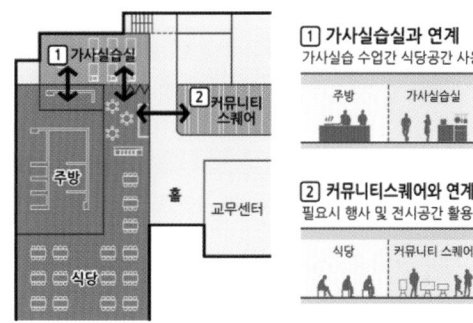

- 가사실습실과 인접배치로 학습공간 연계 및 확장가능

- A library plan to become the central space of a complex plaza facility

- 컴퓨터 교육실-커뮤니티 라운지-홈베이스와 연계하여 정보검색, 토론, 전시 등이 가능한 소통공간 제공

- Various community space planning through open hall

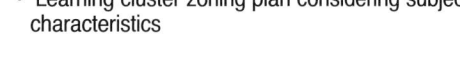

- 홀과 연결된 커뮤니티스퀘어-커뮤니티라운지 계획으로 접근이 편리하며 자유로운 소통이 가능한 열린공간 조성

- Learning cluster zoning plan considering subject characteristics

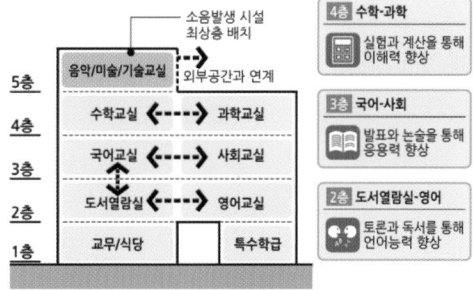

- 교과특성 분석을 통한 협업교과 인접배치로 맞춤형 교육이 가능한 효율적인 학습 클러스트 구성

- Smart classrooms responding to the characteristics of each subject and future-oriented education

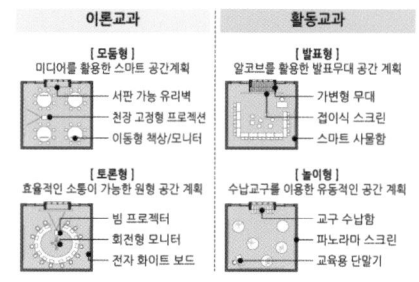

- 교과특성을 반영하여 창의적인 수업을 지원하는 교실유닛 계획

■ Section Plan

- Pleasant interior and exterior rest area plan for students

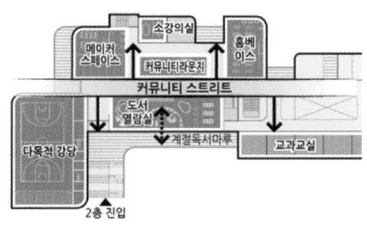

- 내·외부를 유기적으로 연계하여 쾌적한 교육활동 지원 및 편안한 휴식공간 제공

- Various movement lines considering the convenience of users

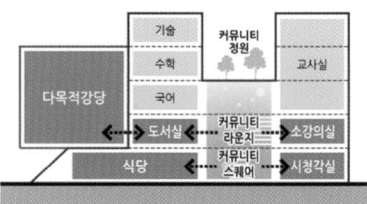

- 시설별 별도출입 계획으로 이용자 편의성 확보

- Appropriate floor height plan considering the characteristics of each room

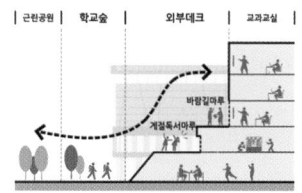
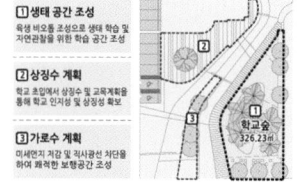

- 용도별 적정 층고계획으로 사용성, 경제성 확보

Winner _ 당선작

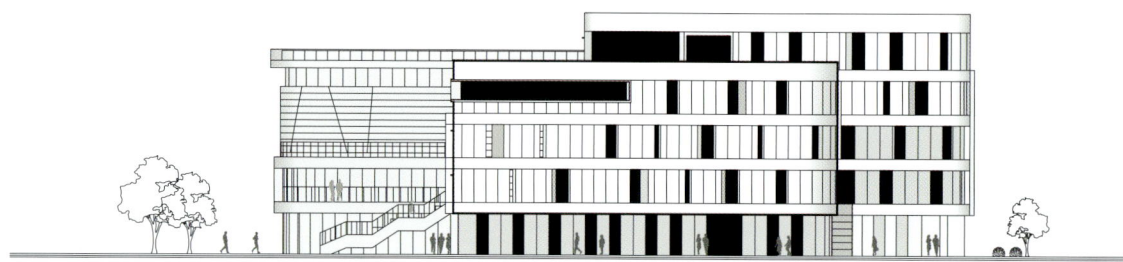

East elevation

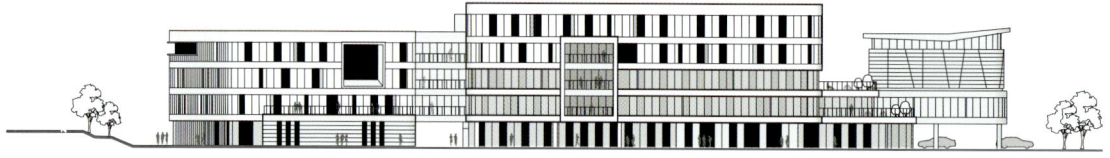

North elevation

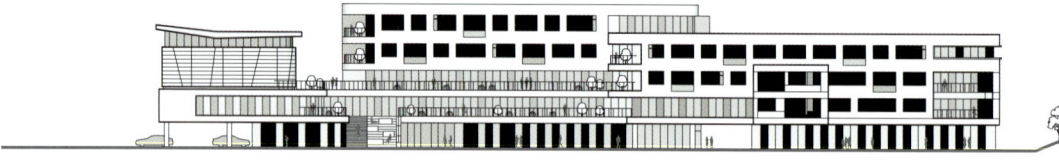

South elevation

West elevation

1. Water tank room
2. Community square
3. Staff lounge
4. Reading room
5. Community lounge
6. Lecture room
7. Classroom
8. Teacher's office
9. Kitchen
10. Restaurant
11. Hall
12. Multipurpose auditorium
13. Makerspace
14. Home base

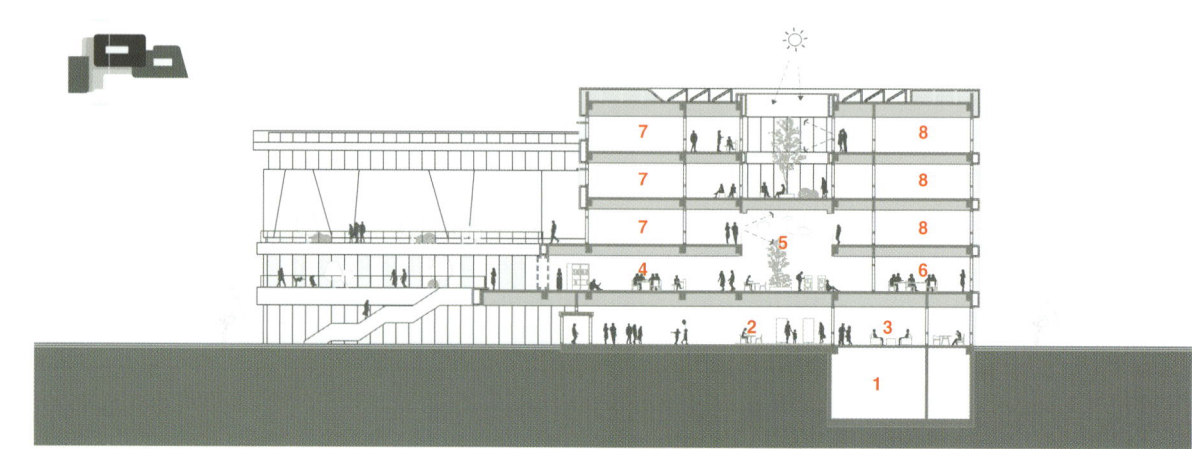

Longitudinal section

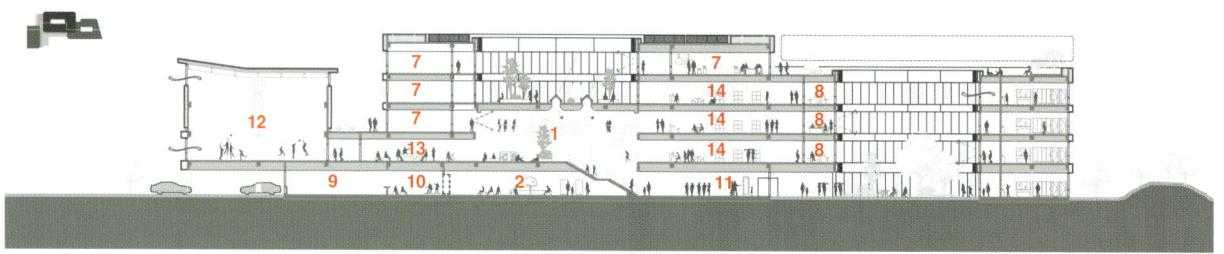

Cross section

Sosabeol 2 Middle School >>

1. Restaurant
2. Kitchen
3. Community square
4. Faculty lounge
5. Education center
6. Administrative office
7. AV room
8. Special classroom
9. School clinic
10. Multipurpose auditorium
11. Reading room
12. Computer training room
13. Home base
14. Classroom
15. Science laboratory
16. Technical laboratory
17. Art room
18. Music room

4th floor plan

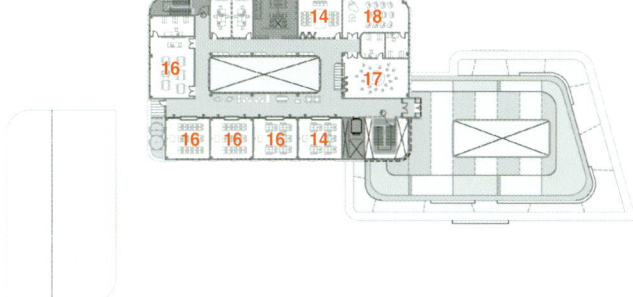

5th floor plan

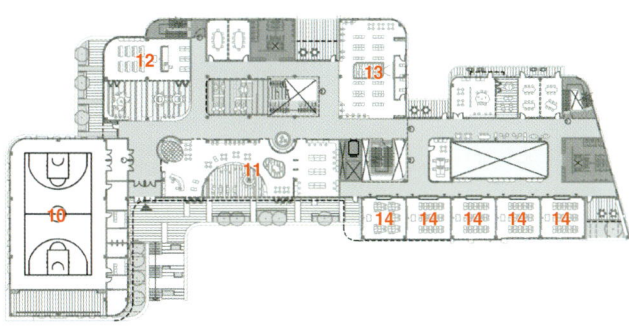

2nd floor plan

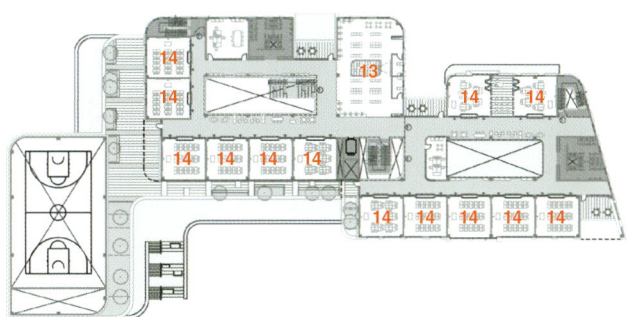

3rd floor plan

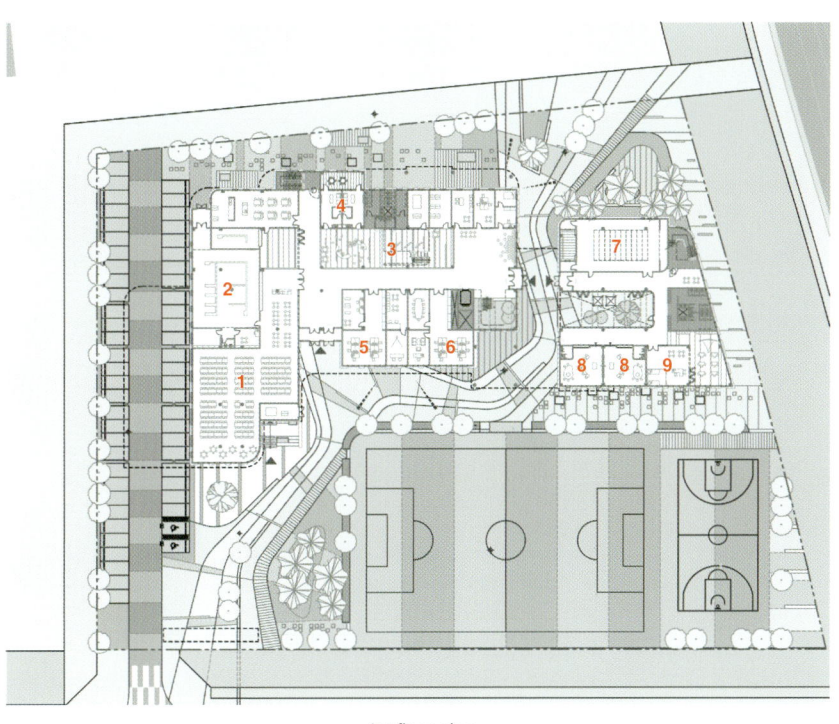

1st floor plan

Winner _ 당선작

Bucheon Okgil Middle & High School Integrated Management School
부천 옥길 중·고 통합 운영학교

Sun and Partners / Yongmin Lee
㈜종합건축사사무소 선기획 / 이용민

Newly built in Bucheon, Gyeonggi-do, Bucheon Okgil Middle and High School Integrated School is a future-oriented and student-centered school in connection with surrounding facilities.
It proposes indoor squares and school forests for various activities such as exchange, learning, play, and rest by transforming from an existing uniform school space into a student-led autonomous space, and it intends to induce the community and create a place that becomes the center of the community.

The site of Bucheon Okgil Middle and High School Integrated School is located between the Okgil District and Beombak District in Bucheon, with Sataemal Byeolbit Park on the right, a natural green space on the south side, and Byeolbit Maru Library about 5 minutes away. Accordingly, the arrangement alternatives are selected in consideration of various school forests, square-type space arrangement in the center, south-facing arrangement of classrooms, safe separation of pedestrians and vehicles, and connection to parks and libraries, and linkage to external spaces and facilities.

In a specialized plan, we will create a GREEN (school forest) that provides a space for students to develop their emotions and relax through the school forests connecting various external spaces, and SMART (smart classroom), a smart and flexible learning space, is designed through the terrace and alcove plan linked to the classroom. A management and safety system considering openness to residents and a parking lot and separate entrances considering open facilities create a SHARING (open facilities). Three-dimensional indoor squares such as a learning center, rest area, and step reading space is created by placing SPACE (blooming square) at the center of various programs.

경기도 부천에 새롭게 들어서는 부천옥길 중·고 통합학교는 주변 시설과 연계하여 미래지향적이고 학생 중심의 학교이다.
기존 획일화된 학교공간에서 벗어나 학생들이 주체가 되는 자율적 공간으로 변모되어 교류와 학습, 놀이, 휴게 등 다양한 활동이 일어나는 실내광장 및 학교 숲을 제안하여 커뮤니티를 유도하고, 지역사회의 중심이 되는 장소로서 함께하는 학교를 제안하였다.

경기도 부천옥길 중·고 통합학교의 사이트는 부천 옥길지구와 범박지구의 사이에 위치해 있으며, 우측에는 사태말별빛공원, 남측에는 자연녹지가 위치해 있으며, 약 5분 거리에 별빛마루도서관이 위치해 있다. 이 점들을 고려해 다양한 학교 숲, 중심에 광장형 공간배치, 교과교실의 남향배치, 안전한 보차분리, 공원 및 도서관과 외부 및 시설 연계를 고려하여 배치대안을 선정하였다.

특화계획으로는 다양한 외부 공간 사이를 이어주는 학교 숲을 통한 학생의 정서함양 및 휴게 공간 제공하는 GREEN[학교 숲]을, 교과교실과 연계한 테라스, 알코브 계획으로 스마트하고 유연한 학습공간으로 SMART[스마트교실]을, 주민개방을 고려한 관리·안전 시스템 및 개방시설을 고려한 주차장 및 별도출입구 계획함으로 SHARING[개방시설]을 그리고 다양한 프로그램의 중심공간에 블루밍스퀘어를 계획하여 러닝센터, 휴게공간, 스텝독서공간 등 입체적인 실내광장 조성하여 SPACE[블루밍스퀘어]을 제안하였다.

Location : 712-2, Okgil-dong, Bucheon-si, Gyeonggi-do, Korea l **Function** : Education & Research facility l **Site area** : 14,007㎡ l **Bldg. area** : 4,264㎡ l **Total floor area** : 14,011㎡ l **Stories** : B1, 4FL l **Structure** : Steel frame, Reinforced concrete l **Finish** : Clay brick, White clay brick, Wood louver, Low-E pair glass

위치 : 경기도 부천시 옥길동 712-2번지 일원 l **용도** : 교육연구시설 l **규모** : 지하1층, 지상4층 l **구조** : 철골, 철근콘크리트 l **마감** : 점토벽돌, 백토벽돌, 목재루버, 로이복층유리 l **건축주** : 경기도 부천 교육지원청 l **설계팀** : 김승태, 김태현, 이창연, 김종섭, 서수민, 이다솜, 김지현, 정상준

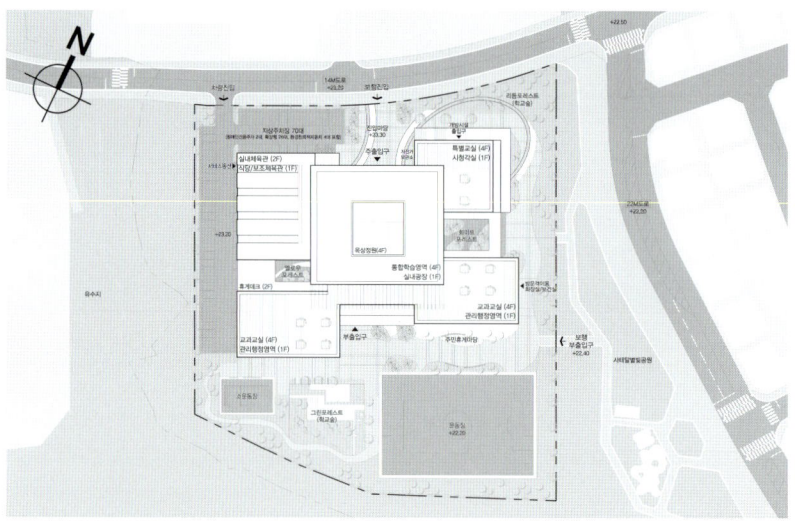

Site plan

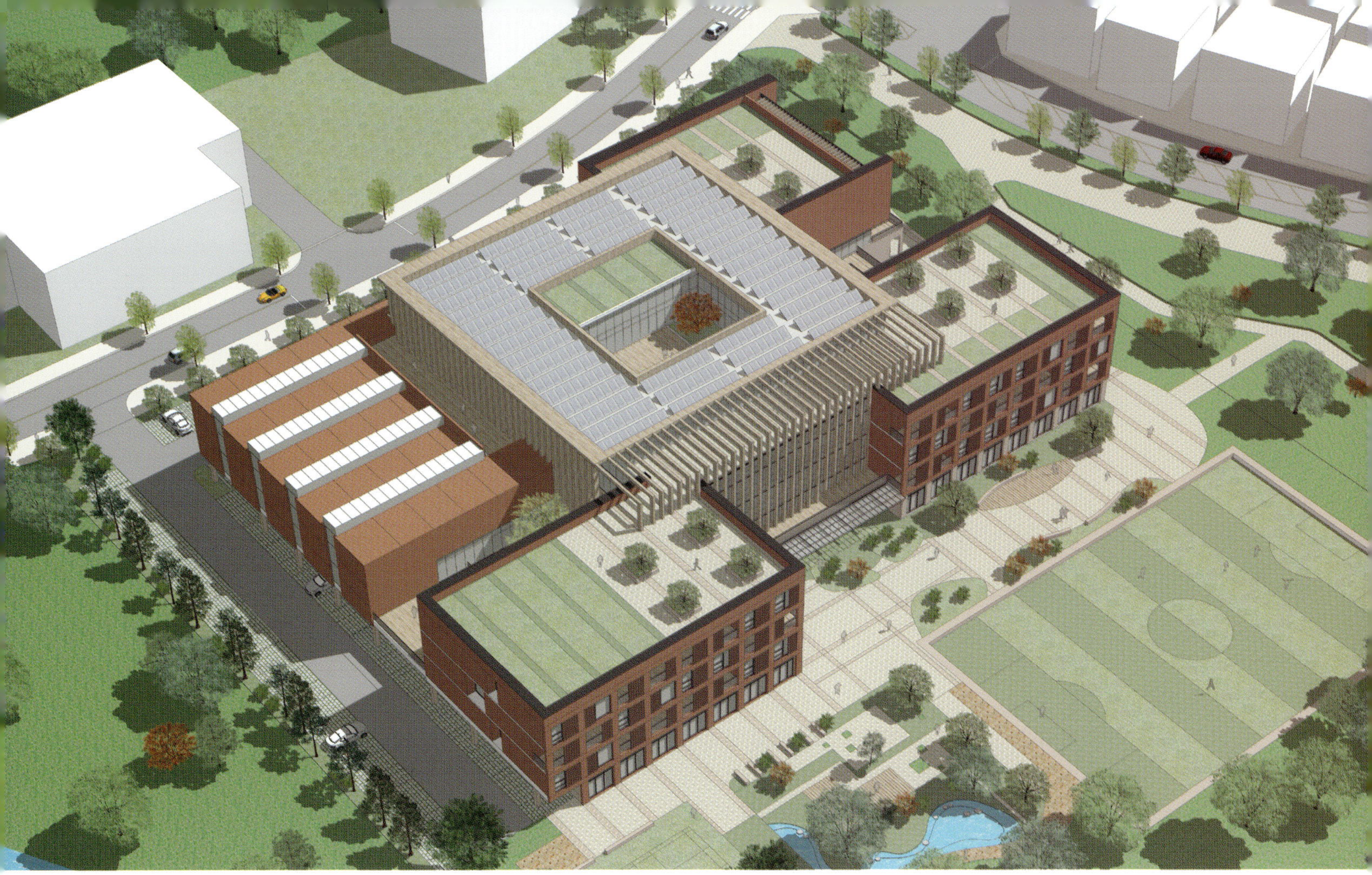

▌Site Plan

- Site plan by area considering use and function
- Various school forest plans linking the inside and the outside
- Classroom site plan considering the surrounding environment

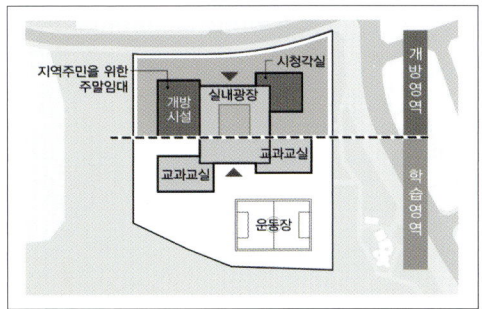

– 학습영역과 개방영역의 명확한 분리로 관리 효율 향상

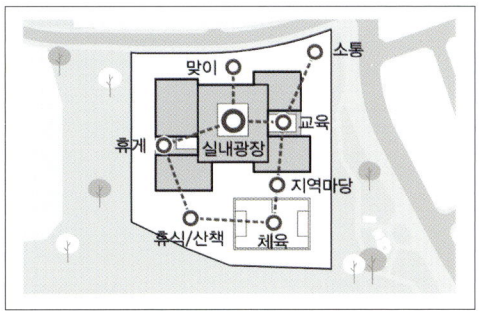

– 내·외부 공간의 유기적인 연계를 통한 미래공간 조성

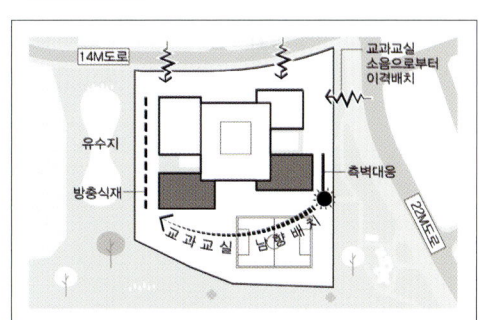

– 교과교실 남향 배치로 일조, 열린조망 확보

Winner _ 당선작

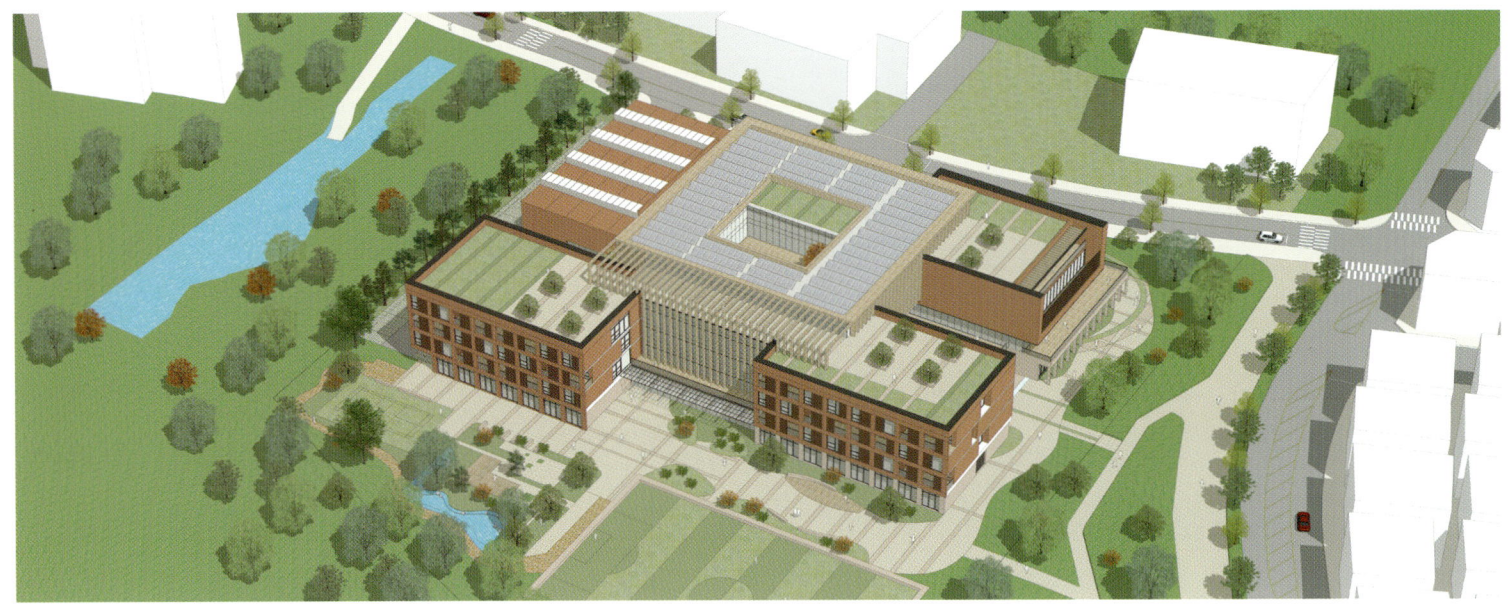

■ Concept Design

· School forest

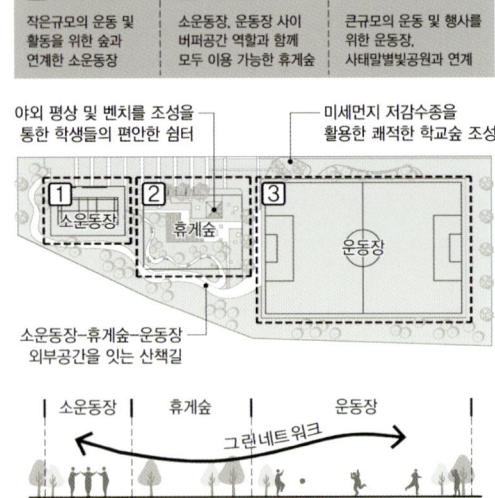

- 다양한 외부공간과 외부공간을 이어주는 학교숲을 통해 학생들의 정서함양 및 휴게공간 제공

· Smart class room

- 테라스, 알코브 계획을 통한 스마트하고 유연한 학습공간

· Sharing

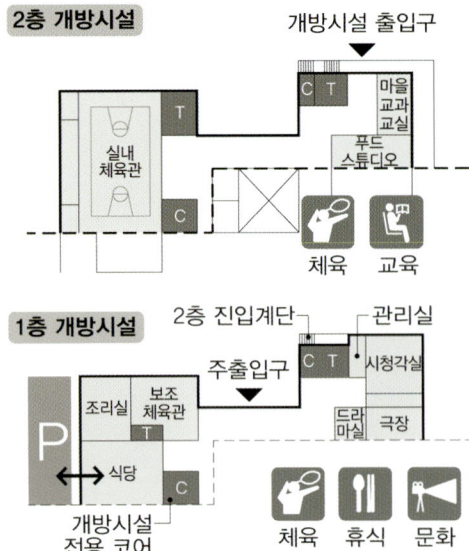

- 주말 주민개방을 고려한 관리·안전 시스템
- 개방시설의 주차장 인접배치 및 별도 출입구 계획

· Blooming square

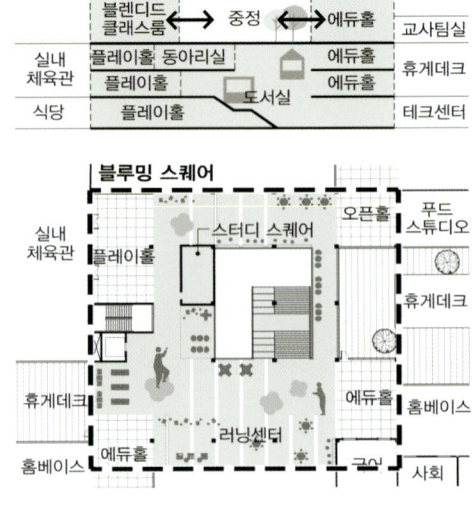

- 다양한 프로그램의 중심공간에 블루밍스퀘어 계획
- 러닝센터, 휴게공간, 스텝독서공간 등 다양한 활동

Bucheon Okgil Middle & High School Integrated Management School >>

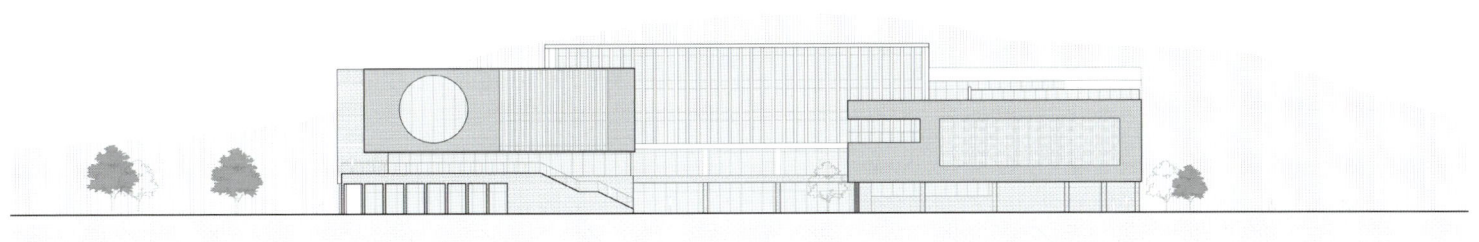

Front elevation

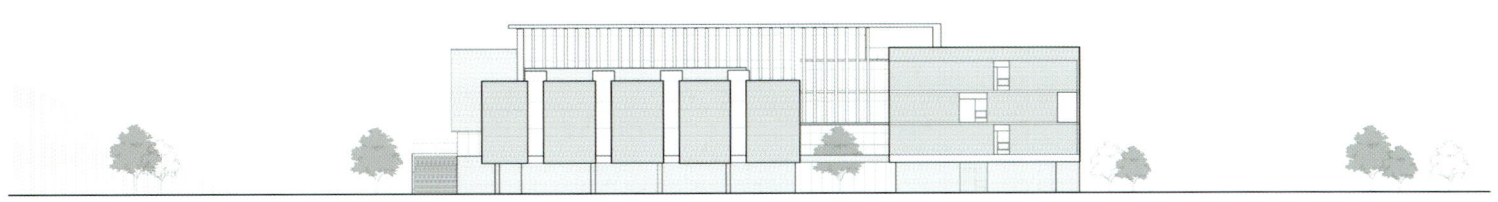

Left elevation

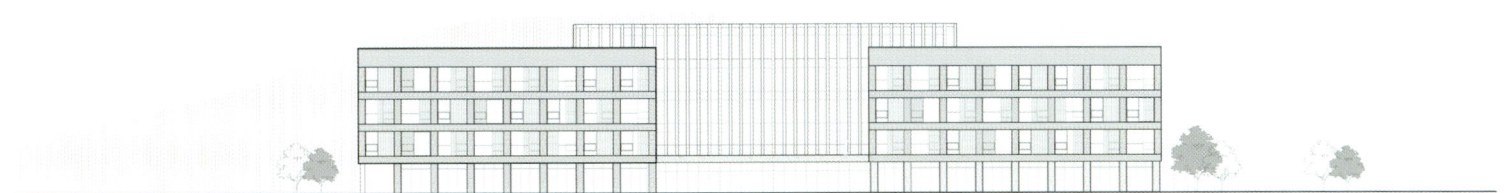

Rear elevation

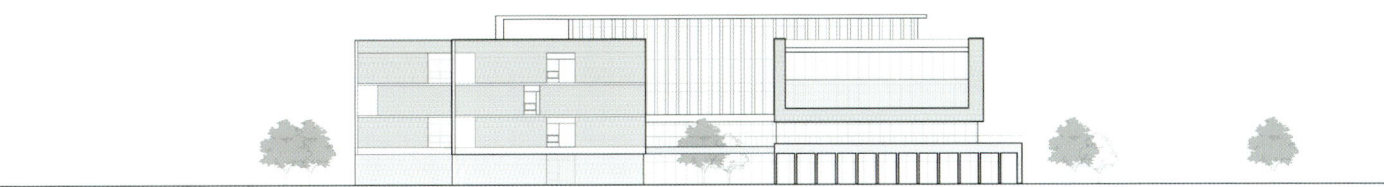

Right elevation

Winner _ 당선작

1. Counseling office
2. Social affairs division classroom
3. Homebase
4. Common classroom
5. Students self-government
6. Drama room
7. AV room
8. Food studio
9. Village classroom
10. Science room
11. Music room
12. Practice room
13. Art room
14. Restaurant
15. Placehall
16. Library
17. Makerspace
18. Tech center
19. Stair hall
20. Toilet
21. Gymnasium
22. Preparation room
23. Club room
24. Edu hall
25. Blended classroom
26. Teacher team room

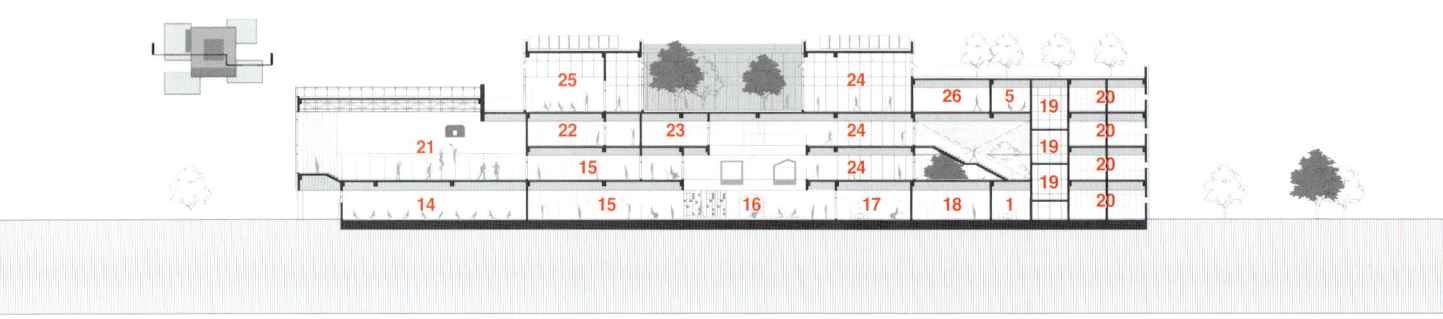

Cross section

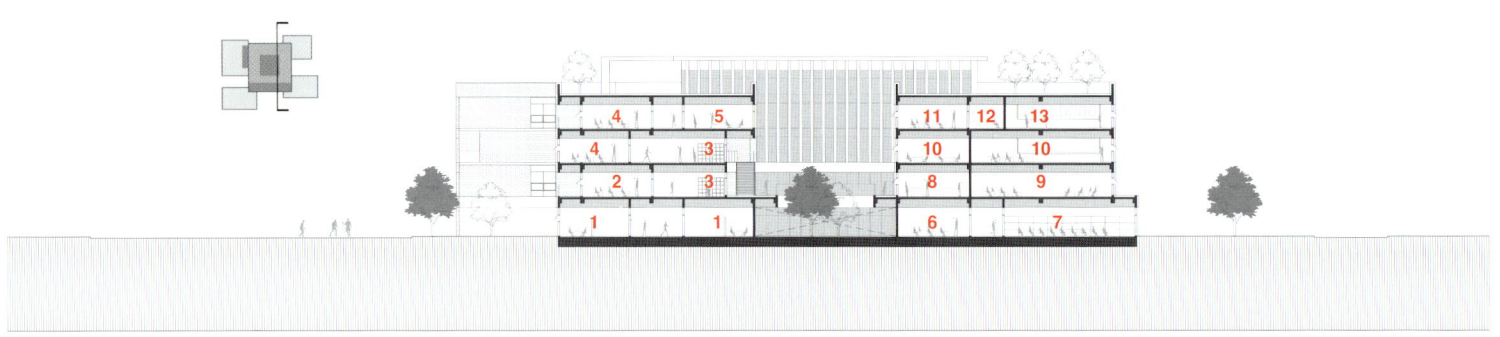

Longitudinal section

Bucheon Okgil Middle & High School Integrated Management School >>

1. Education center
2. Teacher lounge
3. Studio
4. Administration office
5. School clinic
6. Makerspace
7. Tech center
8. Restaurant
9. Auxiliary gym
10. Community hall
11. Drama room
12. Small cinema
13. AV room
14. Gymnasium
15. Classroom
16. Living lab room
17. Village classroom
18. Food studio
19. Changemakeroom
20. Teacher team room
21. Learning community council room
22. Student-led project room
23. Blended classroom
24. Art room
25. Music room

3rd floor plan

4th floor plan

1st floor plan

2nd floor plan

Winner _ 당선작

Janghang Elementary School
장항초등학교

Sun and Partners / Yongmin Lee
㈜종합건축사사무소 선기획 / 이용민

Newly built in the Janghang District in Goyang, Gyeonggi-do, Janghang Elementary School intends to form an education and community cluster in the Janghang District by becoming a community learning nest where nature, students, and residents are connected. It proposes indoor squares and school forests for various activities such as exchange, learning, play, and rest by transforming from an existing uniform school space into a student-led autonomous space. It suggests a 'Green Nest School' that embraces the region by creating a school forest and external space for each user with openness, a theme-type nature learning center flowing into the school, a smart complex square to provide a variety of learning, communication, and creative education space, and a three-dimensional house square.

Located in the center of a neighborhood park and residential area, a 'Green Nest School' is an open green smart school that embraces the local community and nature. It plans an open complex community indoor square that can be used variably where village open facilities, Janghang Elementary School, and local residents are connected. It proposes an eco-friendly themed school forest for each user, such as an independent and organically connected community, resident exchange, play rest, and ecological experience.

The community road of the village is designed by creating a community school forest road that connects the green axis of the city with a village layout plan similar in scale to the northern residential area. As a village connected along the school forest road, a buffer space is placed between the spaces to build an external space system in which various interiors and exteriors are blended according to their characteristics. To do this, the natural flow to a school forest road, a communication forest, a house square and a dream forest is planned to create an eco-friendly learning environment in which the inside and outside are naturally expanded and nature and education are harmonized.

경기도 고양의 장항지구에 새롭게 들어서게 될 장항초등학교는 자연과 학생, 주민이 연결되는 공동체 학습둥지가 되어 장항지구의 교육 및 커뮤니티 클러스터를 형성하고자 하였다. 기존 획일화된 학교공간을 벗어나 학생들이 주체가 되는 자율적 공간으로 교류와 학습, 놀이, 휴게 등 다양한 활동이 일어나는 실내광장 및 학교숲을 제안한다. 개방감 있는 이용자별 학교숲 및 외부공간, 학교 내부로 유입되는 테마형 자연학습장, 다채로운 배움과 소통, 창의적 교육공간을 제공하기 위한 스마트 복합광장, 입체적인 집집광장 조성을 통하여 지역을 품은 '그린 둥지학교'를 제안하였다.

근린공원과 주거지역의 중심에 위치한 '그린 둥지학교'는 지역사회와 자연을 품은 열린 그린스마트학교로, 마을개방시설-장항초-지역주민이 연결되는 가변 활용 가능한 열린 복합형 커뮤니티 실내광장 계획, 독립적이면서 상호 유기적으로 연결되는 커뮤니티, 주민교류, 놀이휴식, 생태체험 등 이용자별 친환경 테마형 학교숲을 제안하였다.

북측 주거지역과 비슷한 스케일을 가진 마을형 배치계획으로 도시의 녹지축을 연결하는 커뮤니티 학교숲길을 조성하여 마을의 커뮤니티 로드를 계획하였으며, 학교숲길을 따라 연결되는 하나의 마을로서 공간 사이에 버퍼공간을 배치하여 성격에 따라 다양한 내외부가 섞이는 외부 공간 체계를 구축하고자 하였다. 이를 위해 학교숲길-소통의숲-집집광장-꿈꿈숲으로의 자연흐름을 계획하여 내외부가 자연스럽게 확장되고 자연과 교육이 한데 섞이는 친환경 학습 환경을 조성하였다.

Location : 662-30, Janghang-dong, Ilsandong-gu, Goyang-si, Gyeonggi-do, Korea I **Function** : Education & Research facility I **Site area** : 14,822㎡ I **Bldg. area** : 5,577㎡ I **Total floor area** : 16,477㎡ I **Stories** : B1, 5FL I **Structure** : Steel frame, Reinforced concrete I **Finish** : Clay brick, Low-E pair glass, Stucco, Color steel plate

위치 : 경기도 고양시 일산동구 장항동 662-30번지 일원 I **용도** : 교육연구시설 I **규모** : 지하1층, 지상5층 I **구조** : 철골, 철근콘크리트 I **마감** : 점토벽돌, 로이복층유리, 스타코, 컬러강판 I **설계팀** : 김승태, 김태현, 이창연, 김종섭, 허장미, 박신영, 박가영, 홍석효

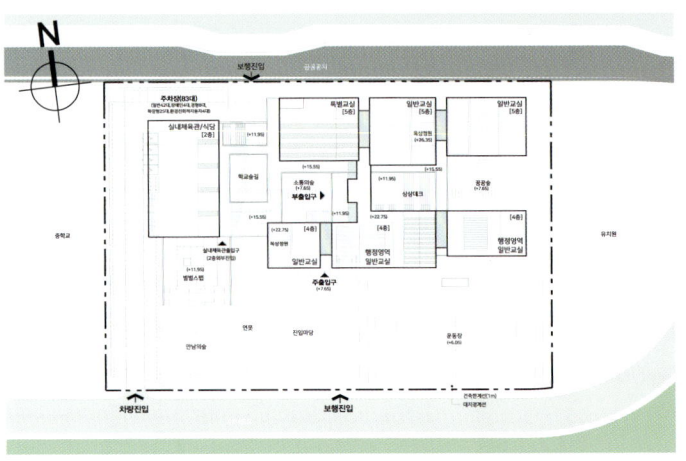

Site plan

Site Plan

- Creating a peasant learning environment

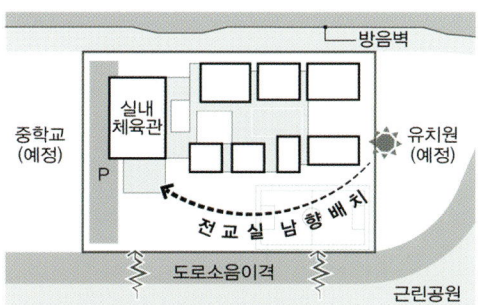

- 도로소음을 고려한 교사동 소음 이격배치
- 전교실 남향배치로 일조 및 공원으로 열린조망 확보

- Open area separation plan for convenient use

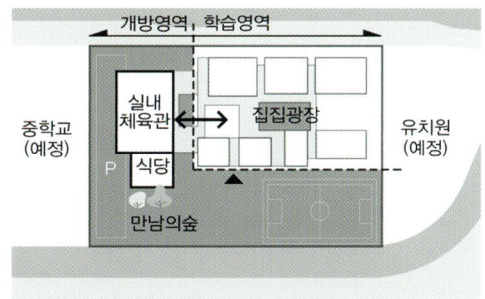

- 주민들의 동선을 고려한 개방시설-집집광장 인접배치
- 주민개방시설 별동배치로 안전한 교육환경 확보

- Creating a 3d green landscape linking the interior · exterior

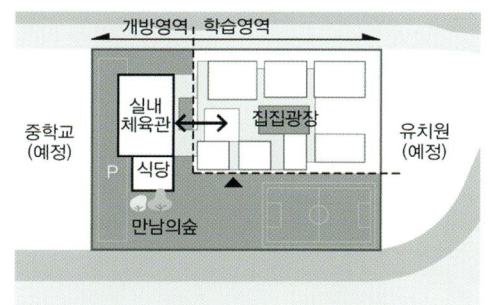

- 내·외부 공간의 유기적인 연계를 통한 미래공간 조성
- 다양한 외부공간을 조성을 통한 그린네트워크 계획

Concept Design

- Education cluster

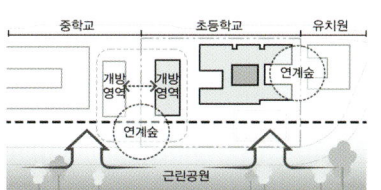

- 고양 장항지구의 교육
- 중, 초, 유치원으로 이어지는 교육 클러스터를 형성 및 개방영역과 학교숲 연계를 통해 지역 커뮤니티장을 형성

- The permeating school forest's

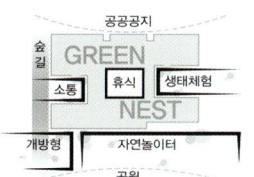

- 휴식과 소통이 있는 학교
- 마당, 중정, 놀이터 등 곳곳에 공원의 학교숲은 다양한 공간에서 녹지를 향유하는 입체적인 배치

- Smart complex plaza

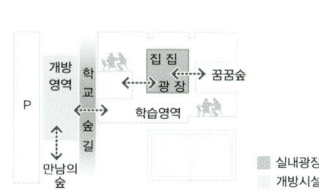

- 협력과 소통의 장, 스마트 기반의 광장형 공간
- 소통과 공유의 공동체 배움공간인 복합광장시설 중심배치 및 광장과 연계된 다양한 레벨의 활동공간 조성

- 3D space plan

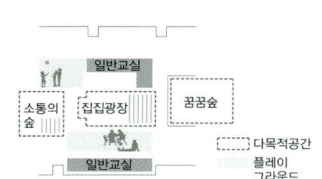

- 내 외부공간이 유기적으로 연결된 미래형 학교
- 일반교실 중심에 실내외 플레이 그라운드로 학습, 놀이 등 다양한 활동이 가능한 창의적 교육 공간 조성

Winner _ 당선작

■ Model

Janghang Elementary School >>

Front elevation

Left elevation

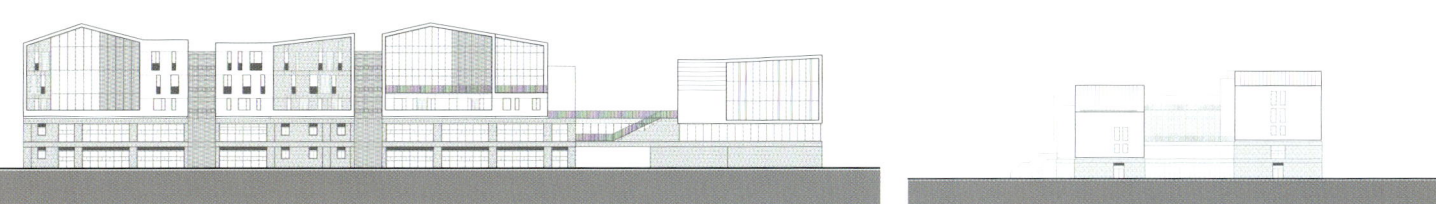

Rear elevation

Right elevation

Winner _ 당선작

1. Assembly hall
2. Kitchen
3. Care classroom
4. Practical course room
5. Art room
6. Science room
7. Classroom
8. Fitting room
9. Library
10. Club room
11. Music room
12. Preparation room
13. Study room
14. Wee class
15. Special class room

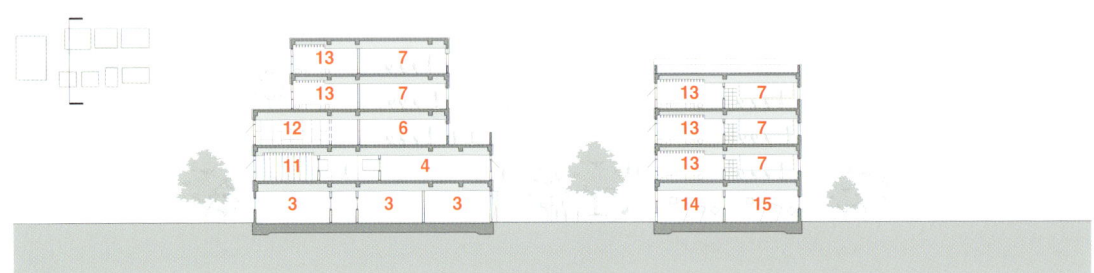

Longitudinal section

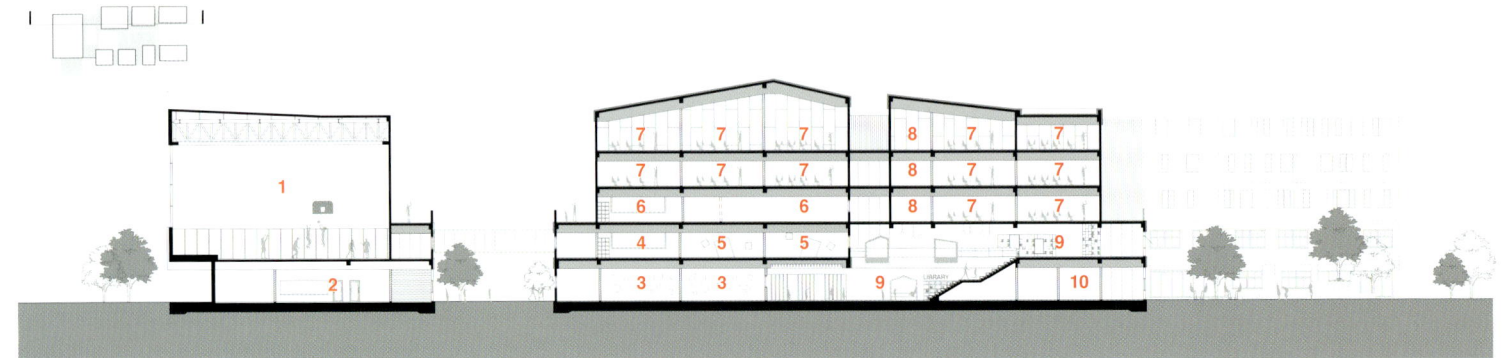

Cross section

Janghang Elementary School >>

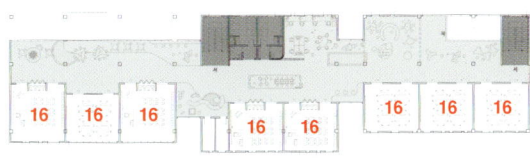

5th floor plan

1. Restaurant
2. Kitchen
3. Care classroom
4. Public education room
5. Convergence education room
6. Special class room
7. School clinic
8. Administration office
9. Education center
10. Studio
11. Assembly hall
12. Music room
13. Practical course room
14. Art room
15. Teacher lounge
16. Class room
17. Science room

3rd floor plan

4th floor plan

1st floor plan

2nd floor plan

Winner _ 당선작

Agricultural Research & Extension Services in Gyeongsangbuk-do
경상북도 농업기술원

dA architecture group / Hyunho Kim + Koma Architects + Kyung Hee University / Eunseok Lee
㈜디에이그룹엔지니어링종합건축사사무소 / 김현호 + ㈜코마건축사사무소 + 경희대학교 / 이은석

Beyond Horizon_Convergence of technology and communication with nature

Gyeongsangbuk-do Agricultural Research & Extension Services aims to be a research and education facility of cutting-edge agricultural technology for the era of convergence and innovation of rapidly changing industrial technology. The site is located in the wide plain of Sangju City, and harmonizes with the topography of the Nakdong River and surrounding mountains. The new Agricultural Research & Extension Services will stay in nature and play a pivotal role in regional agricultural development, and it will become an open place to communicate with the local community through festivals and agricultural education.

Under the slogan of 'Beyond Horizon_Convergence of technology and communication with nature', the facility is based on three concepts with the goal of leaping forward as a next-generation Agricultural Research & Extension Services preparing for a new millennium in Gyeongsangbuk-do, beyond boundaries.
First, with the theme of 'Nature_Restoration of nature', the disconnected flow of nature is restored, and the grid of farmland is introduced into the site to create a harmonious complex with the surrounding context through the organic combination of the topographical context and grid pattern.
Second, with the concept of 'Solid & Void_Aesthetics of solid and void', the horizontality of traditional architecture is borrowed to give the building a strong monumentality, and a Korean-style research facility with traditional aesthetics is planned through void with the motif of an auditorium yard where academics are exchanged.

Lastly, it is intended to realize the 'Future_Convergence Research Complex of Communication and Exchange'. The building is composed of five clusters for the characteristics of each program of education, support, and research, and the three-dimensional connection between facilities creates natural communication and various activities between researchers in one space.

The master plan is to be the meeting place with the farmland as an open space to the public through centralized arrangement. In the large green space in front, an experience yard, an ecological yard, and a grass yard are configured to accommodate various events. In the Interior, three yards of a garden of hospitality, a yard of exchange, and a forest of thoughts are arranged, and they are integrated into one space through piloti. In connection with the open studio, exhibition space, and public relations space, it becomes a space for exchange for both researchers and visitors.

The exterior of the building symbolically expresses convergence and communication with horizontality through repetition of louvers. In particular, the roof focuses on sustainability for the future by applying passive systems such as blocking solar radiation, discharging high heat, and reducing the heat island effect in addition to energy production.
It is hoped that Gyeongsangbuk-do Agricultural Research & Extension Services, where tradition and future are harmonized in the richness of the landscape, becomes a new standard for a Korean-style research facility as a user-centered convergence research complex.

Location : 864, Samdeok-ri, Sabeolguk-myeon, Sangju-si, Gyeongsangbuk-do, Korea I **Function** : Education & Research facility I **Site area** : 121,315m I **Bldg. area** : 32,287m I **Total floor area** : 33,600m I **Stories** : 3FL I **Structure** : Reinforced concrete, Steel frame I **Finish** : Etching glass, Low-E pair glass, Wood louver

위치 : 경상북도 상주시 사벌국면 삼덕리 864번지 일원 I **용도** : 교육연구시설 I **규모** : 지상3층 I **구조** : 철근콘크리트, 철골 I **마감** : 에칭유리, 로이복층유리, 목재루버 I **설계팀** : ㈜디에이그룹엔지니어링종합건축사사무소 / 이태민, 김상, 이근명, 이근택, 최재혁, 이혁, 정연욱, 강대규, 정동혁, 송예한, 김채린 + ㈜코마건축사사무소 / 안옥순, 이주연, 김산희, 배강현

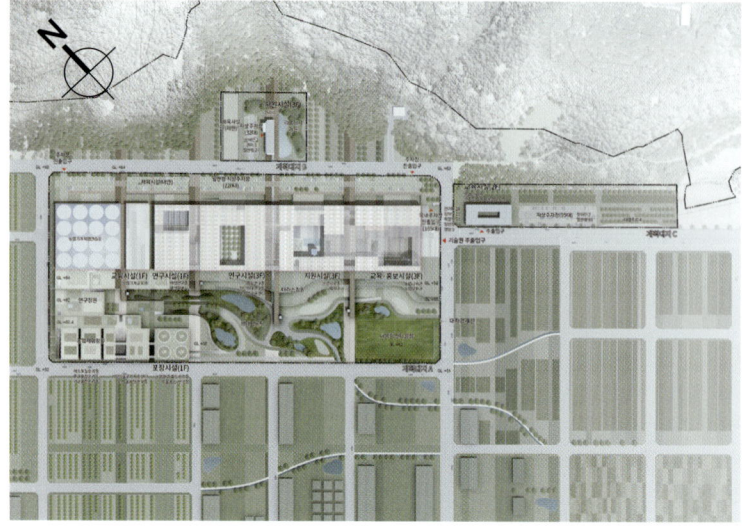

Site plan

Masterplan Concept

- Construction of network through connection and union

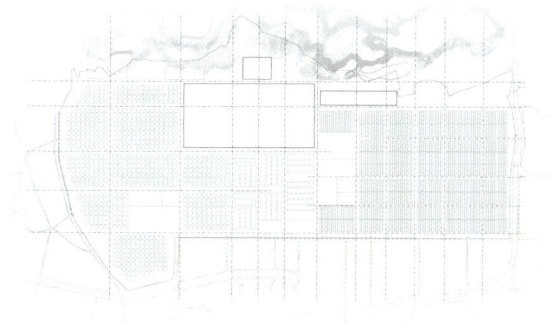

– Grid: 농경지 패턴의 질서

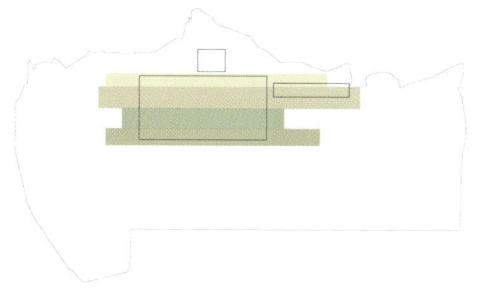

– Band: 건축과 조경 개념의 공존

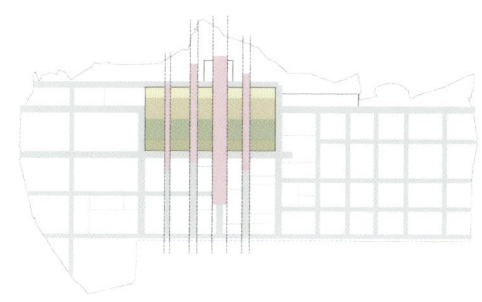

– Extension: 자연의 확장과 유입

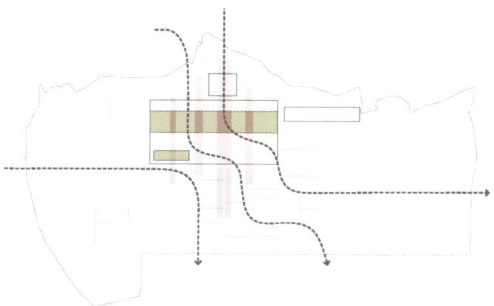

– Convergence: 자연, 건축, 조경의 융합

Beyond Horizon_기술의 융합, 자연과의 소통

경상북도 농업기술원은 급속하게 변화하는 산업 기술의 융합, 혁신의 시대에 맞는 최첨단 농업기술의 연구·교육시설을 목표로 한다. 대지는 상주시의 드넓은 평야에 위치하며, 낙동강과 주변 산세의 지형과 조화를 이루는 장소이다. 새로운 농업기술원은 자연 속에 머물며, 지역 농업발전의 중추적인 역할을 수행할 것이고 도민 축제와 농업교육 등 지역과 소통하는 열린 장소가 될 것이다.

본 농업기술원은 'Beyond Horizon_기술의 융합, 자연과의 소통'을 슬로건으로 내세우며, 경계를 뛰어넘어 경상북도의 새로운 천년을 준비하는 차세대 농업기술원으로의 도약을 목표로 3가지 개념을 바탕으로 계획되었다.
먼저, 'Nature_자연의 회복'의 테마로 단절된 자연의 흐름을 복원하고, 경작지의 그리드를 대지 내로 유입하여 지형의 맥락과 그리드 패턴의 유기적 결합을 통해 주변 맥락과 조화로운 단지를 완성하고자 했다.
두 번째, 'Solid & Void_채움과 비움의 미학'이라는 개념을 바탕으로 전통건축의 수평성을 차용함으로써 건축물에 강한 기념성을 부여하여, 학문이 교류되는 서원의 마당을 모티브로 한 비움을 통해 전통의 미학이 숨 쉬는 한국형 연구시설을 계획하였다.
마지막으로 'Future_소통과 교류의 융·복합 연구 단지'를 구현하고자 하였다. 건물은 교육, 지원, 연구의 프로그램별 성격을 고려한 5개의 클러스터로 구성되어 있으며, 시설 간 입체적인 연결을 통해 하나의 공간에서 연구원 간의 자연스러운 교류와 다양한 활동이 일어날 수 있도록 유도하였다.

마스터플랜은 집중화된 배치를 통해 경작지와 만나는 곳을 공공에 열린 공간으로 계획하였다. 전면의 넓은 녹지공간에는 체험 마당과 생태 마당, 잔디 마당을 구성하여 다양한 이벤트를 수용할 수 있도록 했으며, 건물의 내부는 3개의 마당인 환대의 뜰, 교류의 마당, 생각의 숲을 배치하고 필로티를 통해 하나의 공간으로 통합된다. 이곳은 오픈 스튜디오, 전시공간, 홍보공간 등과 연계하여 연구원과 방문객 모두를 위한 교류의 공간이 된다.

건물의 외피는 루버의 반복을 통한 수평성으로 융합과 소통을 상징적으로 표출하였다. 특히 루프는 에너지 생산과 더불어 일사 차단, 고임열 배출, 열섬효과 저감 등 패시브 시스템을 적용하여 미래를 위한 지속 가능성을 고려하였다.
경관의 풍요로움 속에 전통과 미래가 어우러진 경상북도 농업기술원이 사용자 중심의 융·복합 연구 단지로서 한국형 연구시설의 새로운 기준이 되기를 희망한다.

Model

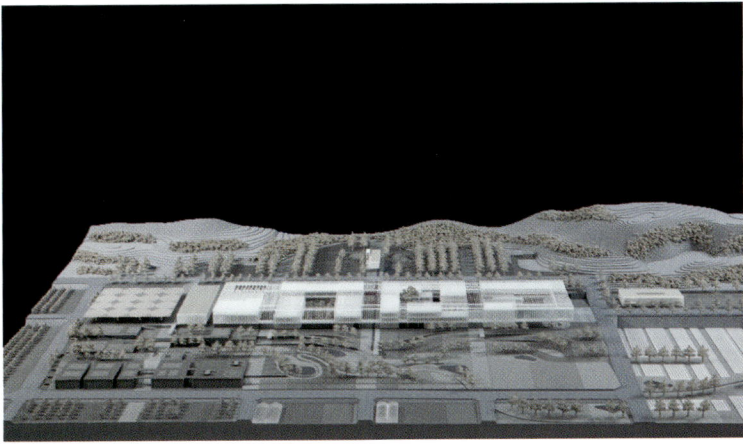
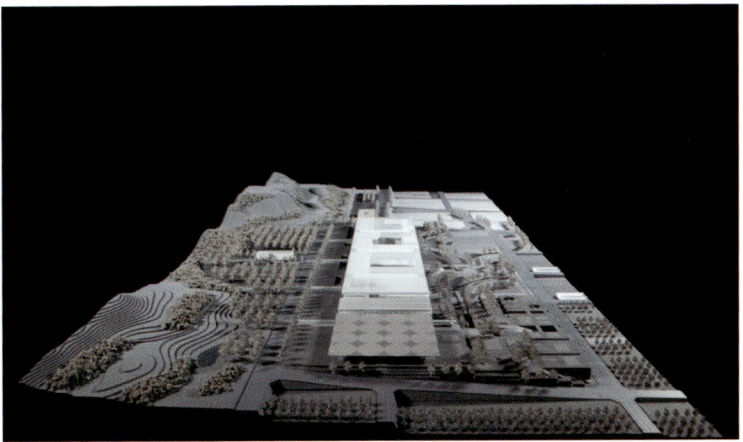

Design Process

- Extension
- Public voids
- Private solids
- Natural context
- Intergration

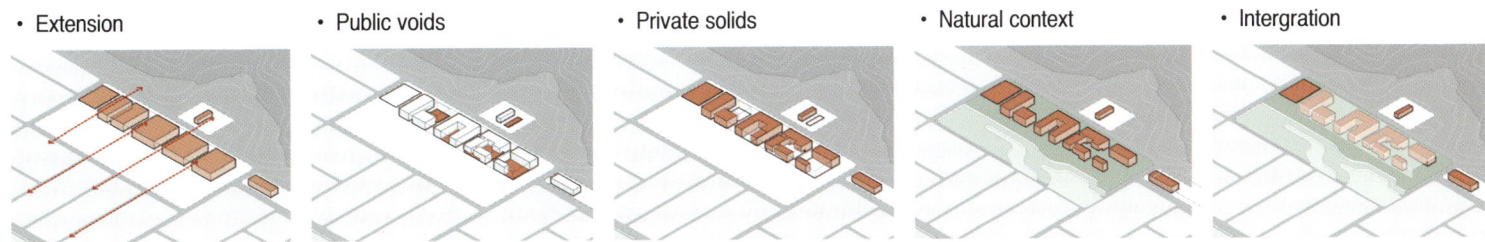

Front elevation

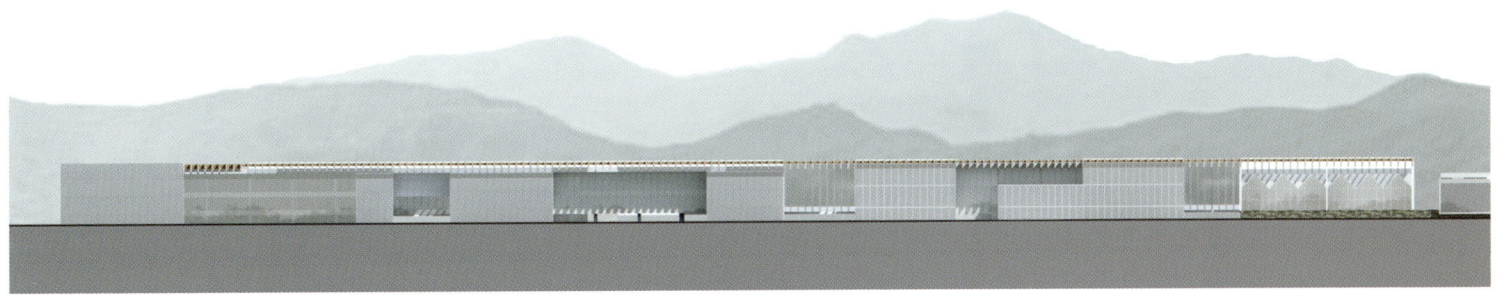

Rear elevation

Winner _ 당선작

1. Mushroom cultivation
2. Rooftop garden
3. Storage
4. Laboratory
5. CA storeroom
6. Analytical laboratory
7. Incubation room
8. Conference room
9. Weight room
10. Office
11. Air handling unit room
12. Lecture room
13. Lounge
14. Accommodation
15. Auditorium
16. Lobby
17. Studio
18. Public facility
19. Director's office
20. Terrace

Section A

Section B

Agricultural Research & Extension Services in Gyeongsangbuk-do >>

▎Circulation Plan

- Flow planning considering connection of facilities and flow of users

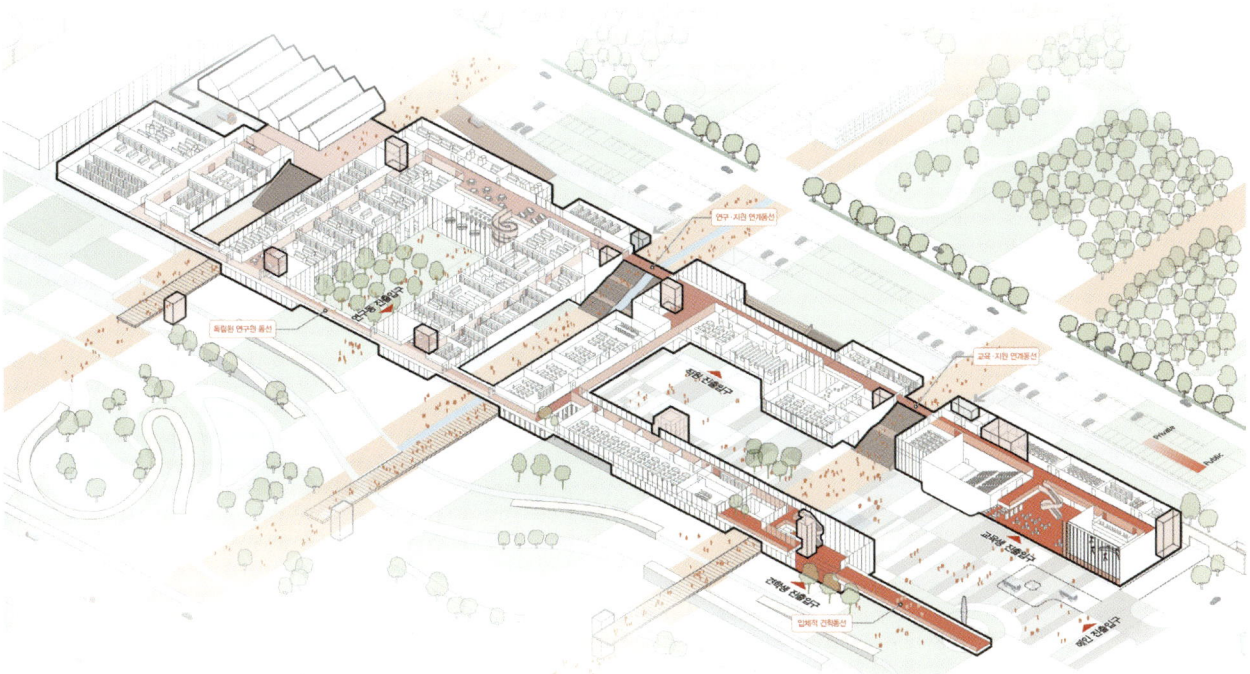

- 경북 농업기술원은 교육, 행정, 연구 기능이 결합된 융복합 연구시설로 다양한 시설 이용자가 존재한다. 연구원, 직원 동선은 동측 주차장에서 각 클러스터에 위치한 코어를 통해 건물로 편리하게 이동할 수 있으며, 각 영역간의 연계는 건물 전면과 후면에 위치한 브릿지를 통해 이루어진다. 명확한 클러스터링을 통해 동선의 혼재를 방지하였고, 시설을 이어주는 동선에는 자연의 풍경과 실내 아트리움 등을 계획, 시각적 개방감과 공간의 입체적 연계를 도모하였다.

Winner _당선작

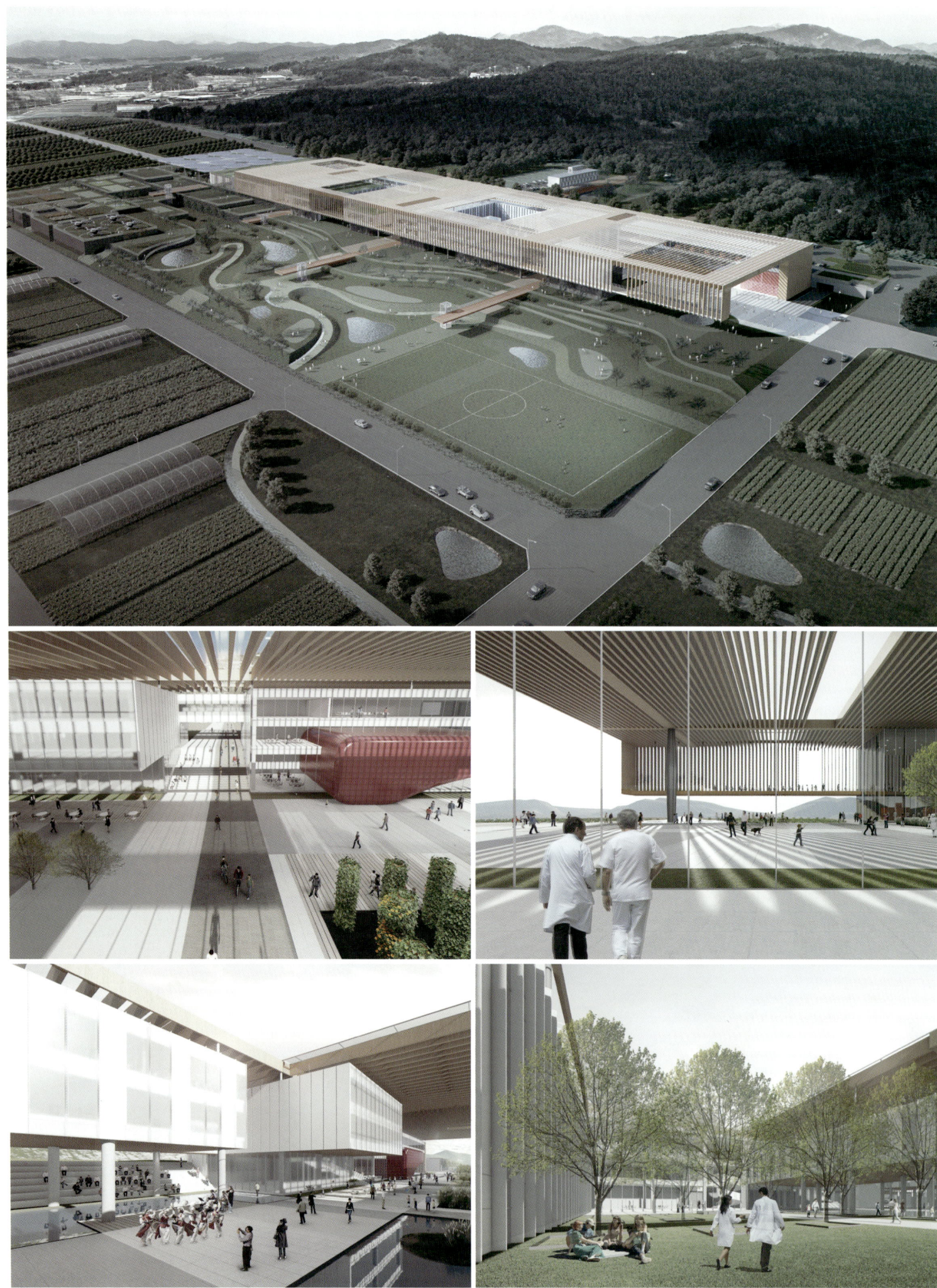

Agricultural Research & Extension Services in Gyeongsangbuk-do >>

3rd floor plan

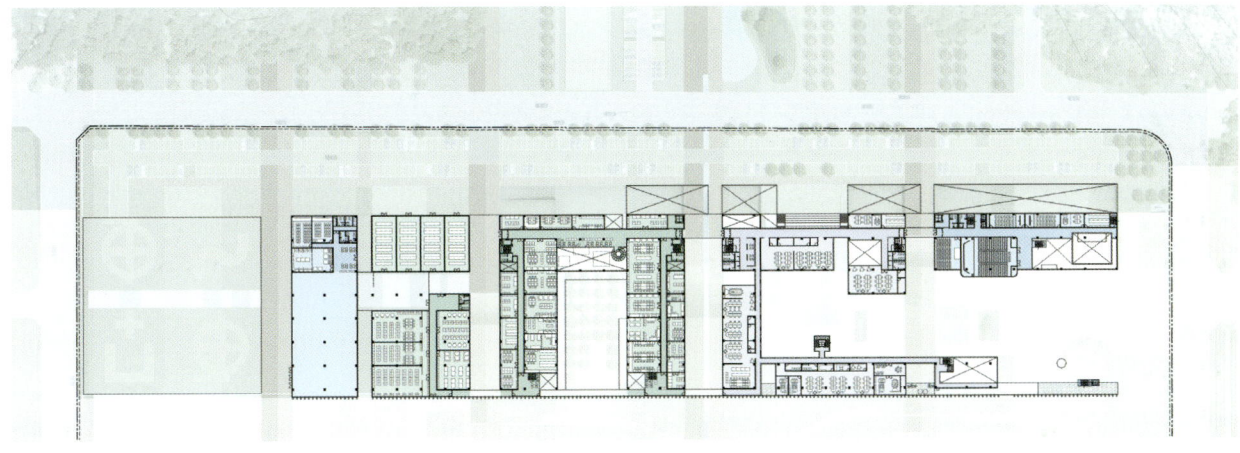

2nd floor plan

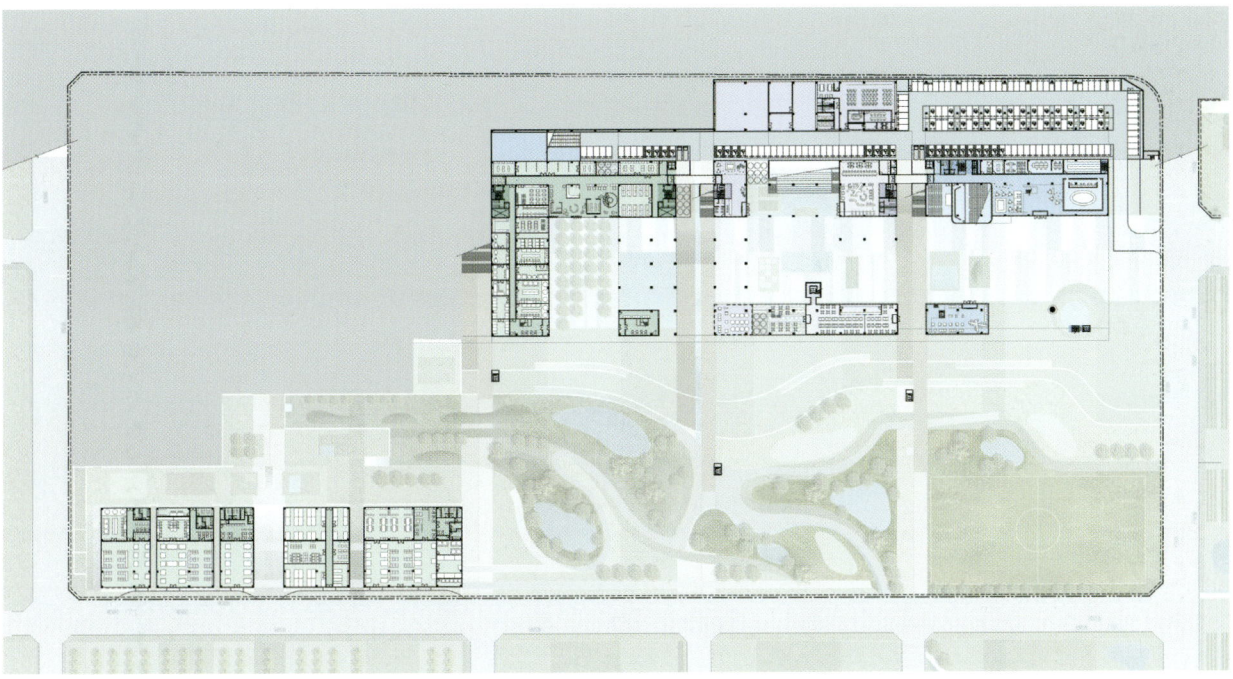

1st floor plan

Winner _ 당선작

Joint Research Centre
공동연구센터

BIG / Bjarke Ingels, João Albuquerque
BIG / 비야케 잉겔스, 주앙 알부케르케

Located at the former EXPO '92 Sevilla site, in Isla de la Cartuja, the new 9,900m² building for the European Commission, ties into the City of Sevilla's goal to become a global benchmark for sustainability by 2025 and the local vision of the eCitySevilla project to decarbonize and transition Isla de la Cartuja to 100% renewable energy sources. The JRC building will house 12 research units and supporting functions as well as public and private outdoor spaces.

"With our design for the Joint Research Centre in Sevilla, more than anything, we have attempted to allow the sustainable performance of the building to drive an architectural aesthetic that not only makes the building perform better but also makes it more inhabitable and more beautiful – a new Andalusian environmental vernacular," said Bjarke Ingels.
Inspired by the shaded plazas and streets of Sevilla, BIG proposes a cloud of pergolas over the entire JRC site, sheltering the plaza, the garden and the building underneath, akin to the shading elements typical to Sevilla. The entire pergola structure is supported by a forest of columns and will be covered in photovoltaics contributing positively to the building's operational footprint.

The periphery of the pergola is lowered to a human scale height creating a variety of spaces underneath it. The research centre adapts to the canopy, creating a series of terraces, shaded outdoor spaces for breakouts, relaxation, and informal meetings with views of the city.

Inside, the functions of the new JRC building are organized with public program and amenities such as dining, a conference center and social spaces on the ground floor, while the offices and research units occupy the upper floors for privacy and security. The collaborative workplaces face the plaza, while the deep-focus workspaces face the garden. The proposed layout is designed to be entirely flexible and adaptable according to any future needs of the JRC.

In addition to the energy harvested from PVs and technologies applied across the building, the canopy will feature integrated rainwater collection technologies and a particular focus on Biophilia. The passive design of the building through its shallow floorplate and constant shading under the pergola cloud enables natural cross ventilation and ideal light qualities, reducing the energy consumption typically used on artificial lightening, air conditioning and mechanical ventilation.

The design prioritizes locally sourced materials, such as limestone, wood and ceramic tiling. The building structure is low-carbon concrete, reducing up to 30% of typical CO_2 emissions, while the pergola cloud is made from recycled steel. Gardens, greenery and water elements in the outdoor environment seek to reduce/eliminate the heat island effect and create a comfortable microclimate.

Location : Isla de la Cartuja, Sevilla, Spain I **Function** : Education facility I **Total floor area** : 9,900m I **Client** : Joint Research Centre, European Commission I **Project manager** : Angel Barreno Gutiérrez I **Project leader** : Stefani Fachini de Araujo I **Design team** : Hanna Ida Johansson, Nir Leshem, Gonzalo Coronado, Jose Gómez Carbonell, Miquel Perez, Luca Fabbri, Matthew Reger, Elena Ceribelli, Pietro Saccardi, Raphaël Logan, Saina Abdollahzadeh I **Landscape** : Giulia Frittoli

위치 : 스페인 세비야 카르투하 I 용도 : 교육시설

카르투하 섬의 이전 엑스포 1992 세비야 부지에 위치하는 유럽연합 집행위원회를 위한 새로운 건물은 2025년까지 지속 가능성에 대한 글로벌 벤치마크가 되겠다는 세비야 시의 목표와 카르투하 섬을 100% 재생 가능한 에너지원으로 탈탄소화하고 전환하려는 e시티세비야 프로젝트의 지역 비전과 연결된다. 공동연구센터 건물에는 12개의 연구 유닛과 지원 기능, 공공 및 개인 야외 공간이 들어선다.

세비야의 공동연구센터를 위한 설계를 통해 무엇보다도 건물의 지속 가능성으로 건물의 전체 성능을 향상시킬 뿐만 아니라 거주하기 좋고 아름답게 만드는 건축 미학을 드러내고자 했으며, 이는 새로운 안달루시아의 환경적 건축 언어이다.
세비야의 그늘진 광장과 거리에서 영감을 받은 BIG는 공동연구센터 부지 전체에 퍼걸러 구름을 제안하여 광장, 정원 및 그 아래 건물을 보호하며, 이는 세비야의 전형적인 차양 요소와 같다. 전체 퍼걸러 구조는 기둥 숲으로 받쳐지고, 건물의 운영 공간에 긍정적으로 기여하는 광전지로 덮일 것이다.

퍼걸러의 주변은 휴먼 스케일 높이로 낮아져 그 아래에 다양한 공간이 만들어진다. 연구센터는 캐노피를 설치하여 일련의 테라스와 그늘진 야외 공간을 형성해 휴식을 취하거나 도시 전망을 보며 비공식 회의를 할 수 있다.

새로운 공동연구센터 건물의 내부 기능은 지상층에 공공 프로그램, 식사 등을 위한 편의 시설, 회의 센터, 사교 공간이 구성되고, 상층부에 개인정보 보호와 보안을 위해 사무실과 연구 부서가 배치된다. 협업 업무 공간은 광장을 향하고, 심층연구 업무 공간은 정원을 향한다. 제안된 배치는 공동연구센터의 향후 요구 사항에 따라 완전히 유연하고 적응할 수 있도록 설계되었다.

PV에서 얻은 에너지와 건물 전체에 적용된 기술 외에도, 캐노피는 통합 빗물수집 기술과 바이오필리아를 집중적으로 적용할 것이다. 퍼걸러 구름 아래의 얕은 바닥판과 지속적인 차양을 통한 건물의 패시브 설계는 쾌적한 자연 통풍과 채광을 받아들이며 인공 조명, 에어컨, 기계 환기에 일반적으로 사용되는 에너지 소비를 줄인다.

디자인은 석회석, 목재, 세라믹 타일과 같은 현지 재료를 우선적으로 사용한다. 건물 구조는 저탄소 콘크리트로 일반적인 CO_2 배출량을 최대 30%까지 감소시키는 동시에, 퍼걸러 구름은 재활용 강철로 만들어진다. 야외 환경의 정원, 녹지, 수공간은 열섬 효과를 줄이거나 없애고 쾌적한 미기후를 조성한다.

■ Design Process

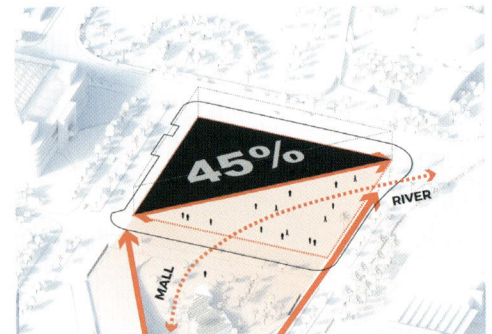

SEAMLESS PUBLIC REALM: FROM TORRE SEVILLA TO THE RIVER

BIG proposes the new Joint Research Centre to run diagonally across the orthogonal site which creates the opportunity to connect the JRC directly to the 'Jardin Americano' river-front promenade and the Torre Sevilla mall in a seamless continuous public space that is both plaza and promenade

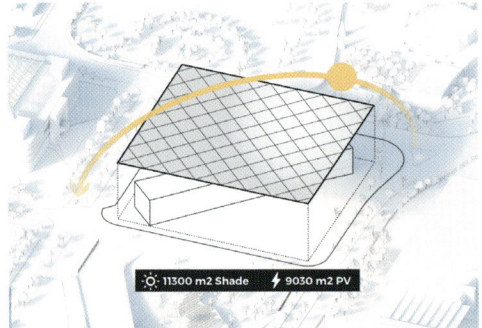

10.000M2 OF SOLAR ENERGY

The JRC will provide 4 times more solar cells than initially planned. We propose to cover the entire site with an array of solar panels to reduce the operational carbon footprint of the building

The sustainable performance of the building drives the architectural aesthetic of the JRC that not only makes the building perform better but also makes it more inhabitable, comfortable and beautiful - BIG's attempt at a new Andalucian environmental vernacular

Winner _ 당선작

Joint Research Centre >>

Zoning

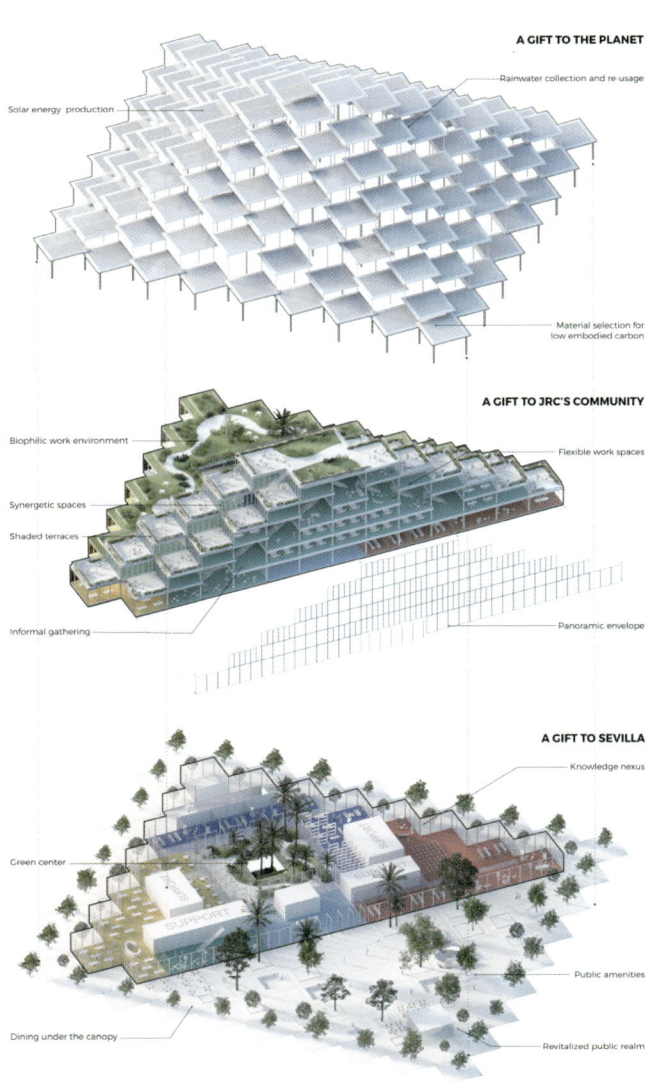

MCI Campus

MCI 캠퍼스

Henning Larsen Architects / Henning Larsen
헤닝 라센 아키텍츠 / 헤닝 라센

Founded in 1995, MCI quickly outgrew its central Innsbruck campus; the university's many faculties today are scattered throughout Innsbruck. The new structure, located next to the SoWi University and the Innsbrucker Hofgarten, will craft a unified campus for MCI for the first time in the school's history.

Bordered by the city to the south and east and by the historical Hofgarten to the north and west (and the Alps all around), the building is designed to have no back or front. Multi-story entries are carved into each facade to break the scale of the building in relation to its surroundings. These pockets are, in turn, planted with gardens to match the identity of their neighbour.

"We found a wealth of inspiration in the landscapes and geography of Tirol," explains Associate Design Director Lucas Ziegler. "…especially in the way changing light alters the way you see the mountains throughout the day. The facade is an interpretation of this – the depth, verticality, and angularity of the facade gives the building an appearance that changes throughout the day and the seasons."

Classrooms and lecture halls populate the outer edge of the ground level, framing a fluid interior space with a large community stair in the centre that not only links the three levels of 'learning' spaces, but also serves as a community space itself. These learning floors are designed to be open and flexible, with nearly as much 'unprogrammed' space for students to study, socialise, and rest as there is actual classroom space.

"In learning projects today, there's much less emphasis on the traditional model of dedicated classrooms connected by empty corridors – more fluid spaces reflect a better understanding of how people actually learn and process new information," explains Lucas Ziegler. "Our design not only supports the variety of ways people learn but is also an inviting place for people to meet and stay."

The building's upper floors are divided in two sections, one containing offices for MCI faculty, administration, and students, the other containing laboratories and research spaces. The design is dense and highly efficient, with four cores that serve not just as vertical circulation, but also as social hubs within the large floorplates.

Location : Innsbruck, Austria | **Function** : Education facility | **Total floor area** : 35,000m

위치 : 오스트리아 인스부르크 | 용도 : 교육시설

Site plan

1995년에 설립된 MCI는 인스부르크 중앙 캠퍼스보다 빠르게 발전했다. 오늘날 이 대학교의 많은 교수진은 인스부르크 전역에서 활동하고 있다. SoWi 대학과 인스부르커 호프가르텐 옆에 위치한 신축 건물은 학교 역사상 처음으로 MCI를 위한 통합 캠퍼스를 이룰 것이다.

남쪽과 동쪽은 도시와, 북쪽과 서쪽은 역사적인 호프가르텐과 접하고 주변에는 알프스 산맥이 있는 이 건물은 전면과 후면이 없도록 설계되었다. 다층 입구는 각 파사드에 주변 환경과 어울리도록 만들어져 있으며, 마을의 정체성에 맞게 정원이 갖춰져 있다.

"우리는 티롤의 풍경과 지리에서 풍부한 영감을 얻었다. 변화하는 빛으로 하루 종일 산을 바라보는 방식을 바꾸는 방법을 파사드에 표현했다. 파사드의 깊이, 수직성, 경사도는 하루 종일 그리고 계절에 따라 변화하는 건물의 모습을 보여준다"라고 설계 책임자 루카스 지글러는 설명한다.

강의실과 강당은 3층 규모의 '학습' 공간을 연결할 뿐만 아니라 커뮤니티 공간 자체의 역할을 하는 중앙의 대형 커뮤니티 계단으로 유동적인 내부 공간을 구성하여 지상층의 바깥쪽 가장자리를 채운다. 이러한 학습 층은 실제 강의실 공간만큼 학생들이 공부하고, 교제하고, 휴식을 취할 수 있는 '프로그래밍되지 않은' 공간이 있어 개방적이고 유연하게 설계되었다.

"오늘날 학습 프로젝트에서는 빈 복도로 연결된 전용 강의실의 전형적인 모델을 고려하지 않는다. 유동적인 공간이 많을수록 학생들이 실제로 새로운 정보를 배우고 처리하는 방법에 대한 이해도가 높아진다. 우리의 디자인은 학생들이 배우는 다양한 방법을 지원할 뿐만 아니라 학생들이 만나고 머물 수 있는 매력적인 장소가 된다"라고 루카스 지글러는 설명한다.

건물의 상층부는 두 부분으로 나뉘는데, 바로 MCI 교수진, 행정 직원, 학생을 위한 사무실과 실험실 및 연구 공간이 있다. 디자인은 밀도가 높고 매우 효율적이며, 4개의 코어가 수직 순환뿐만 아니라 큰 바닥판 내 사회적 허브 역할을 한다.

Winner _ 당선작

Design Process

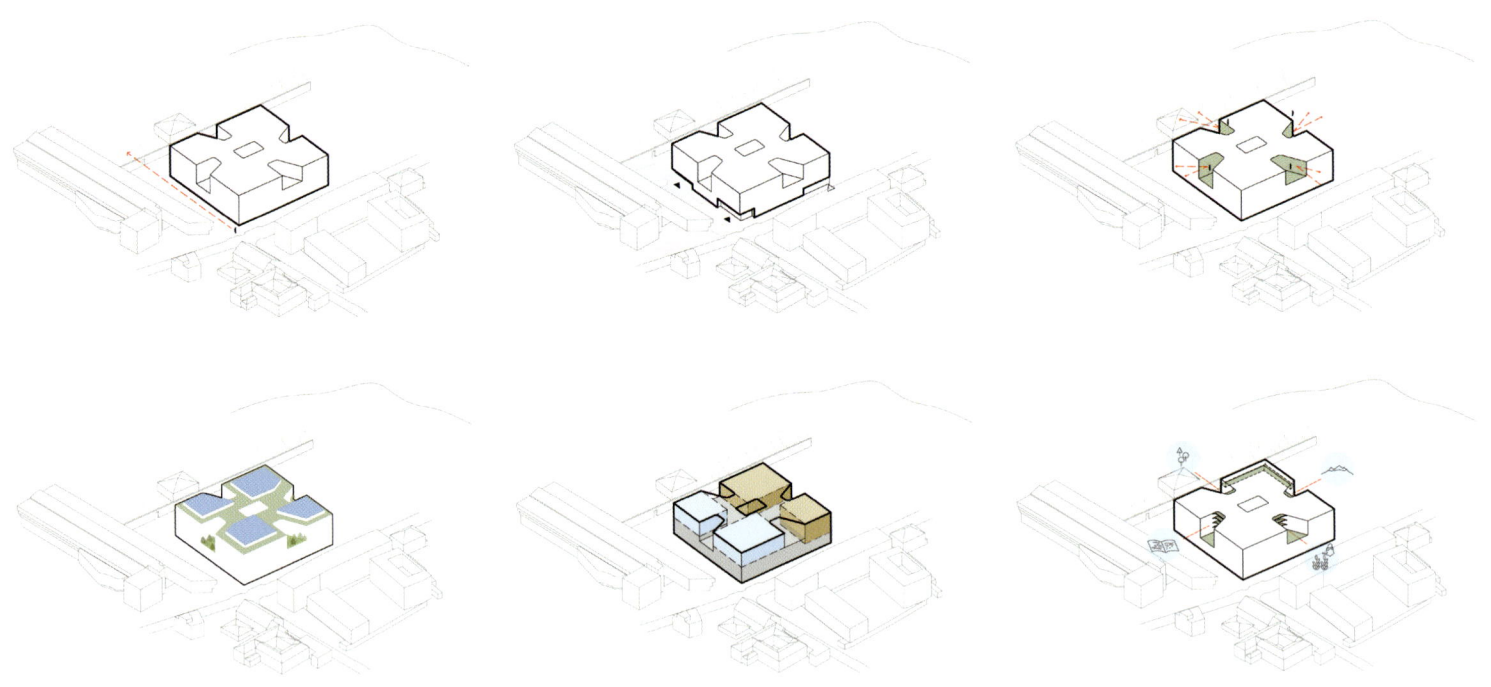

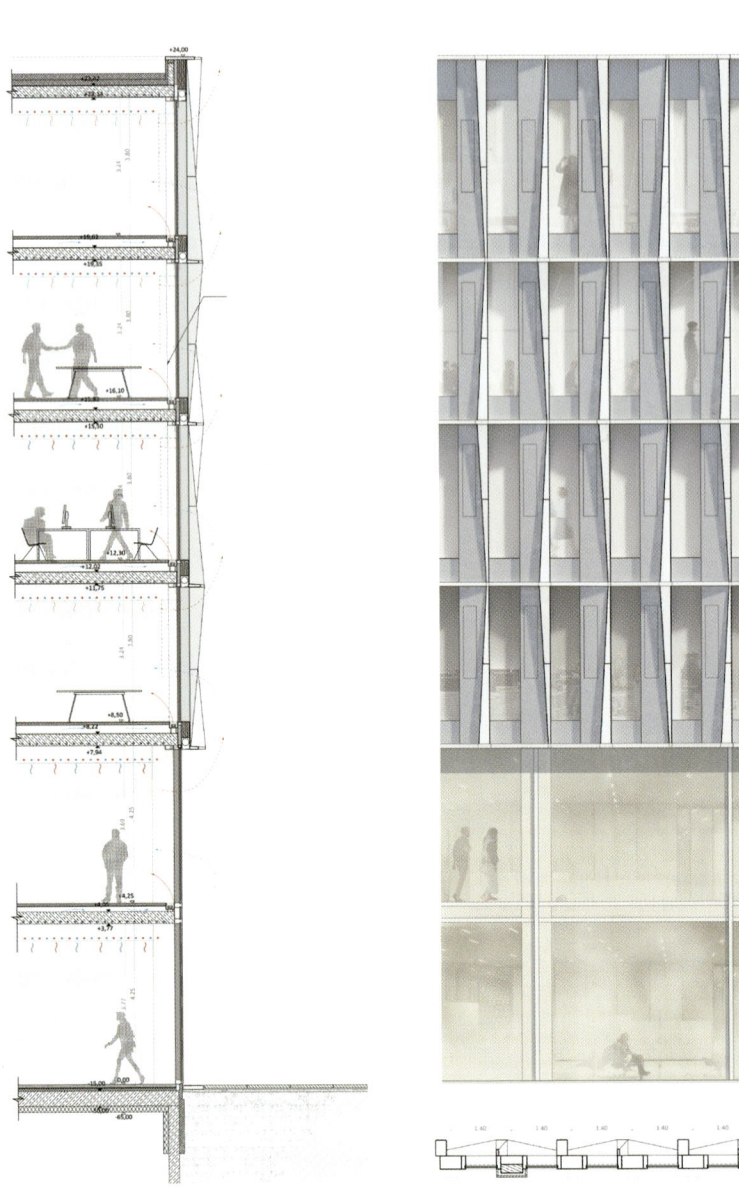

Detail - Elevation

MCI Campus >>

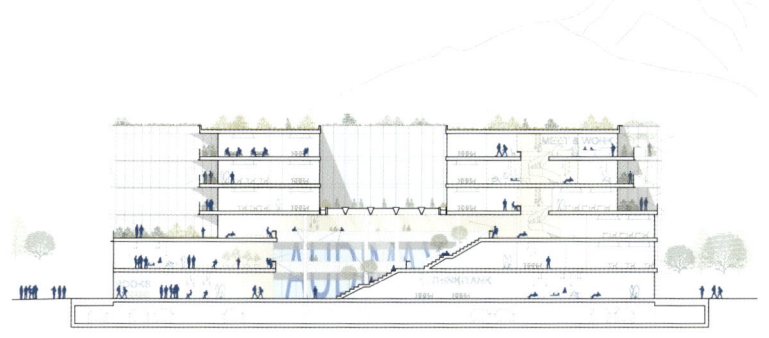

Section I

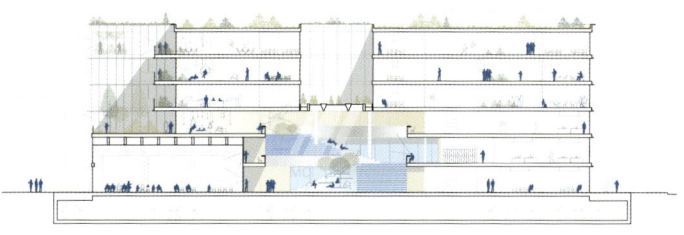

Section II

Singil District 12 Social Welfare Facility & Kindergarten
신길12구역 사회복지시설 및 유치원

D.LIM architects / Yeonghwan Lim, Sunhyun Kim
㈜디림건축사사무소 / 임영환, 김선현

Between the apartment and the convent

"De Mihi Animas Cetera Tolle," is written on the website of the Salesian Sisters on the eastern side of the site. The Shinmi apartment complex, which is about to be reconstructed, is located on the west side of the site. The Singil District 12 complex is a social welfare facility that will be built between a convent whose vocation is to sacrifice and serve, and an apartment complex that will be transformed with reconstruction. The existence of the facilities is clear since the union favors the community facility and the kindergarten. However, in order to minimize damage to the apartment in the near future, the floor level of the first floor of the kindergarten is already set at +24.00, which is higher than the road, and half of the south side of the site should be left empty to place the kindergarten.

The goals of the convent and the social welfare facility are almost the same. They include Youth activity assistance, early childhood education, multicultural care, and welfare counseling. The exterior material of the building was borrowed from the red bricks of the convent. It aims to inform that it is a public facility with the same purpose as the convent, and to combine the yards of the convent and the social welfare facility.

Public pedestrian promenade through the site

Public facilities are yards for neighbors. In the reality that a large yard in front of the building cannot be arranged, a public pedestrian promenade with a width of 10m through the site is first planned. The kindergarten and the welfare facility are naturally separated, and the front yard of the site and the large back yard are connected. Basically, all circulations start from public pedestrian promenade. In addition, a sunken garden is directly connected to the living culture center on the first basement floor, and the kindergarten can be accessed directly from a sidewalk. The entrance to the parking lot is placed at the lowest level of the road, and you can safely enter the kindergarten from the highest level.

The front yard of the social welfare facility, and the back yard of the kindergarten

The front yard and the back yard face each other around the central public pedestrian promenade. The hierarchy of the yard is three-dimensional because it is divided into three access circulations according to the level of the road. The front yard and back yard are not disconnected thanks to the public pedestrian promenade that connects the front and back of the site with a wide and gentle slope. Rather, the building blocks the uncomfortable sunlight in the early morning and late afternoon and creates a comfortable resting shade. The back yard of the kindergarten becomes the front yard of the daycare center, and on weekends, it can be used as a shelter for residents and as a flea market event space.

Location : 4966, Singil-dong, Yeongdeungpo-gu, Seoul, Koera | **Function** : Welfare, Education & Research facility | **Site area** : 2,890m² | **Bldg. area** : 1,399m² | **Total floor area** : 8,252m² | **Stories** : B2, 4FL | **Structure** : Reinforced concrete | **Finish** : Clay brick, Color steel plate, Clear low-E pair glass

위치 : 서울시 영등포구 신길동 4966 | **용도** : 노유자, 교육연구시설 | **규모** : 지하2층, 지상4층 | **구조** : 철근콘크리트 | **마감** : 점토벽돌, 컬러강판, 투명로이복층유리 | **건축주** : 영등포구청 | **설계팀** : 최정호, 박수현, 박재우, 김수진, 이진성

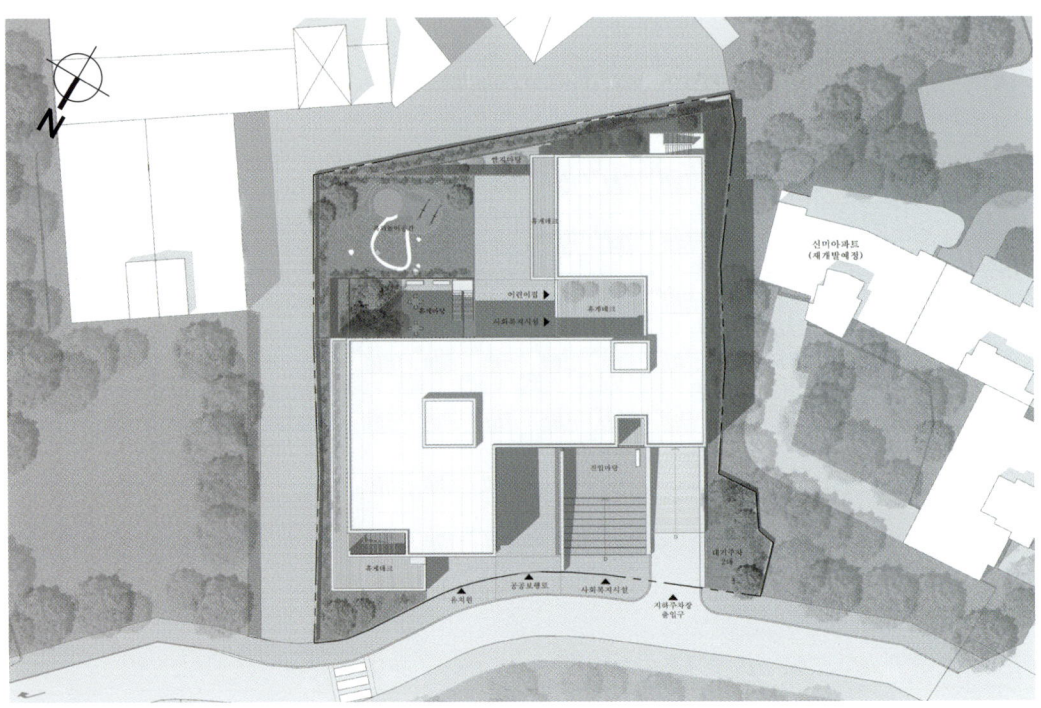

Site plan

▮ Diagram

- Guide

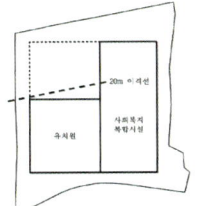

- 유치원의 2시간 일조 채광을 만족시키기 위해 20m 이격 선을 고려한 배치이며, 유치원과 사회복지 시설이 한 동으로 제안된 지침안이다.

- Public footpath

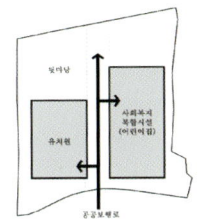

- 지역주민과 유치원, 사회복지시설 이용객이 함께 사용하는 공개공지이며, 대지를 관통해 모든 시설을 연결하는 주동선이다.

- Separation of facilities

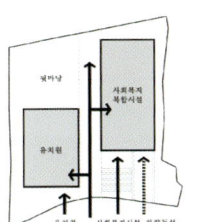

- 각 시설별 주출입구는 공공보행로를 따라 자연스럽게 연결되며, 도로경사에 맞추어 각각 독립된 출입동선을 확보한다.

- Completion

- 가장 이용 빈도가 높은 생활문화센터를 지하 1층에 두고, 대지 내 3개의 선큰 가든으로 자연채광 및 환기를 통한 열린 지하공간을 계획한다.

아파트와 수녀원사이

"나에게 영혼을 주고 나머지는 다 가져가라" 대지 동쪽에 면한 살레시오 수녀원 홈페이지 글귀다. 대지의 서쪽으로는 재건축을 목전에 둔 신미아파트 단지가 있다. 신길12구역 복합시설은 희생과 봉사가 천직인 수녀원과 재건축을 통해 새롭게 변화할 아파트 단지 사이에 들어설 사회복지시설이다. 다행히 조합도 반기는 커뮤니티시설과 유치원이 들어서기 때문에 시설의 존립에 대한 걱정은 없다. 하지만, 조만간 들어 아파트에 피해를 최소화하기 위해 유치원의 1층 바닥 레벨은 도로보다 높은 +24.00으로 이미 정해져 있고 남측 대지의 절반을 비워두고 유치원을 배치해야 한다.

수녀원과 사회복지시설의 목표는 거의 동일하다. 청소년활동 보조, 유아교육, 다문화 보살핌, 복지상담. 건물의 외장재료는 수녀원의 붉은 벽돌을 차용했다. 수녀원과 같은 목적을 가진 공공시설임을 알리고 수녀원과 사회복지시설의 마당을 하나로 묶어내기 위함이다.

대지를 관통하는 공공의 보행로

공공시설은 이웃 주민의 마당이다. 건물 앞 넓은 마당을 둘 수 없는 현실에서 대지를 관통하는 폭 10미터의 공공보행로를 우선 계획했다. 유치원과 복지시설을 자연스럽게 분리하고 대지 앞마당과 넓은 뒷마당을 연결한다. 기본적으로 모든 동선은 공공보행로에서 시작된다. 더불어 지하 1층의 생활문화센터로 바로 연결되는 선큰 가든이 있고, 유치원도 인도에서 바로 접근할 수 있다. 도로의 가장 낮은 레벨에 주차장 출입구를 두고 가장 높은 레벨에서 안전하게 유치원으로 들어간다.

사회복지시설의 앞마당, 유치원의 뒷마당

중앙의 공공보행로를 중심으로 앞마당과 뒷마당이 마주한다. 도로의 레벨에 맞추어 세 개의 진입 동선으로 분리했기 때문에 마당의 위계가 입체적이고, 넓고 완만한 경사로 대지의 앞과 뒤를 연결하는 공공의 보행로 덕분에 앞마당과 뒷마당은 단절되지 않는다. 오히려 이른 아침과 늦은 오후의 불편한 햇빛을 건물이 막아주며 편안한 휴식 그늘을 만들어준다. 유치원의 뒷마당은 어린이집의 앞마당이 되고, 주말에는 주민들을 위한 휴식 공간이자 프리마켓 행사장으로 사용할 수 있다.

Winner _ 당선작

Front elevation

Left elevation

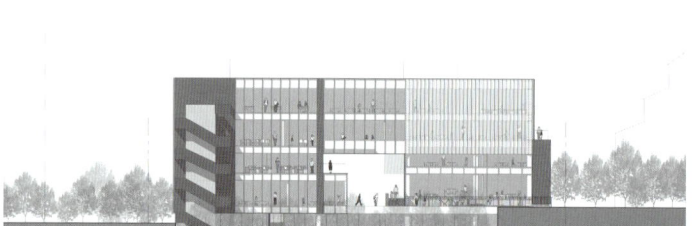

Rear elevation

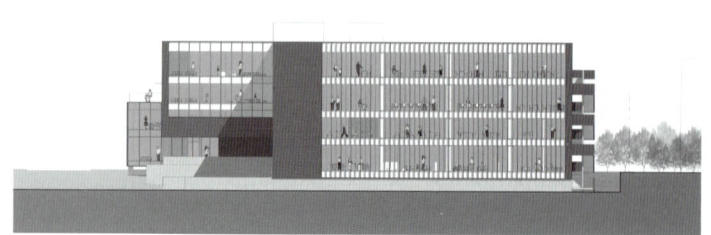

Right elevation

1. Machine / Electrical room
2. Parking
3. Community space
4. Kitchen
5. Hall
6. Wind break room
7. Playing room
8. Nursery room
9. Group counseling room
10. Personal counseling room
11. Part-time nursery room
12. Meeting room
13. Office
14. Sky lounge
15. Lounge deck
16. Activity space
17. Private room
18. Education room

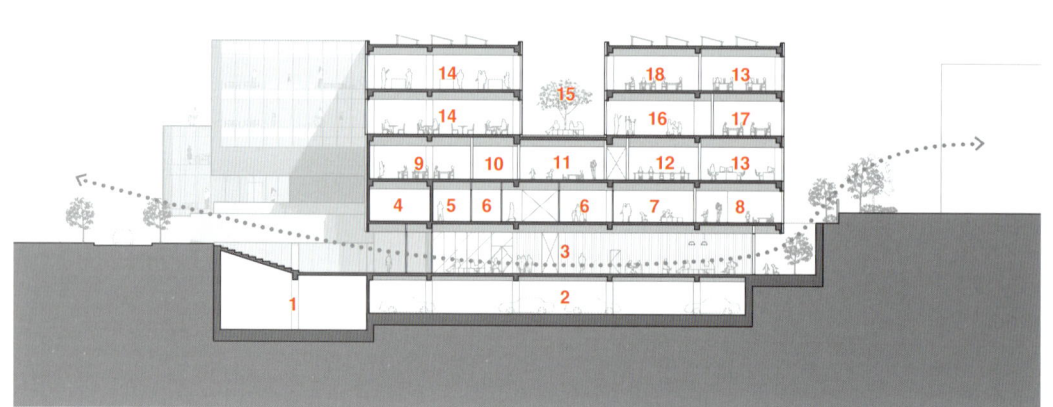

Section

Singil District 12 Social Welfare Facility & Kindergarten >>

3rd floor plan

1. Educare classroom
2. Director office
3. Health room
4. Hall
5. Playing room
6. Special classroom
7. Kitchen
8. Teacher's room
9. Wind break room
10. Nursery room (0-year-old)
11. Nursery room (1-year-old)
12. Nursery room (2-year-old)
13. Nursery room (3,4-year-old)
14. Nursery room (5-year-old)
15. Administration office / Library
16. Learning material room
17. Teacher's lounge
18. Classroom
19. Group counseling room / Small lecture room
20. Personal counseling room
21. Lobby
22. Women lounge
23. Part-time nursery room
24. Warehouse / Storage room
25. Parents & Teacher counseling room
26. Center director office
27. Office
28. Library
29. Study room
30. Sky lounge
31. Powder room
32. Private room
33. Activity space

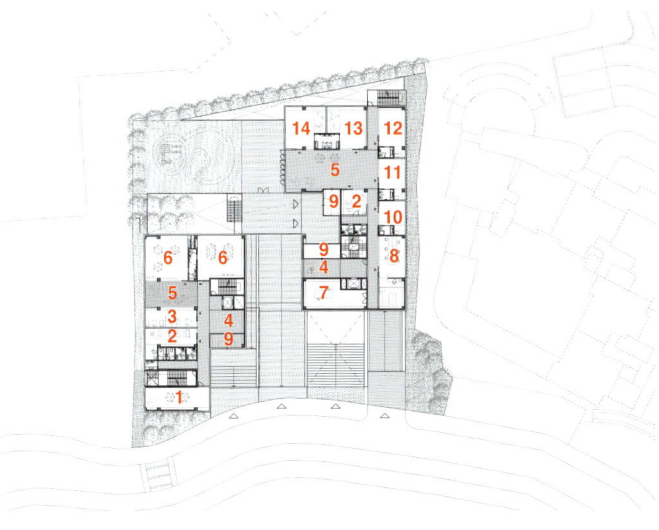

1st floor plan

2nd floor plan

Singil District 12 Social Welfare Facility & Kindergarten

신길12구역 사회복지시설 및 유치원

An Architects / Jonghwan Ahn + studio METAA / Uijeong Woo
㈜건축사사무소 안 / 안종환 + ㈜건축사사무소 메타 / 우의정

The lost village street becomes a children's alley
The coexisting street seeks to revive the memories that have been lost through the 'children's alley,' which recycles the alleys that have been lost due to redevelopment by large-scale capital, and to restore relationships with neighbors that have been disconnected.

Arrangement plan
Surrounded by various facilities of the Salesian Sisters, which have been located in the south for a long time, and the apartment house on the west side of the new site, the facility requires a lot of area due to the complexities of facilities, but according to the nature of the front life street and the conditions of various programs, it emphasizes harmony with the surrounding facilities to create a natural urban landscape.

Children's alley
In order to enter various facilities from the limited front road and the children's playground inside is perceived as an open environment, a children's alley is created that draws the street image into the interior. The street connects to the front road and focuses on the publicness of the facility as a pleasant alleyway where the public flow naturally connects to the inside.

Principle of floor plan
The main rooms of the program are basically arranged to face south so that the environment can be advantageously located. Administrative management areas such as stairs, elevators, and toilets are arranged facing a children's alley across the center of the site, and a multi-layered space system is established so that the public flow can flow flexibly through the public areas between them. In addition, the continuity of the small hall is highlighted so that it is not perceived as a uniform space with a middle corridor.

Formative creation
In order to build a facility with many functions on the irregularly shaped site, we create a strategy to expand the contact surface with the public area by segmenting the mass. Although the masses are separated under the necessity, we attempt a formative combination to emphasize the formation of a single image and give a sense of unity. This way, we propose a small architecture that is less overwhelming and not huge.

Elevation design
The composition of the mass and the elevation of the two buildings uses the image of a block that children play with, and the facade is designed to be friendly to children. The opening like a hole in the kindergarten mass adds lightness by forming a facade as if stacked blocks are removed, and the social welfare facility uses brown louvers to harmonize with the red bricks of the adjacent Salesian Cultural Center. The louvers are designed to have different facades by controlling the density, and privacy is ensured and sunlight is controlled by arranging the louvers over the openings according to the purpose of the room. The facade surrounding the children's alley between the social welfare facility and the kindergarten uses red bricks to give it a warm atmosphere like a neighborhood alley. The children's alley extends naturally into the outdoor play area inside the site and the facade of the Salesian Cultural Center.

Location : 4966, Singil-dong, Yeongdeungpo-gu, Seoul, Koera I **Function** : Welfare, Education & Research facility I **Site area** : 2,890m² I **Bidg. area** : 1,724m² I **Total floor area** : 8,452m² I **Stories** : B2, 4FL I **Structure** : Reinforced concrete I **Finish** : Synthetic resins plastering, Finished brick, Aluminium louver

위치 : 서울시 영등포구 신길동 4966 I **용도** : 노유자, 교육연구시설 I **규모** : 지하2층, 지상4층 I **구조** : 철근콘크리트 I **마감** : 합성수지미장, 치장벽돌, 알루미늄루버 I **건축주** : 영등포구청 I **설계팀** : 신아름, 오기백, 최석, 이민철, 최지수

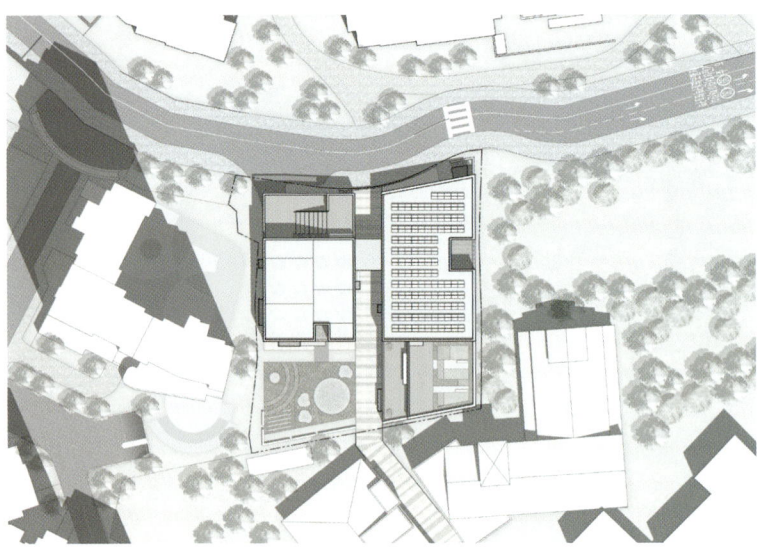

Site plan

사라진 마을길이 어린이 골목으로
공생가로는 대규모 자본에 의한 재개발로 사라진 골목을 새롭게 조성하는 '어린이 골목'으로 인하여 사라진 기억을 되살리며 단절된 이웃과의 관계를 회복시키고자 한다.

배치계획
남측에 오래 전부터 자리잡고 있는 살레시오 수녀회의 여러 시설과 새롭게 들어서는 대지 서측의 공동주택에 의해 둘러싸이는 형상의 본 시설은 시설의 복합화로 인하여 제법 많은 용적을 요구하지만 전면 생활가로의 성격과 다양한 프로그램의 조건에 따라 시설을 흐름이 자연스러운 도시경관을 조성하도록 주변시설과의 조화를 강조한다.

어린이 골목
제한적인 전면도로에서 다양한 시설의 진입을 고려하고 안쪽에 마련되는 어린이 놀이터가 열린 환경으로 인식되기 위하여 가로의 이미지를 내부로 끌어들이는 어린이 골목을 조성한다. 이 길은 전면도로와 이어지며 공공의 흐름이 자연스럽게 내부로 연결되는 쾌적한 골목길로서 이 시설의 공공성을 강조한다.

평면구성 원칙
프로그램 주요실은 남향 배치를 원칙으로 환경이 유리한 영역에 자리잡도록 하고 대지 중앙을 가로지르는 어린이 골목을 맞대고 계단실, 엘리베이터, 화장실 등의 행정관리영역을 배치하며 그 사이의 공공영역을 통하여 공공의 흐름이 유연하게 이어지도록 다층의 켜를 갖는 공간체계를 수립하도록 한다. 그러면서도 공간의 구성이 지나치게 중복도의 획일적인 공간으로 인식되지 않게 하기 위하여 작은 홀의 연속성을 강조한다.

조형의 생성
부정형의 대지에 많은 기능을 담는 시설이 들어서기 위해 매스를 분절하여 공공영역과의 접촉면을 넓히는 전략을 세운다. 그리고 필요에 의해 매스를 분리하긴 하지만 하나의 이미지를 형성하는 것을 강조하기 위하여 조형적 결합을 시도하고 통일감을 부여한다. 이러한 방식으로 부담이 적고 거대하지 않은 작은 건축을 제안한다.

입면 디자인
두 건물의 매스와 입면의 구성은 아이들이 가지고 노는 블록의 이미지로 어린이들에게 친근한 입면을 계획한다. 유치원 매스에 구멍처럼 난 개구부는 쌓은 블록을 빼낸 듯한 입면을 형성해 경쾌함을 더하고, 사회복지시설은 브라운 계열의 루버를 사용하여 인접한 살레시오 문화원의 적벽돌과 위화감이 없게 조성한다. 루버는 밀도의 조절로 입면을 다채롭게 구성할 뿐만 아니라, 실의 용도에 따라 개구부 위에 루버를 배치함으로써 프라이버시 확보와 일조 조절을 동시에 기능한다. 사회복지 복합시설과 유치원 사이의 어린이 골목을 둘러싼 입면은 적벽돌을 사용하여 동네의 골목과 같은 따뜻한 분위기를 부여한다. 어린이 골목은 대지 안쪽 옥외 놀이공간과 살레시오 문화원의 입면으로 흐르듯이 연장된다.

2nd Prize _ 2등작

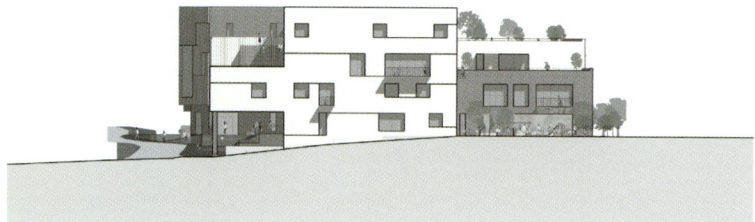

Front elevation

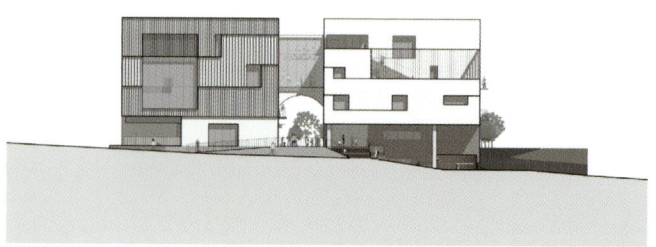

Left elevation

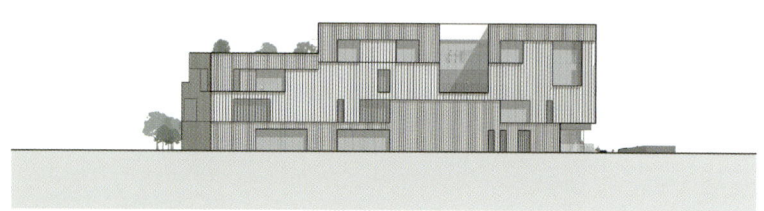

Rear elevation

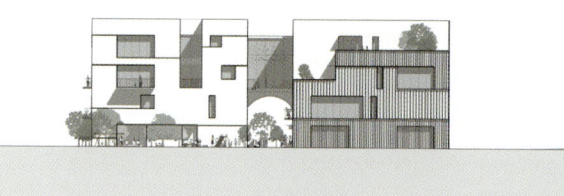

Right elevation

Singil District 12 Social Welfare Facility & Kindergarten >>

Elevation Design

- Block

– 두 건물의 매스와 입면의 구성은 아이들이 가지고 노는 블록의 이미지로 어린이들에게 친근한 입면을 계획한다.

- Kindergarten-punching window

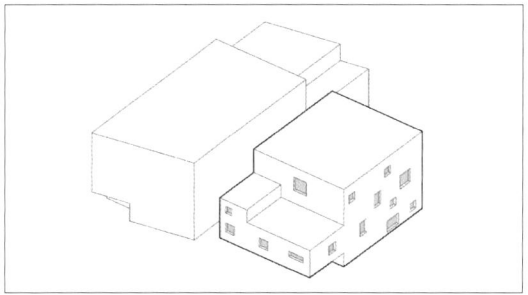

– 매스에 구멍처럼 난 개구부는 쌓여있는 블록을 하나하나 빼낸듯한 입면을 형성해 경쾌함을 더한다.

- Social welfare organization-vertical louver

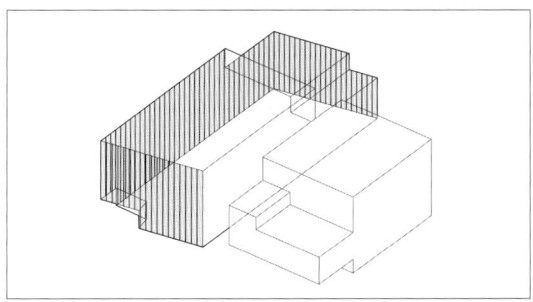

– 브라운 계열의 루버를 사용하여 인접한 살레시오 문화원의 적벽돌과 위화감이 없게 조성하고 일조조절의 기능도 함께 한다.

- Child alley-red brick

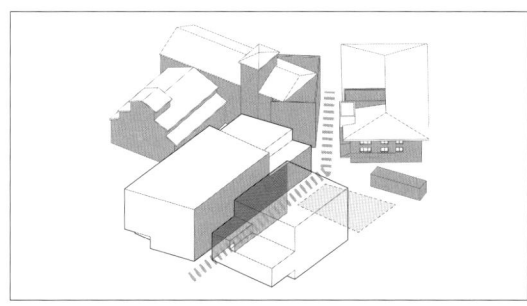

– 어린이 골목을 둘러싼 입면은 적벽돌을 사용하여 동네의 골목과 같은 따뜻한 분위기를 부여하고 살레시오 문화원으로 연장된다.

Composition of Section

- Daycare center
- Woman friendly center
- Kindergarten

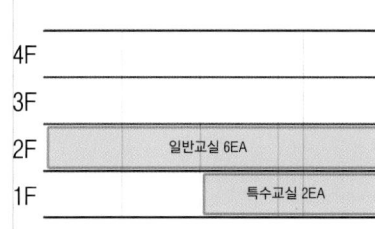

– 이 공간에는 다양한 용도의 시설이 복합적으로 함께하기에 층수에 따르는 위치의 적정성이 매우 중요하다. 이 중 유치원과 어린이집 그리고 여성친화센터의 조성이 특히 중요하다. 유치원과 어린이집은 저층에 위치하면서 외부의 어린이놀이터와 연결이 전제되며, 라운지와 결합하는 여성친화센터는 안정적인 운영이 가능하도록 한다.

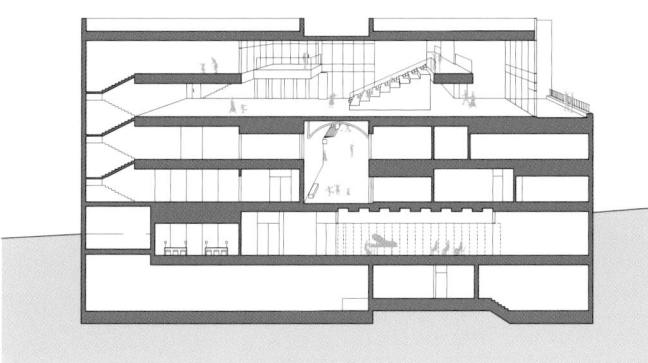

Cross section

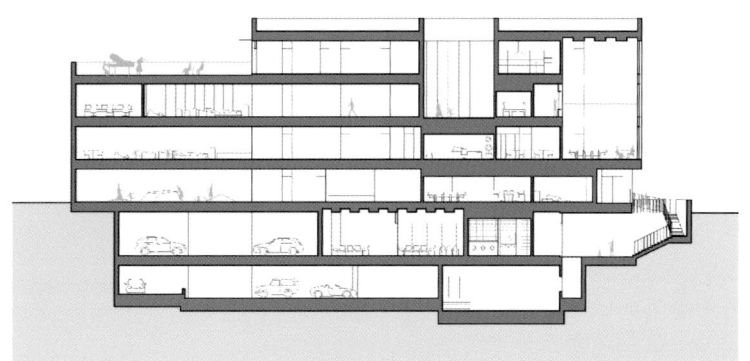

Longitudinal section

2nd Prize _ 2등작

Singil District 12 Social Welfare Facility & Kindergarten

1. Children's playground
2. Special classroom
3. Administration office
4. Library
5. Learning material room
6. Hall
7. Health room
8. Director office
9. Nursery room
10. Kitchen
11. Food warehouse
12. Book warehouse
13. Teacher's room
14. Part-time nursery room
15. Deck
16. Classroom
17. Kids cafe
18. Children library
19. Tea making room / Lounge
20. Toy warehouse
21. Group counseling room
22. Office
23. Personal counseling room
24. Center director office
25. Mamdeundeun center
26. Lactation room
27. Assembly hall
28. Playing room
29. Lounge
30. Activity space
31. Private room
32. Powder room
33. Study room
34. Cafe shelter
35. Small kitchen
36. Restaurant
37. Laundry room
38. Material store
39. Education room
40. Kitchen ancillary room

3rd floor plan

4th floor plan

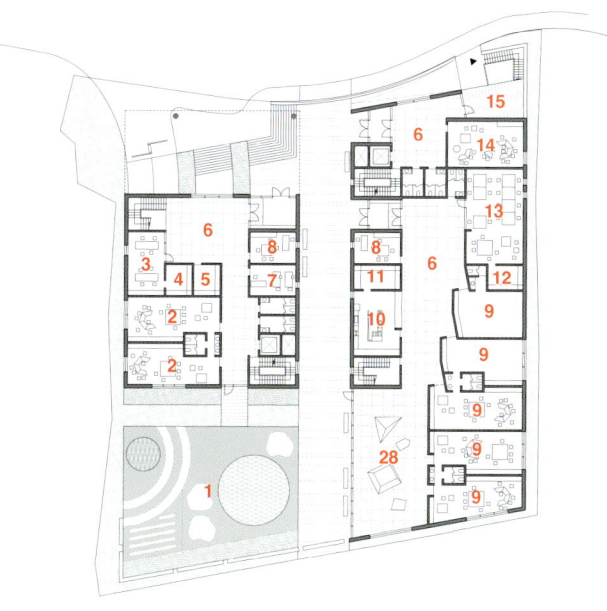

1st floor plan

2nd floor plan

Wanju-gun Agricultural Technology Center
완주군 농업기술센터

REGAON Architecture&Planners Co., LTD / Hyunjo Lee
㈜리가온건축사사무소 / 이현조

In the '2050 Carbon Neutral Declaration', 'circular economy' emerged as a prerequisite for carbon zero. The circular economy is a concept that symbolizes carbon zero by transforming into a sustainable economic structure that recycles resources through management and regeneration. With the motif of 'resource circulation', which symbolizes carbon zero, the arrangement of functional spaces for work, research, production, and education of the Agricultural Technology Center is transformed from a linear spatial structure around the existing main road to a 'circular spatial structure' that can efficiently use the complex. This takes it one step further as a low-carbon, zero-energy facility. It is hoped that the 'Wanju-gun Agricultural Technology Center' will become a symbolic advanced example of the carbon-neutral pilot project.

Considering the characteristics of the remodeling project, it is reviewed so that the facilities for realizing carbon zero can be placed in the optimal space by analyzing the facility hierarchy and circulation in the existing complex. It is proposed to create a circular walkway with a new master plan that connects facilities within the complex. It is intended to increase the efficiency of research and work through the circular shielding road and eco-cycling that is made up of solar power facilities connected to the central street of the research complex, accommodate various events such as rest, field trips, and experiences, and improve pedestrian safety and use efficiency of the site.

'2050 탄소중립 선언'에서, 탄소중립의 선행과제로 '순환 경제'가 대두되었다. 순환경제란 관리와 재생을 통해 자원을 재활용하는 지속적 경제구조로 바꾸어 나가는 것으로 탄소중립을 상징하는 개념이다. 탄소중립을 상징하는 '자원순환'을 모티브로, 농업기술센터의 업무/연구/생산/교육 등의 기능 공간 배치를 기존 메인도로 중심의 선형적 공간구조에서 단지를 효율적으로 활용할 수 있는 '순환적 공간구조'로 탈바꿈하고자 한다. 이를 통해 저탄소 제로에너지 시설로서 한 단계 더 성장하며, 탄소중립 시범사업의 상징적인 선진사례 '완주군 농업기술센터'가 되기를 기대한다.

리모델링 사업의 특성을 고려하여 기존 단지 내 시설 위계와 동선 분석을 통해 탄소중립 실현을 위한 시설물이 최적의 공간에 배치될 수 있도록 검토하고, 단지 내 시설을 연결하는 새로운 마스터 플랜을 제안하여 순환 산책로 조성을 제안하였다. 연구단지의 중심가로와 연계된 태양광 시설로 이루어진 순환형 우천로, 에코 사이클링을 통해 연구 및 업무의 효율을 높이고 휴게, 견학 및 체험 등 다양한 이벤트를 수용하며 보행 안전성과 부지의 이용 효율을 향상시키고자 하였다.

Location : 945-70, Samgi-ri, Gosan-myeon, Wanju-gun, Jeollabuk-do, Korea I **Function** : Research facility I **Site area** : 32,196㎡ I **Total floor area** : 3,941㎡ I **Stories** : 2FL I **Structure** : Reinforced concrete I **Finish** : Aluminium composite panel, Aluminum vertical louver, Low-E triple glass

위치 : 전라북도 완주군 고산면 삼기리 945-70번지 일원 I **용도** : 연구시설 I **규모** : 지상2층 I **구조** : 철근콘크리트 I **마감** : 알루미늄복합패널, 알루미늄수직루버, 로이삼중유리 I **설계팀** : 김용준, 윤용상, 유세란, 강민지, 김민성

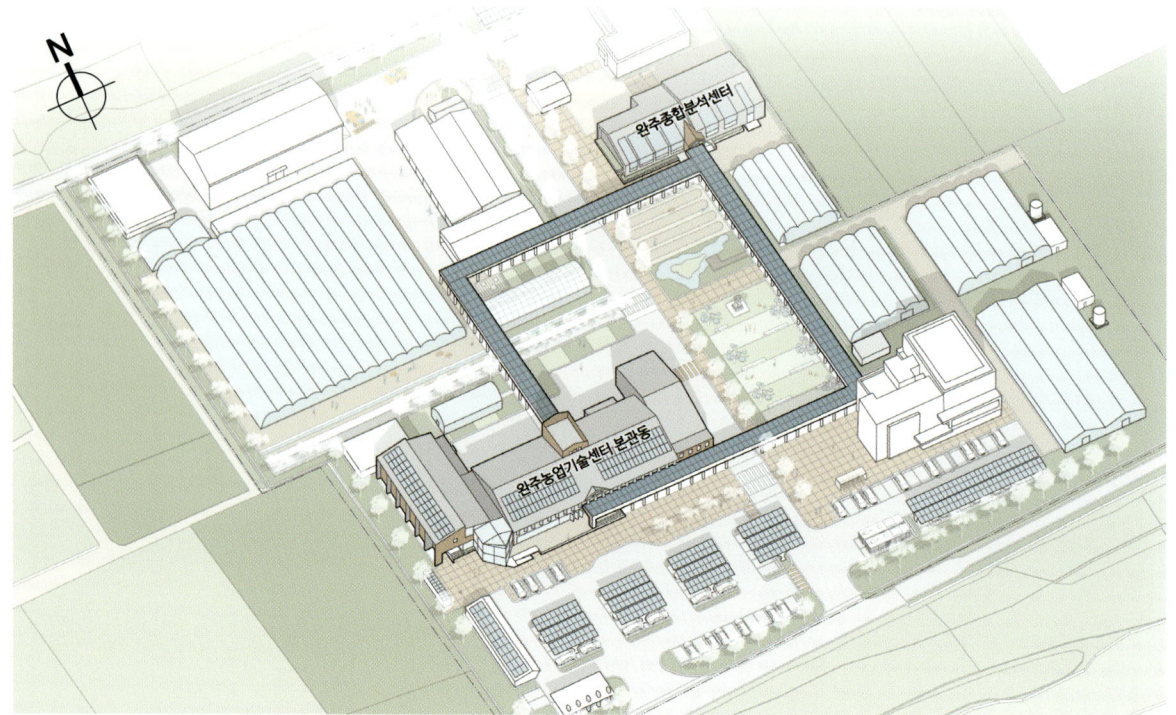

Site plan

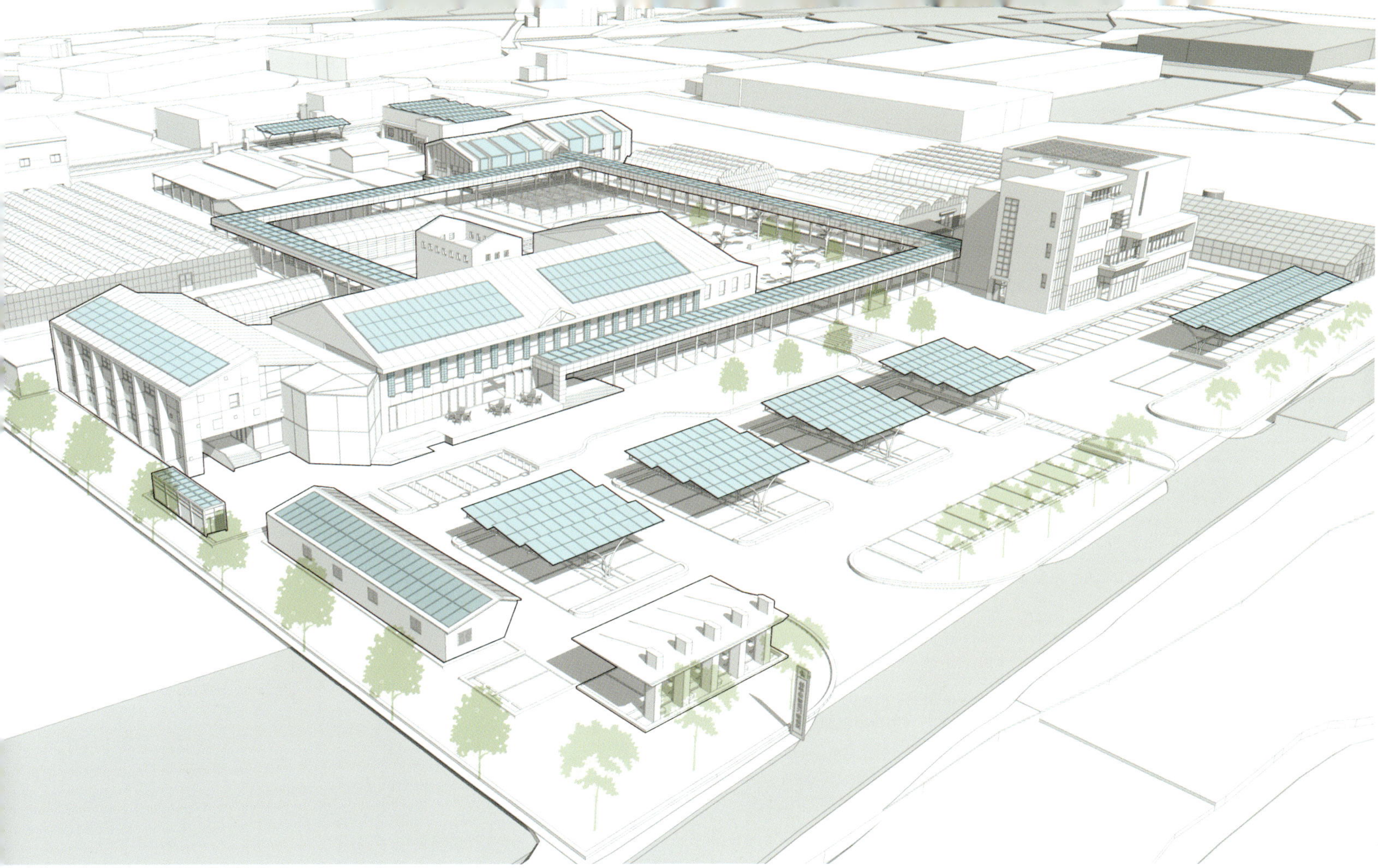

Concept

탄소중립 목표를 구현하는 농업기술센터의 녹색성장고리
ECO - CYCLE RING
[에코-사이클링]

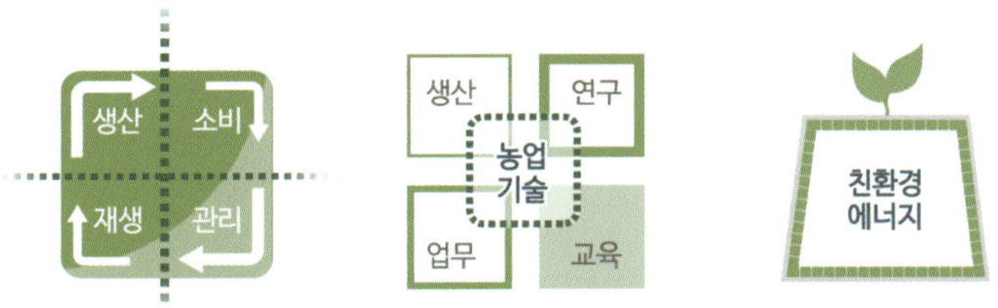

SYMBOL
탄소중립을 위한
자원의 순환을 상징

FUCTION
시설간 연계로
연구/업무 시너지 강화

ENERGY
저탄소, 제로에너지
연구단지 조성

Process

- Hierarchy setting

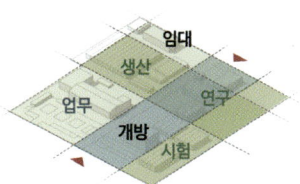

- Outside space

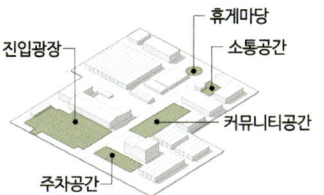

- Layout & Remodeling

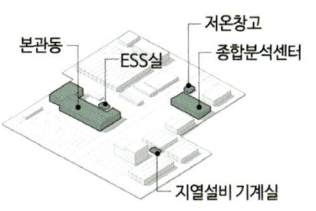

- Energy production & Saving facility plan

Winner _ 당선작

Master Plan

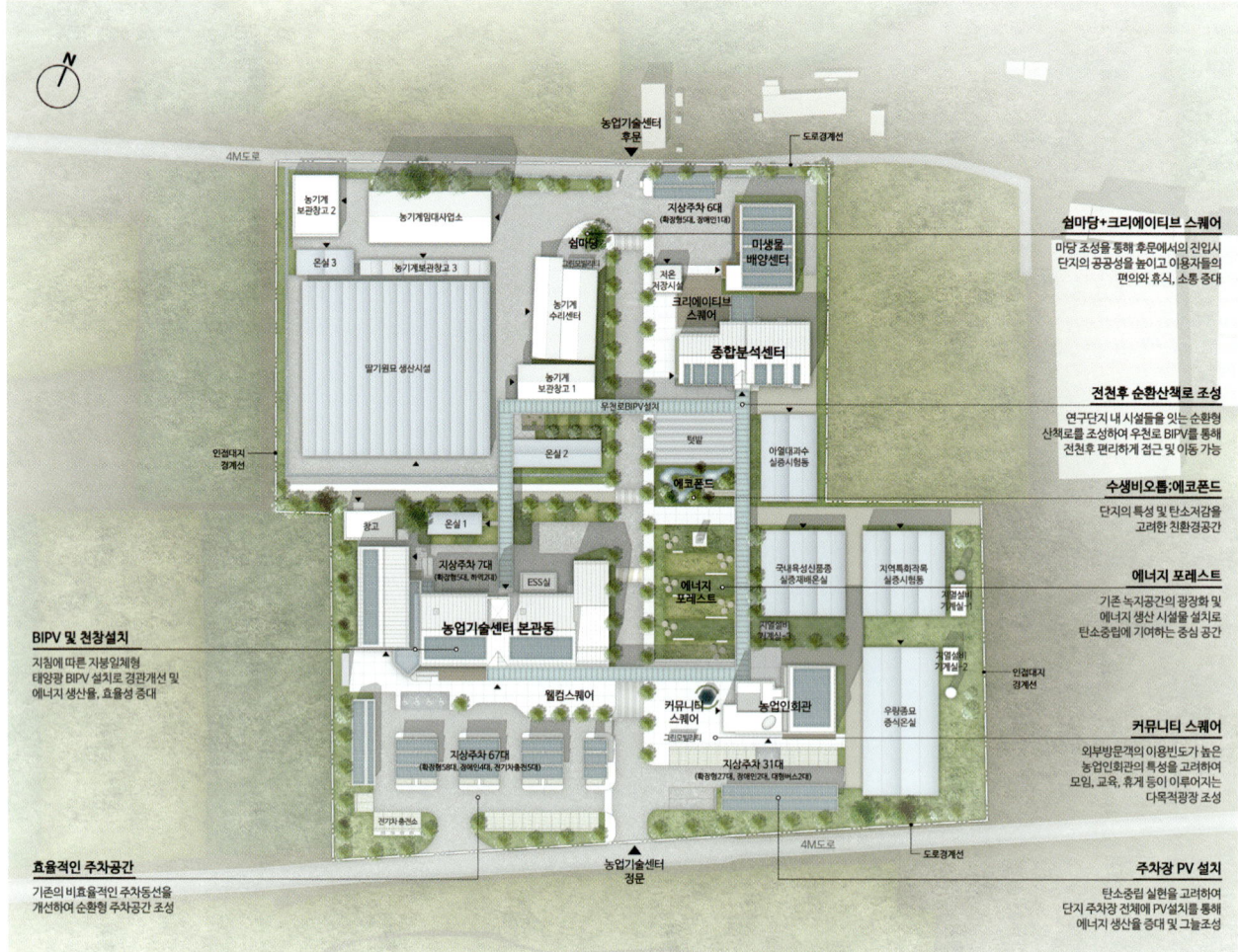

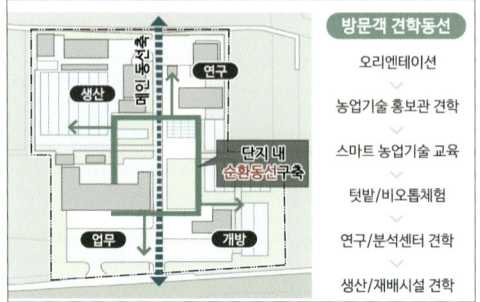

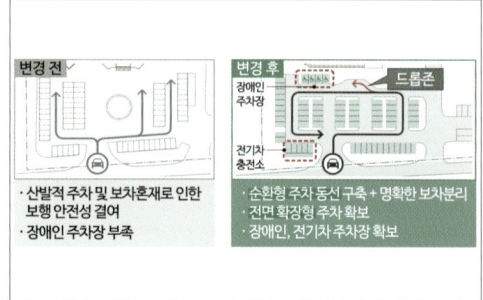

Elevation Plan

- Agricultural technology center main building

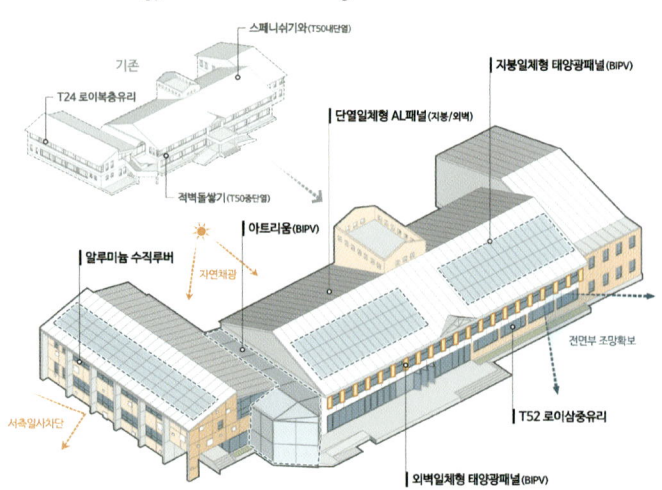

- Wanju comprehensive analysis center

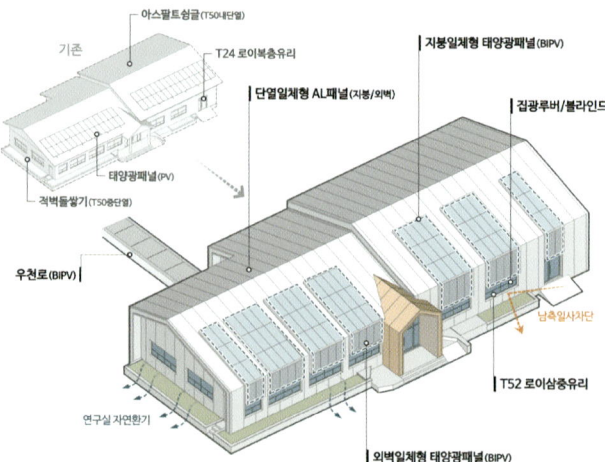

Wanju-gun Agricultural Technology Center >>

1. Kitchen
2. Restaurant
3. Seminar room
4. Community lounge
5. Toilet
6. Wind break room
7. Crop general hospital
8. Pesticide residue analysis room
9. Livestock examination room
10. Agricultural field support office
11. General laboratory
12. Creative hall
13. Tissue culture room
14. Rest hall
15. Eco hall
16. Alcohol theme TF team
17. Department of eco-friendly agriculture and Livestock
18. Business group
19. Technology dissemination department
20. Resource development department
21. Small meeting room
22. Director's office
23. Document room

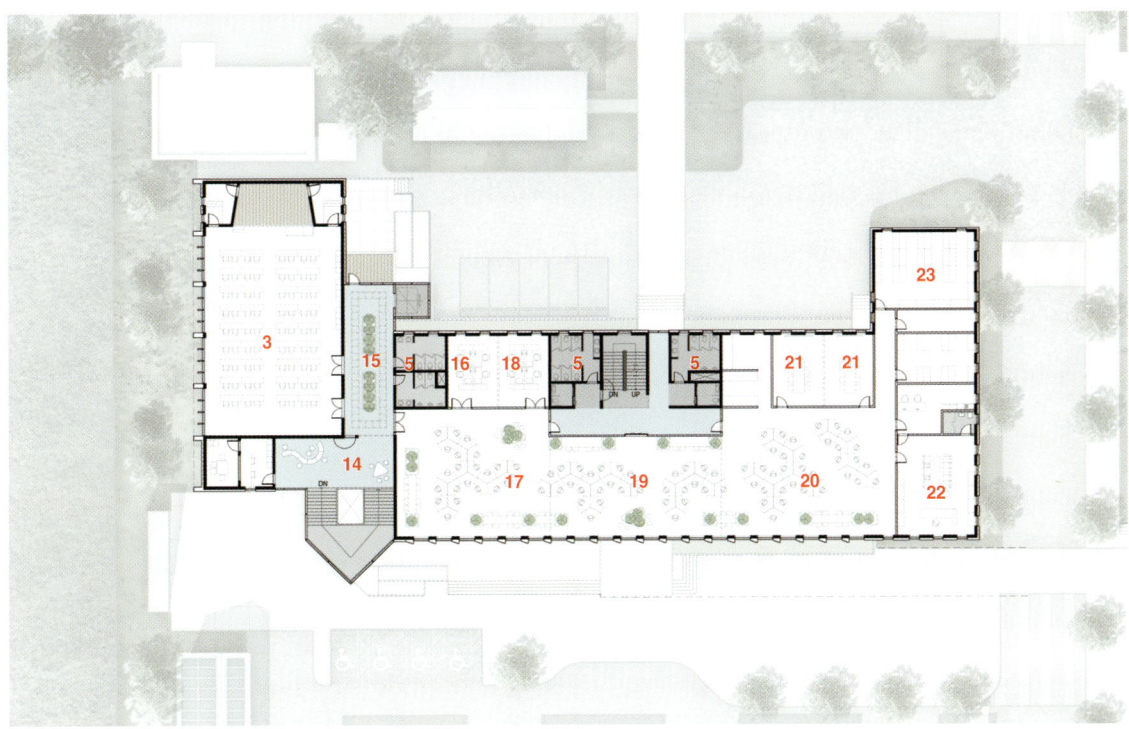

2nd floor plan

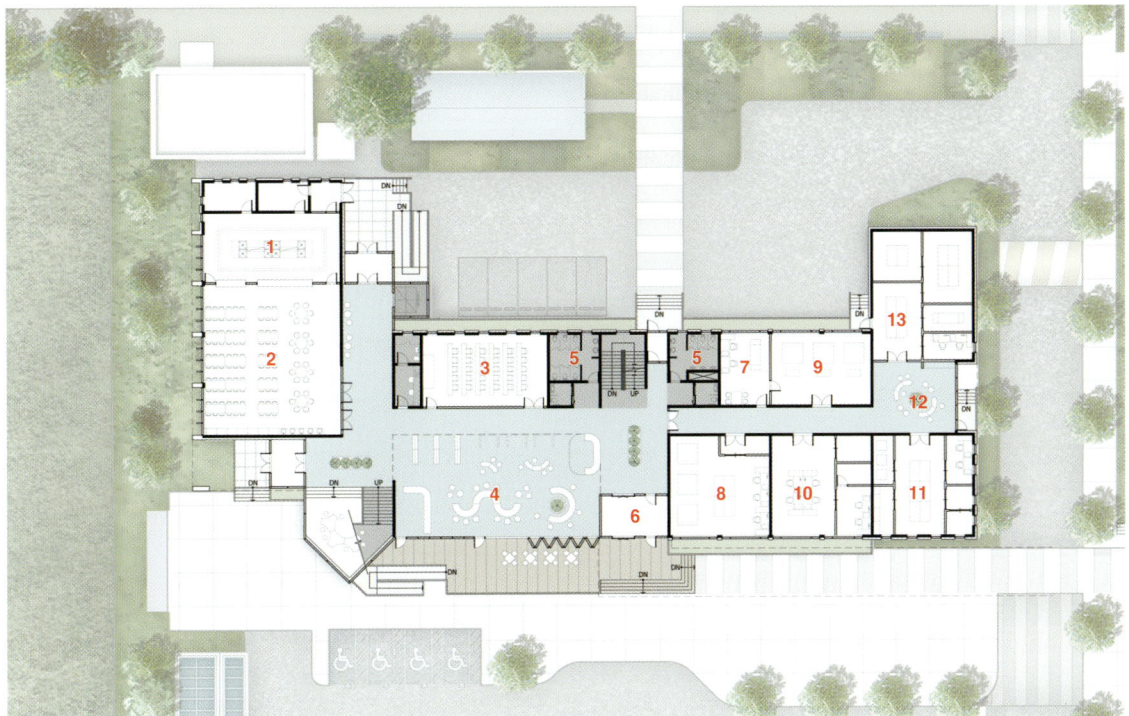

1st floor plan

| 108 | Jinghe New City Culture & Art Centre
징허 신도시 문화& 예술 센터 |
| 114 | Construction Project of Third KINTEX Exhibition Hall
킨텍스 제3전시장 |
| 120 | National Digital Heritage Center
국립디지털문화유산센터 |
| 126 | National Jeongdong Theater_Winner
국립 정동극장_1등작 |
| 132 | National Jeongdong Theater_2nd Prize
국립 정동극장_2등작 |
| 138 | House for Film & Media in Stuttgart
슈투트가르트 영화 & 미디어 하우스 |
| 142 | Montreal Holocaust Museum
몬트리올 홀로코스트 뮤지엄 |
| 146 | Turku Music Centre
투르쿠 음악 센터 |
| 154 | Wuxi Art Museum
우시 미술관 |
| 158 | Philharmonic Hall
필하모닉 홀 |
| 162 | Jinju Public Science Museum
진주공립 전문과학관 |
| 166 | Saskatoon New Central Library
새스커툰 뉴 센트럴 라이브러리 |
| 172 | Terezín Ghetto Museum
테레진 게토 박물관 |
| 178 | National Intangible Heritage Center Miryang Branch
국립무형유산원 밀양 분원 |
| 186 | Namdo Righteous Army History Museum
남도의병역사박물관 |

Jinghe New City Culture & Art Centre
징허 신도시 문화 & 예술 센터

Zaha Hadid Architects / Patrik Schumacher
자하 하디드 아키텍츠 / 패트릭 슈마허

Jinghe New City is growing as a science and technology hub north of Xi'an in China's Shaanxi province. Supported by new scientific research institutes and driven by environmental considerations, the city is becoming a centre for developing industries focussing on new energy and materials, artificial intelligence and aerospace.

Echoing the meandering valleys carved by the Jinghe River through the mountains and landscapes of Shaanxi province, the Jinghe New City Culture & Art Centre is located within the Jinghe Bay Academician Science & Technology Innovation district of the city.

The centre's design intertwines with the city's existing urban masterplan to connect the new multimedia library to the north of Jinghe Avenue with the new performing arts theatre, multi-function halls, studios and exhibition galleries to the south via elevated courtyards, gardens and paths that span the avenue's eight lanes of traffic below.

With gently sloping ramps providing a gateway to the district's network of elevated public walkways, the centre weaves through the city to link its commercial and residential districts with the parks and river to the south, while also bringing the city's residents into the heart of the building and providing direct access to the planned metro station.

Organized as a series of flowing volumes, layers and surfaces interconnecting with courtyards and landscapes, the design defines a sequence of interior and exterior cultural and recreational spaces for its community.

The multimedia library's terraces overlook its full-height atrium with diffusing skylights to provide a variety of public reading zones for individual and collective research. The library will integrate print publications together with immersive virtual reality technologies that expand the boundaries of learning and enrich the exchange of knowledge.

Located on the southern side of the avenue, the performing arts theatre accommodates 450 people and can be adapted for many types of events. The multi-function hall, studios and galleries are stacked and arranged around the theatre to share public areas designed to enhance accessibility and inter-disciplinary collaboration.

Solar irradiation analysis and responsive site planning optimise the centre's use of natural ventilation and daylight in the mild temperate climate of Jinghe New City. Incorporating photovoltaic panels for on-site power generation together with rainwater collection, the centre's construction will prioritise locally-produced materials with a high recycled content to achieve a 3-star certification in China's Green Building program.

Location : Shaanxi province, China I **Function** : Cultural facility I **Commercial director** : Charles Walker I **Project director** : Satoshi Ohashi I **Associate director** : Yang Jingwen I **Project architect** : Nan Jiang I **Project team** : Sanxing Zhao, Lianyuan Ye, Shaofei Zhang, Qiyue Li, Shuchen Dong, Yuan Feng, Congyue Wang, Yuling Ma, Yanran Lu I **Sustainability** : Carlos Bausa Martinez, Bahaa Alnassrallah, Aditya Ambare

위치 : 중국 시안시 I 용도 : 문화시설

Site Analysis

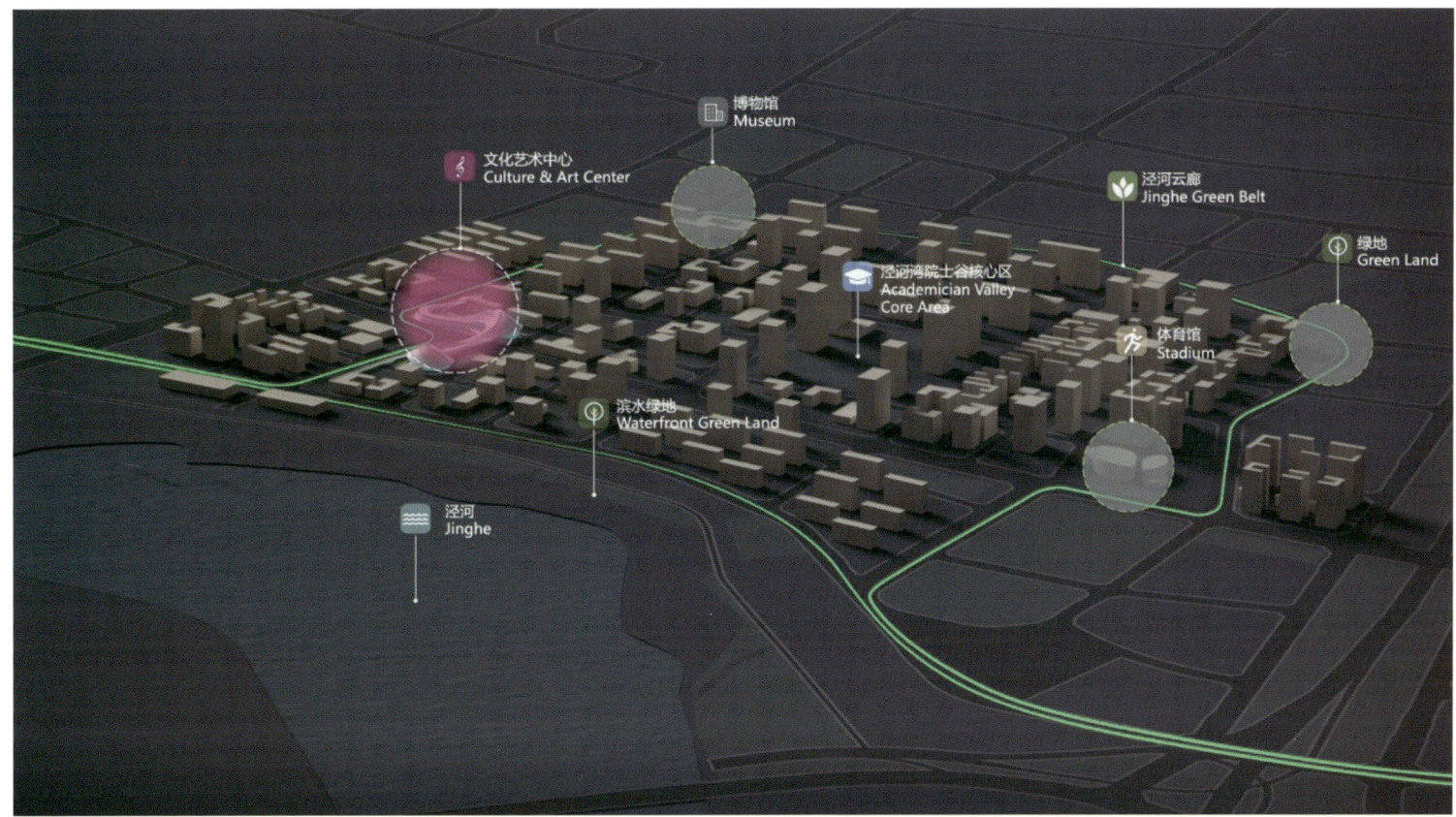

Circulation Plan

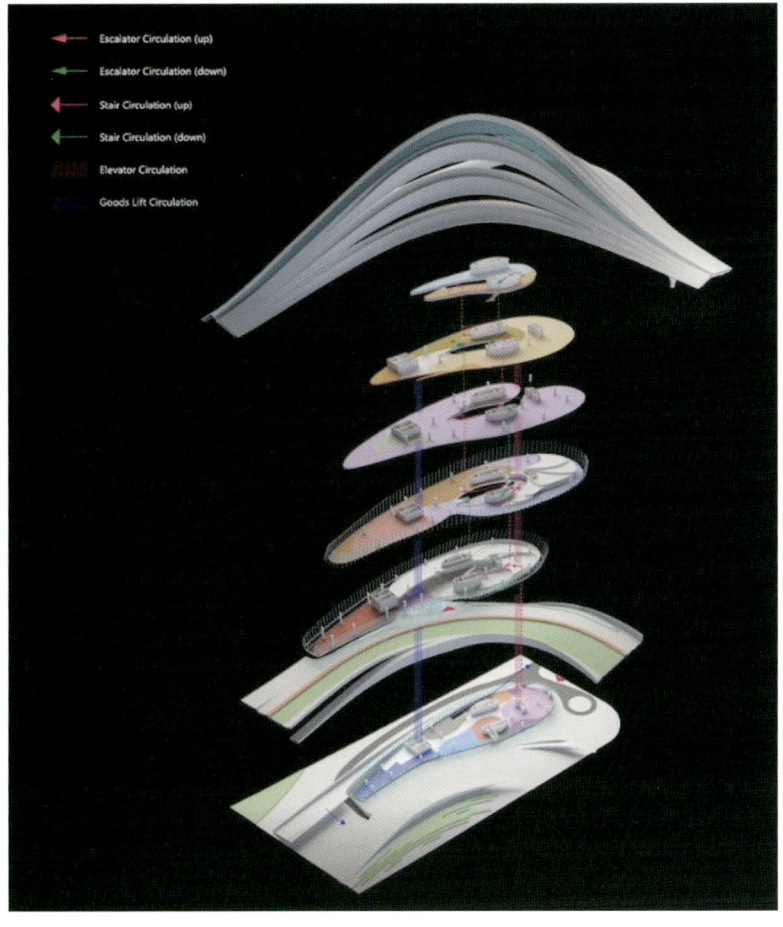
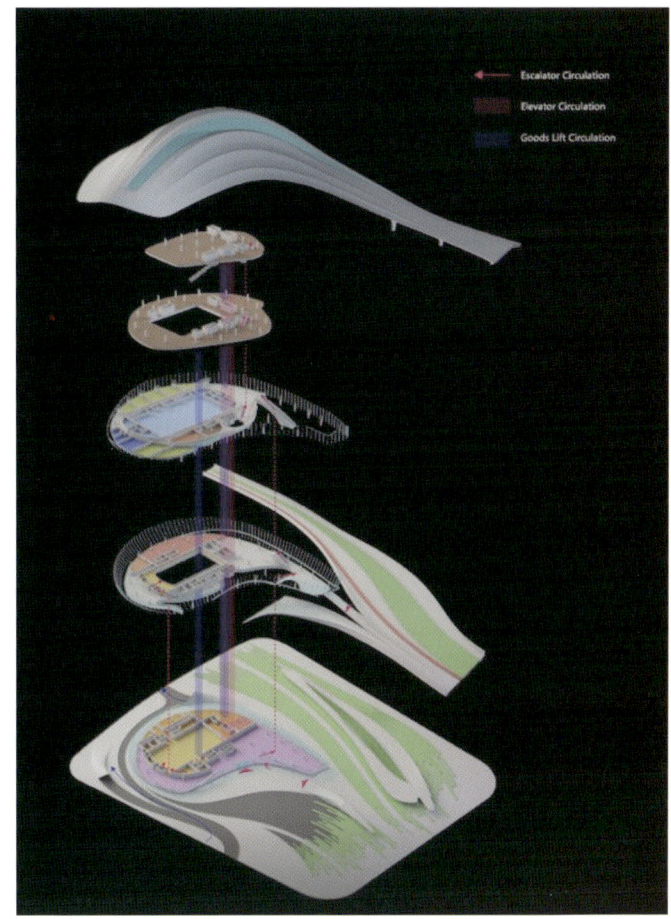

Winner _ 당선작

중국 산시성 시안 북쪽에 있는 징허 신도시는 과학기술허브로 성장하고 있다. 새로운 과학 연구기관의 지원과 환경적 고려 사항을 통한 추진을 바탕으로, 이 도시는 새로운 에너지 및 재료, 인공 지능, 항공 우주에 주력하는 산업 개발의 중심지가 되고 있다. 징허강이 산시성의 산과 풍경을 가로지르면서 생긴 구불구불한 계곡이 느껴지는 징허 신도시 문화예술센터는 징허베이 과학기술혁신 지구 내에 위치하고 있다.
이 센터의 디자인은 기존의 도시 마스터플랜과 얽혀서 아래 에비뉴의 8차선을 가로지르는 융기된 중정, 정원, 보행로를 통해 징허 애비뉴 북쪽의 새로운 멀티미디어 도서관과 남쪽의 새로운 공연장, 다목적홀, 스튜디오, 전시 갤러리를 연결한다.
완만한 경사로에 융기된 공공 보도 네트워크의 관문이 있으므로 이 센터는 도시를 가로지르며 상업 및 주거 지역을 남쪽의 공원 및 강과 연결하면서 도시 주민들을 건물의 중심부로 안내하며 계획된 지하철역에 바로 접근할 수 있다.
중정, 풍경과 연결되어 흐르는 볼륨, 층, 표면으로 구성된 디자인은 커뮤니티를 위한 일련의 내외부 문화여가공간을 정의한다.

멀티미디어 도서관의 테라스에서 개인 및 공동 연구를 위한 다양한 공공 열람 구역을 갖춘 확산 채광창이 있는 전체 높이의 아트리움을 내려다볼 수 있다. 도서관은 학습의 경계를 확장하고 지식의 교환을 풍부하게 하는 몰입형 가상 현실 기술과 함께 인쇄 출판물을 제공할 것이다.
애비뉴 남쪽에 위치한 공연장은 450명을 수용할 수 있으며 다양한 유형의 행사에 적합하다. 다목적홀, 스튜디오, 갤러리는 공연장을 중심으로 적층되고 배치되어 접근성과 학제 간 협업을 향상시키기 위해 설계된 공공 영역을 공유한다.
일사량 분석과 반응형 대지 계획은 징허 신도시의 온화한 온대 기후에서 센터의 자연 환기와 일광 사용을 최적화한다. 빗물 수집과 함께 발전용 태양광 패널을 결합하고, 재활용 가능한 현지 생산 자재를 우선적으로 사용하여 중국의 그린 빌딩 프로그램에서 3성급 인증을 획득할 것이다.

Jinghe New City Culture & Art Centre >>

Section Plan

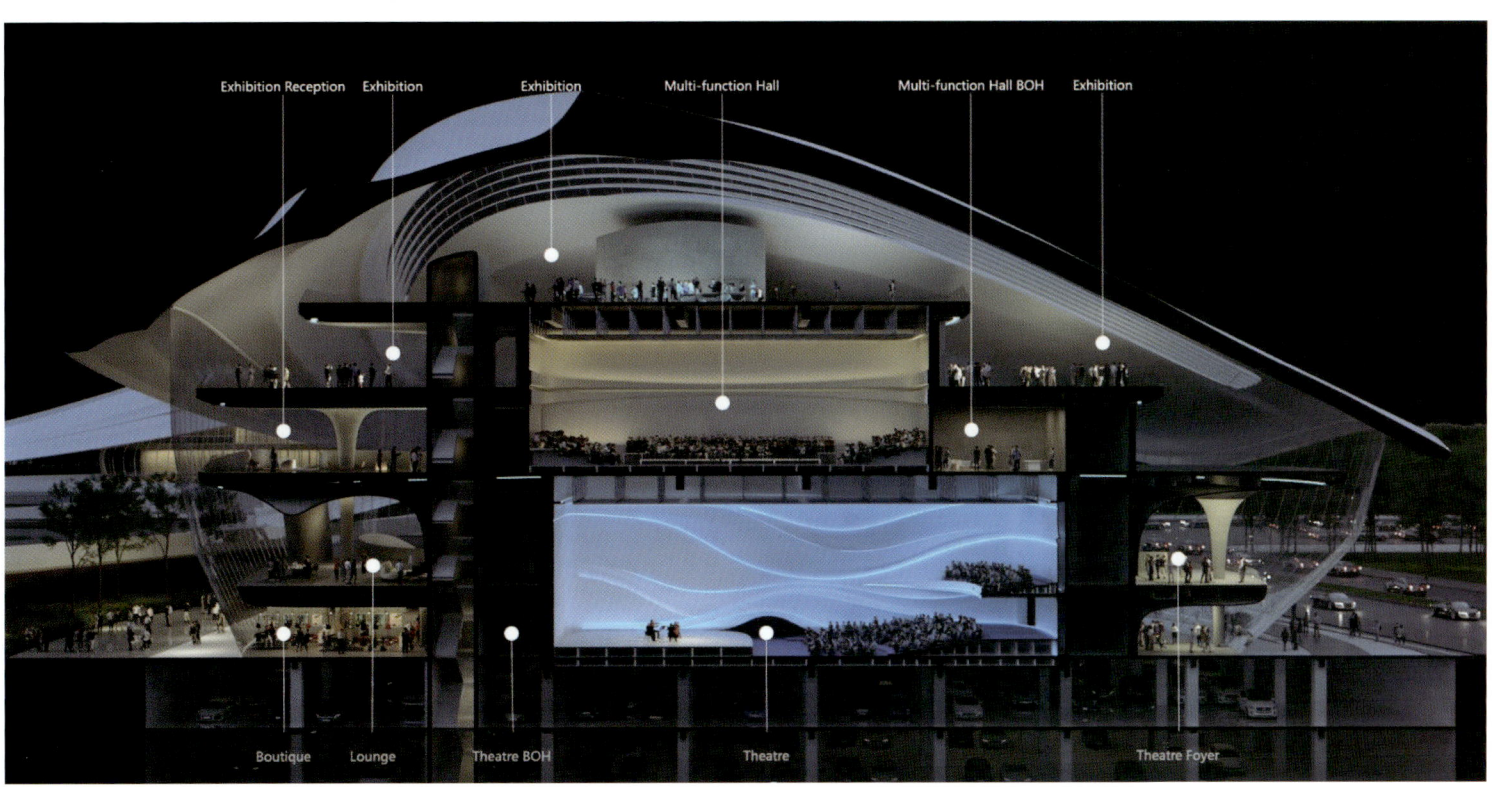

Program I

Lobby
Lockers
Storage&Reading
24h Book Return
Office
BOH
Parking
Social
Study Room
Digital Reading
Kid's Club
Parent&Child
Lounge
Commercial
Cafe
Restaurant
Retail
Develop
Robot Librarian
Augmented Reality
Screening / DIY
Roof
Augmented Reality
Study Room
Lounge
Parent&Child
Restaurant
Lounge
Lobby
IP Retail
MEP
Robot Librarian
Parking Entrance
Management
Digital Reading
Open-stack
Reading
Kid's Club
Reading
Cafe
Entrance
Parking Exit
Lockers
Lobby
24h Book Return
BOH
F&B
Screening / DIY
| F5
Lounge / Augmented Reality
| F4
Study Room / Digital Reading
| F3
Open-stack
| F2
Dining / Playground / Storage
| F1
IP Retail / Reading/ Landscape
| B1
F&B / Workshop / Management
| B2&B3
Parking

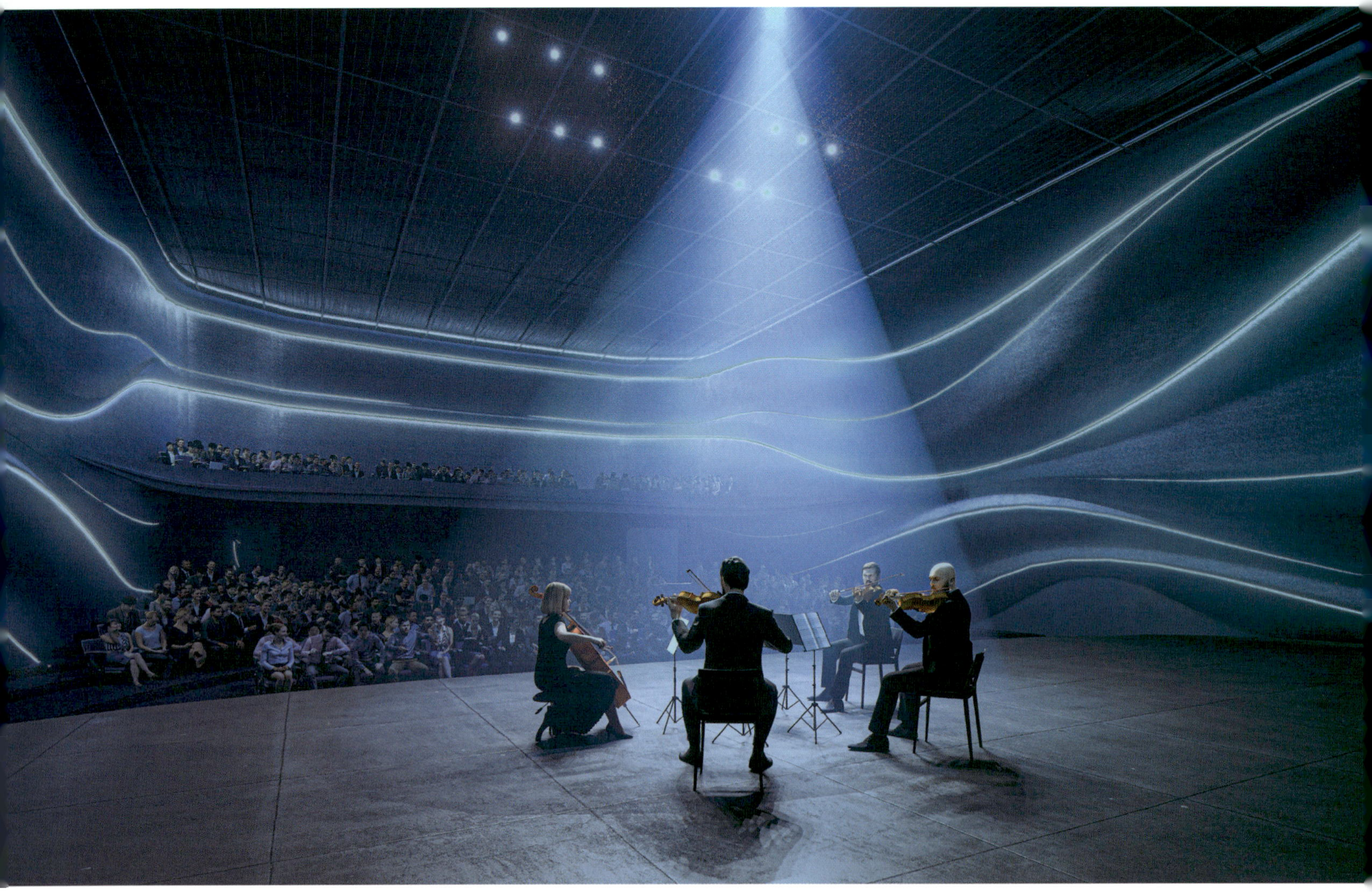

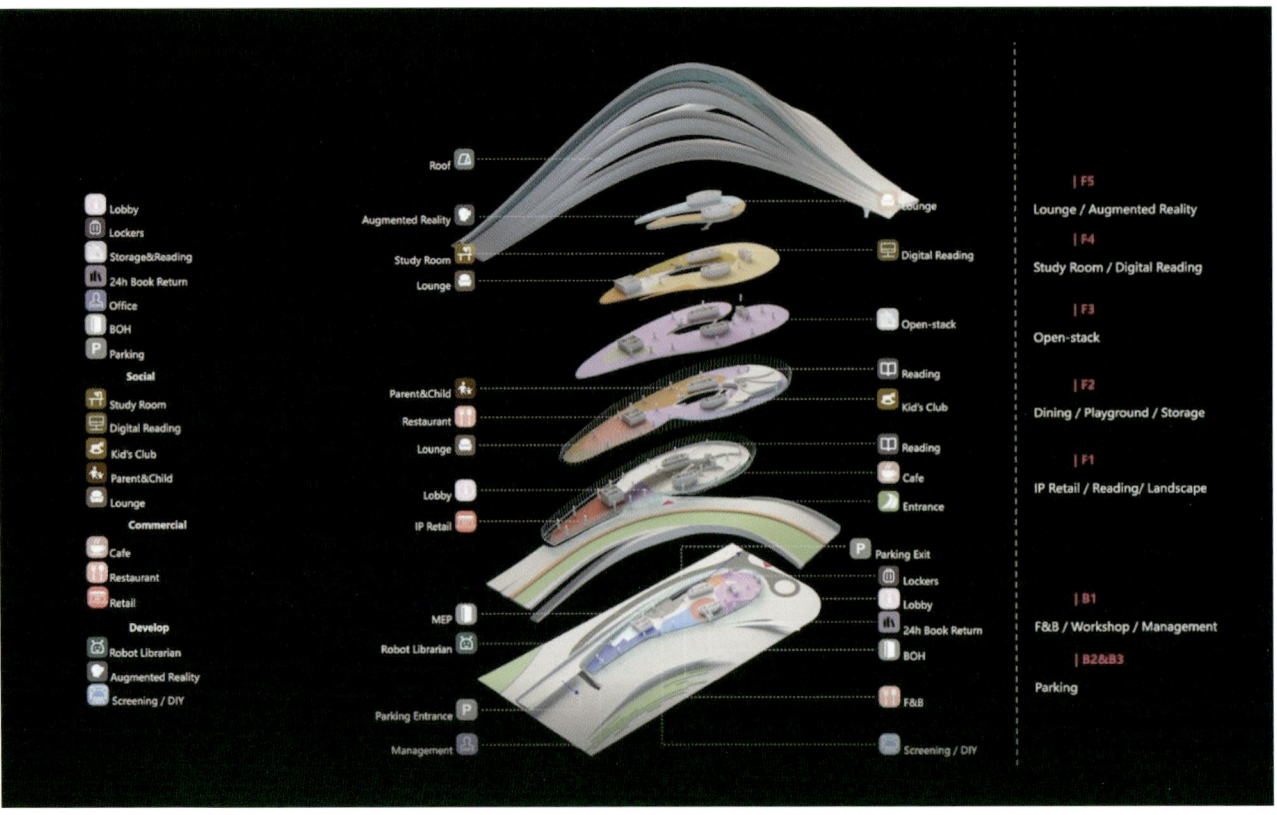

Program Ⅱ

| F5
Exhibition
| F4
Muti-function Hall / Exhibition
| F3
Muti-function Hall / Studio / Office
| F2
Theatre / Restaurant / Landscape
| F1
Theatre / Landscape
| B1&B2
Parking
Roof
Exhibition
Exhibition
Muti-function Hall BOH
Office
Restaurant
Retail
Cafe
Parking Exit
Theatre BOH
Theatre
Parking Entrance
Double-height Muti-function Hall
Training / Studio
Muti-function Hall
Cafe
Theatre Auditorium
Entrance
Retail
Lobby
Entrance
Basic
Lobby
Office
Parking
Social
Theatre
Exhibition
Training / Studio
Commercial
Cafe
Restaurant
Retail
Develop
Muti-function Hall

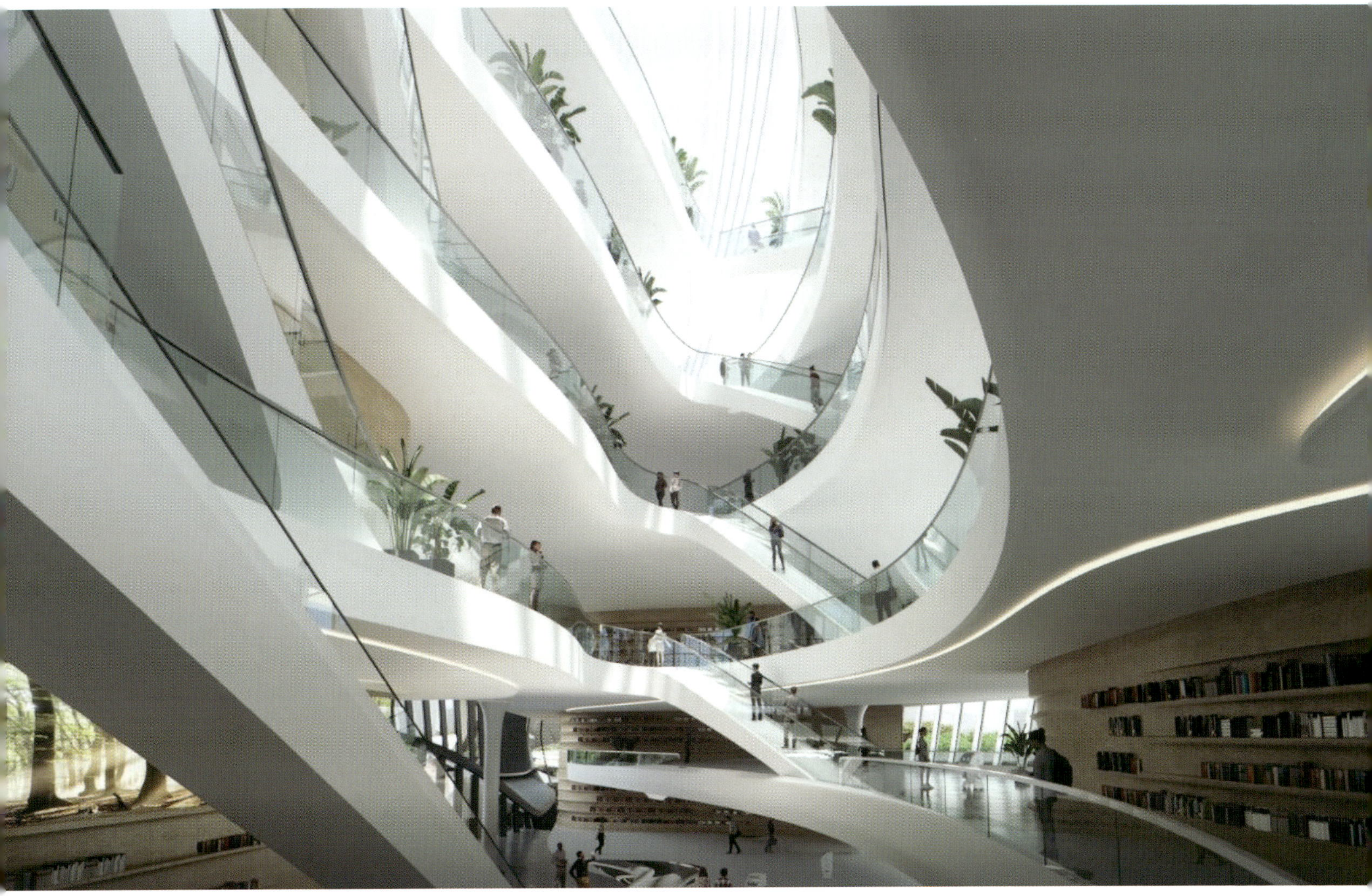

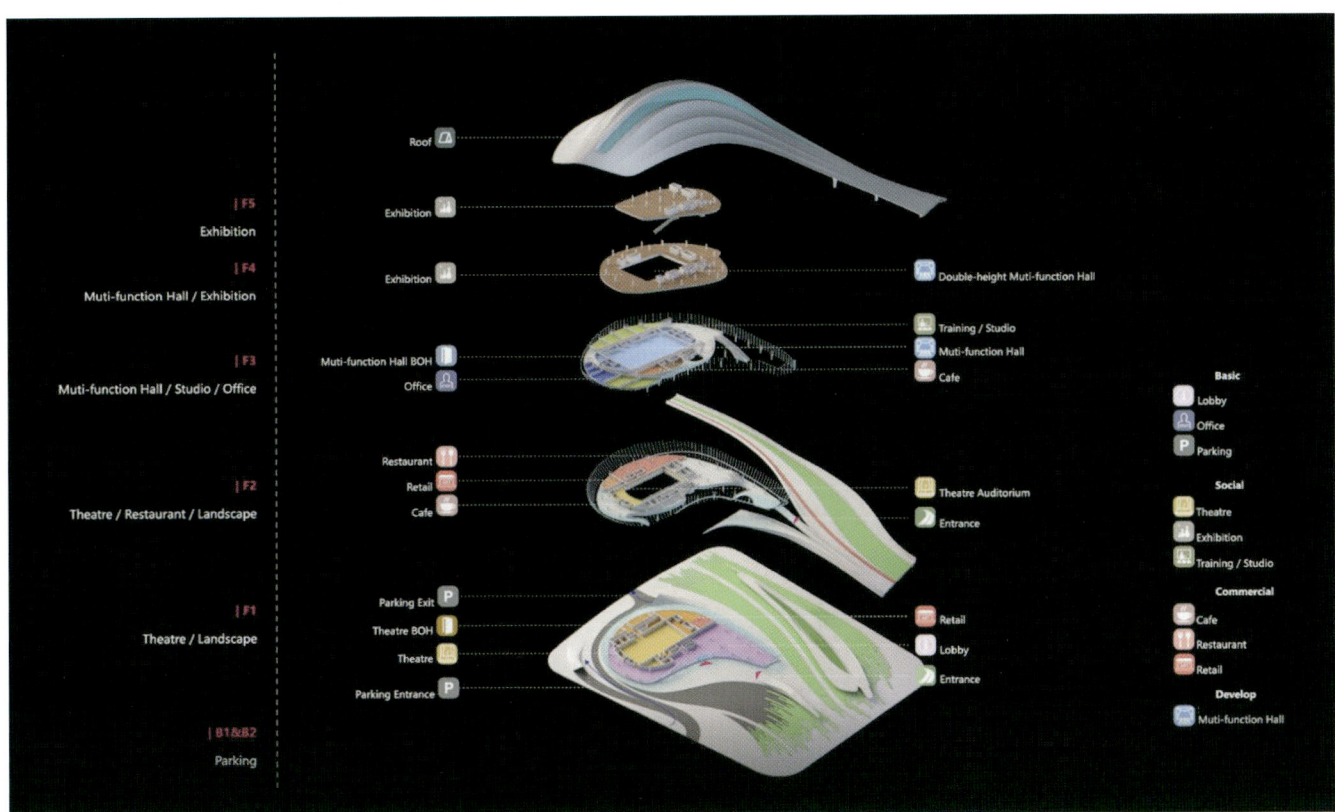

Winner _ 당선작

Construction Project of Third KINTEX Exhibition Hall
킨텍스 제3전시장

HAEAHN Architecture, Inc. / Taeman Kim + SANGJI Architecture / Dongyoon Heo
㈜해안종합건축사사무소 / 김태만 + ㈜상지엔지니어링건축사사무소 / 허동윤

The Third KINTEX Exhibition Hall Construction Project aims to evolve the entire complex into a global business platform by planning Exhibition Halls 3A and 3B that can expand the existing functions in harmony with KINTEX Exhibition Halls 1 and 2 and creating an integrated passage connecting all exhibition halls. By accepting the concept of Exhibition Halls 1 and 2 with flowers and butterflies as motifs, the existing Exhibition Hall 1 is linked with Exhibition Hall 3A and completed on a scale of 100,000m², and Exhibition Hall 2 is integrated with Exhibition Hall 3B to expand its size and function. The integrated connecting passage is set as the central walking axis in the north-south direction to easily and clearly combine the all facilities. The main direction of the plan is to consider this connecting passage as a special internal/external space along with three-dimensional circulations to make it a new central space for KINTEX and to present a vision to serve as the main axis of growth for future expansion. Accordingly, the arrangement and mass design are made so that Exhibition Halls 1, 2, and 3 are integrated and harmonized with the same motif, and the integrated connecting passage is designed in a flexible and symbolic form by applying a shell structure to expand the concept of 'Urban Concourse' and serve as the hub of the city for the connection of the inside and outside. Each exhibition hall is created as a smart exhibition hall to enable flexible operation in response to the changes in the paradigm of the MICE industry by giving characteristics to each area. Urban Concourse is an open concourse in Goyang City that weaves the entire exhibition hall and connects it with the outside pedestrian plaza to become a space for communication between the city and citizens. It provides easy movement and rest for exhibition visitors, and a green shelter for citizens to visit comfortably at any time. It will become a new symbolic space of KINTEX and a multipurpose cultural space that is reborn as a city attraction. "KINTEX THE GRAND", completed through three-stage expansion, is a complex MICE cluster that encompasses conferences, exhibitions, and culture with a multipurpose exhibition facility of 180,000m² and a complex cultural space that is an Urban Concourse. It will become a strong center of the domestic MICE industry and a venue for the global MICE industry.

Location : 217-59, 217-60, Kintex-ro, Ilsanseo-gu, Goyang-si, Gyeonggi-do, Korea I **Function** : Cultural & Assembly facility I **Site area** : 182,115m I **Bldg. area** : 342,906m I **Total floor area** : 301,628m I **Stories** : B2, 3FL I **Structure** : Reinforced concrete, Steel frame, Steel framed reinforced concrete I **Finish** : Metal panel, Metal louver, Low-E pair glass

위치 : 경기도 고양시 일산서구 킨텍스로 217-59, 217-60 일원 I **용도** : 문화 및 집회시설 I **규모** : 지하2층, 지상3층 I **구조** : 철근콘크리트, 철골, 철골철근콘크리트 I **마감** : 금속패널, 금속루버, 로이복층유리 I **설계팀** : ㈜해안종합건축사사무소 / 오현석, 심경아, 이찬주, 신상진, 홍준호, 엄수려, 이한새, 김찬, 조한울, 김도훈, 서미경, 김남훈, 노승민, 민태영, 박우정, 이행숙, 김성겸, 김영용, 황수미, 김용원 + ㈜상지엔지니어링건축사사무소 / 양성원, 조한중, 이한수

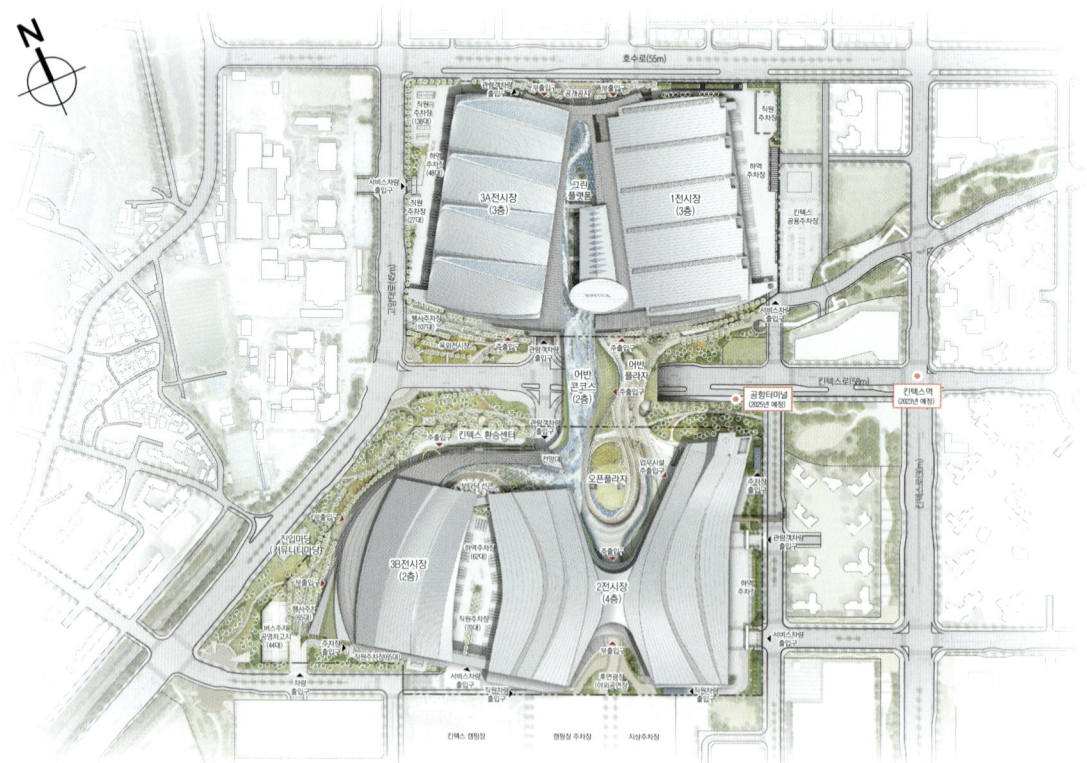

Site plan

■ Mass Process

기존	조화	경관	통합
	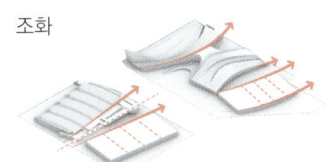	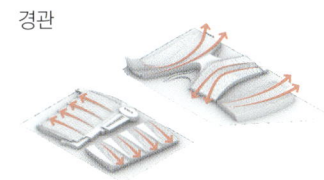	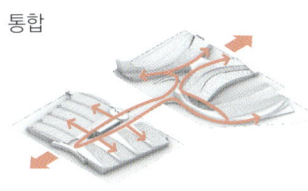

킨텍스 제3전시장 건립사업은 킨텍스 1, 2전시장과 조화를 이루면서 기존 기능을 확장시킬 수 있는 3A, 3B전시장을 계획하고 모든 전시장을 잇는 통합연결통로를 만들어 단지 전체를 글로벌 비즈니스 플랫폼으로 진화시키는 목표를 담고 있다. 꽃과 나비 모티브인 1, 2전시장 개념을 받아들여 기존 1전시장은 3A전시장과 연계시켜 100,000㎡ 규모로 완결하고, 2전시장은 3B전시장과 통합시켜 규모와 기능을 확장시켰다. 통합연결통로는 남북 방향의 중심보행축으로 설정하여 쉽고 명확하게 전체 시설이 하나로 연결되게 했다. 이 연결통로를 입체적 동선과 함께 특별한 내·외부 공간으로 계획하여 킨텍스의 새로운 중심공간으로 만들고 추후 확장까지 고려한 성장의 메인축의 역할을 할 수 있도록 비전을 제시한 것이 계획안의 주된 방향이다. 이에 1, 2, 3전시장이 같은 모티브로 통합되어 조화를 이루도록 배치 및 매스 디자인을 했고 통합연결통로는 '어반콘코스'로 개념을 확장시켜 내외부가 연결되는 도시의 허브역할을 할 수 있도록 쉘구조를 적용하여 유연하고 상징적인 형태로 디자인했다. 각 전시장은 구역별 특성을 부여하고 마이스 산업 패러다임의 변화에 따라 유연한 운영이 가능하도록 스마트 전시관으로 조성했다. 어반콘코스는 전체 전시장을 엮는 동시에 고양시의 열린 콘코스로서, 외부 보행 광장과 연결하여 도시와 시민의 소통 공간이 되게 했다. 전시 관람객에게는 쉬운 이동과 휴식의 장소가 되고 시민들에게는 언제나 편안하게 방문할 수 있는 녹색 쉼터가 된다. 킨텍스의 새로운 상징공간이자 도시 명소로 거듭나는 다목적 문화공간이 되는 것이다. 3단계 확장을 통해 완성되는 '킨텍스 더 그랜드'는 180,000㎡ 규모의 다목적 전시시설과 어반콘코스인 복합문화공간을 갖춘 회의, 전시, 문화를 아우르는 복합 마이스 클러스터로 국내 마이스 산업의 확고한 중심이며 글로벌 마이스 산업의 장이 될 것이다.

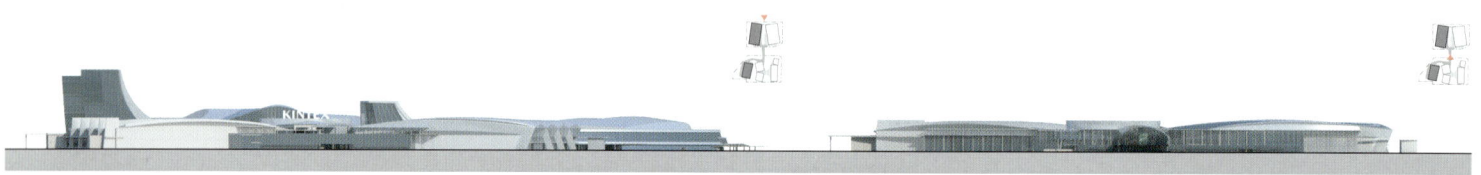

Elevation I - 3A+1

Elevation II - 3B+2

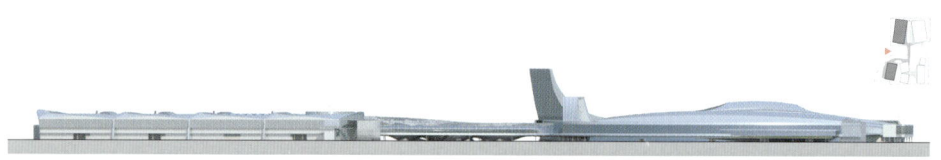

Elevation III

Urban Concourse Plan

- Program concept

- Program layout

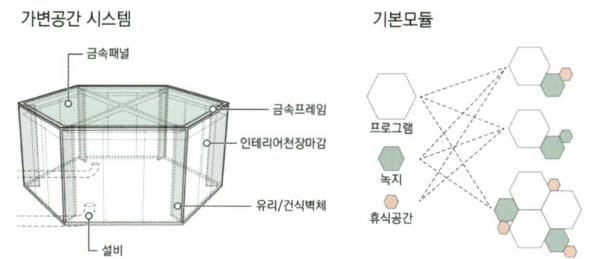

365일 계속되는 시설로 특정 프로그램을 도입하지 않아도 다양한 단위의 공간을 조합하여 1인, 2인, 가족단위들이 영위할 수 있는 다양한 이벤트 공간을 제시했다.

"Daily Convenience"
킨텍스 주변 거주자들의 일상을 고려한 시설

"Culture Anchor"
F&B와 식료품, POP-UP시설이 결합된 데일리 앵커

"Completed Play Ground"
키즈존, 플리마켓, 푸드트럭등 온 가족이 즐기는 시설

"Business & Exhibition Backup"
식음, 휴식 등 킨텍스 방문객의 편의를 고려한 시설

Winner _ 당선작

▌ Circulation Plan

- 3A exhibition hall
- Integrated passageway
- 3B exhibition hall
- Office layout for intensive management

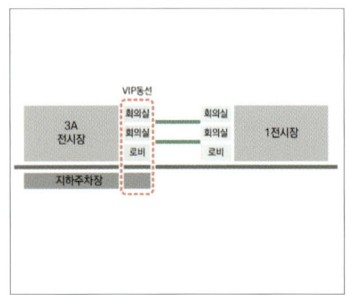
- 대규모 전시공간과 업무 및 회의 집중시설

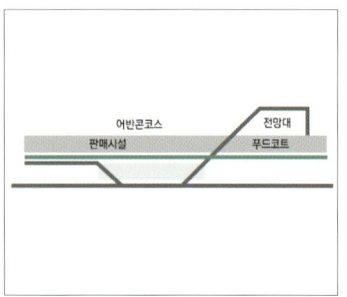
- 자연과 함께 전체를 연결하는 수평 보행통로

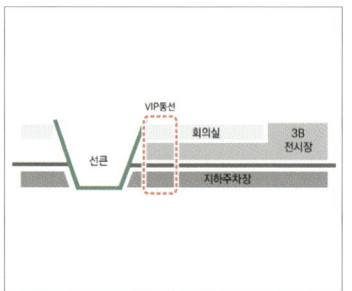
- 지하주차장 및 선큰과 연계된 각 층별 진출입공간 구성

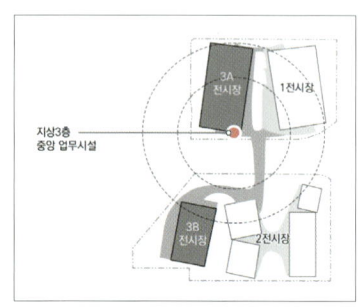
- 전체적인 유지관리 가능한 위치에 중앙 업무시설 배치

▌ Floor Plan

- Arrangement of appropriate service facilities for each unit of exhibition hall
- Intensive arrangement of conference facilities in 3A exhibition hall
- Integrated exhibition composition in Exhibition Hall 1 and Exhibition Hall 2
- Urban Concourse that can be operated separately

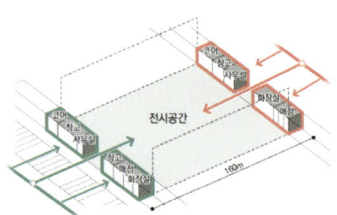
- 방문객공간과 서비스공간의 명확한 분리 및 화장실, 사무실 등의 적정 시설 배치계획

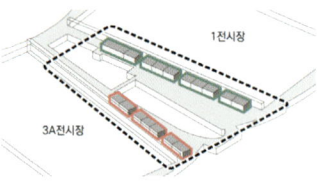
- 2, 3층에 위치한 회의, 컨벤션 시설을 기존 회의공간과 연계 배치하여 이용의 효율성 증대

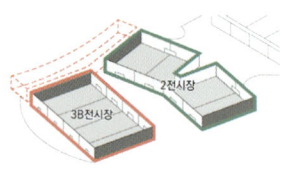
- 2전시장과 3B전시장의 적극적인 연결을 통해 다양한 행사에 상호 보완가능한 전시 구성

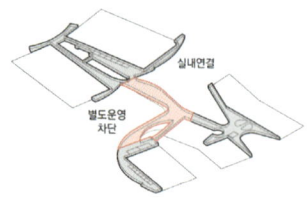
- 킨텍스 전체를 아우르는 연결통로를 실내로 계획하여 사계절 제약 없이 별도 운영이 가능한 공간으로 계획

Construction Project of Third KINTEX Exhibition Hall

▍ Program & Zoning

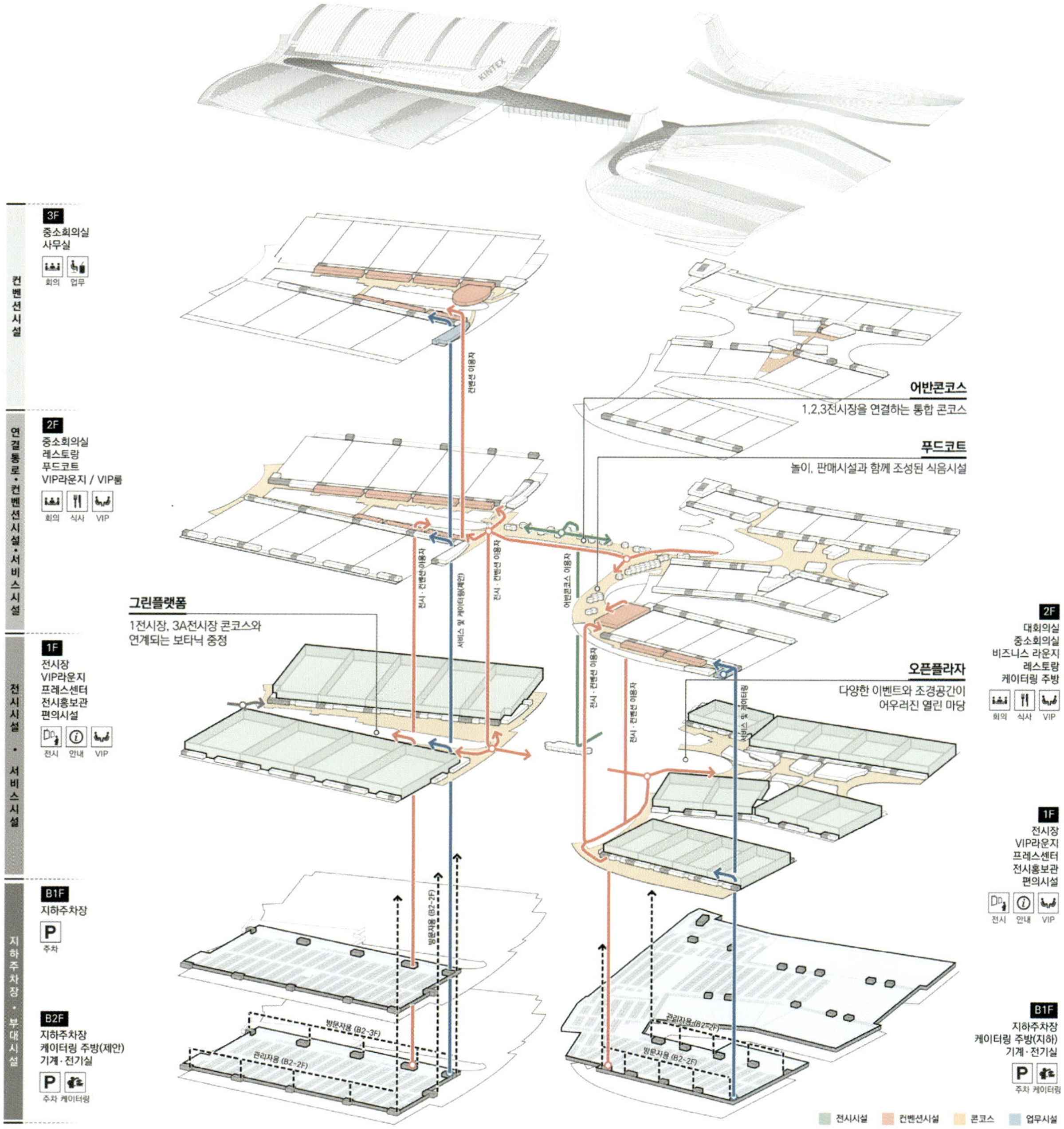

National Digital Heritage Center
국립디지털문화유산센터

HAHN International Architects / Jongruhl Hahn + GilBartolome Architects / Pablo Gil Martinez, Jaime Bartoloome Yllera
㈜한종률도시건축건축사사무소 / 한종률 + 길바르톨로메 아키텍츠 / 파블로 길 마르티네즈, 하이메 바르톨로메 일레라

A spatial and media instrument

Our proposal is to offer a 'Spatial Instrument' to be 'played' by the people running the building; institutions public and private, managers, curators, creatives, technologists, maintenance and operations personnel, and finally its users.

A contemporary and technologically advanced tool that will develop contents as per the plan established in 'Plan and exhibition report', but also that will need to adapt to future technological developments and changes in the exhibitions and other activities that will be held on the building.

The building is more than an exhibition space; it is also a place of creation and production of contents. In addition, the building is the re-presentation of a new institution- the National Digital Heritage Centre that needs to communicate what is the past, the present and the future of the Republic of Korea. Therefore the building needs to communicate values that go beyond the use of its square meters of space, it should also be a 'media instrument of communication'.

The NDHC building is a contextualized piece within the museum complex, adding value to the complex through the technology employed in the building envelope.

공간과 매체의 도구

디지털문화유산센터(NDHC)는 직원부터 창작자, 방문하는 사람들이 '연주'하는 '공간적인 악기가 될 것이다. 공모자료에 따르면 현대적이고 기술적으로 진보된 도구는 컨텐츠를 발전시킨다. 더불어 NDHC는 향후 건물에서 일어날 활동들과 전시에 따른 변화와 미래의 기술적인 발전에 적응할 것이다. 건축은 오랜 시간에 걸친 변화에 탄력적으로 대응한다. 이는 Liquid Society(경계없는 사회)에 긍정적인 기여를 하고, 조화롭게 한다.

NDHC는 다양한 공간과 영역에 따라 간단하지만 상당히 유연하게 설계되었다. 또한 박물관 단지의 맥락에 어울리며, 건물 자체가 기술적으로 가치를 더해준다. 건물 앞 광장은 공공공간으로 잠재적으로 발표회나 이벤트, 전시 등을 열 수 있다. 미디어파사드는 북쪽과 동쪽으로 큰 스크린이 될 것이다. 건물의 주출입은 북쪽 정면의 사이로 접근하여 비나 과도한 햇빛을 막아주는 열린 공간이 된다.

Location : (Moon S-1) cultural facility site, (S-1 living area) Sejong-dong, Sejong-si, Korea I **Function** : Cultural & Assembly facility I **Site area** : 8,189㎡ I **Bldg. area** : 1,607㎡ I **Total floor area** : 8,951㎡ I **Stories** : B2, 3FL I **Structure** : Reinforced concrete, Steel truss I **Finish** : Glass curtain wall

위치 : 세종시 세종동(S-1생활권) 문화시설용지(문S-1) 내 I 용도 : 문화 및 집회시설 I 규모 : 지하2층, 지상3층 I 구조 : 철근콘크리트, 철골트러스 I 마감 : 유리커튼월 I 설계팀 : ㈜한종률도시건축건축사사무소 / 김창원, 이규호, 김수정, 김유경

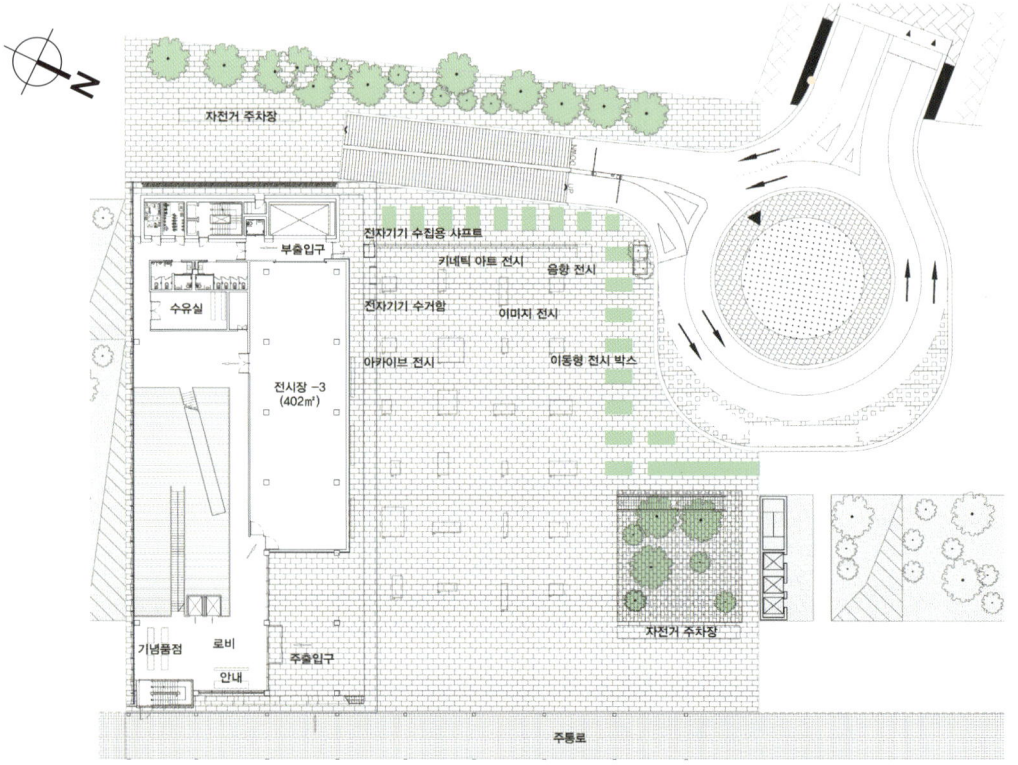

Site plan

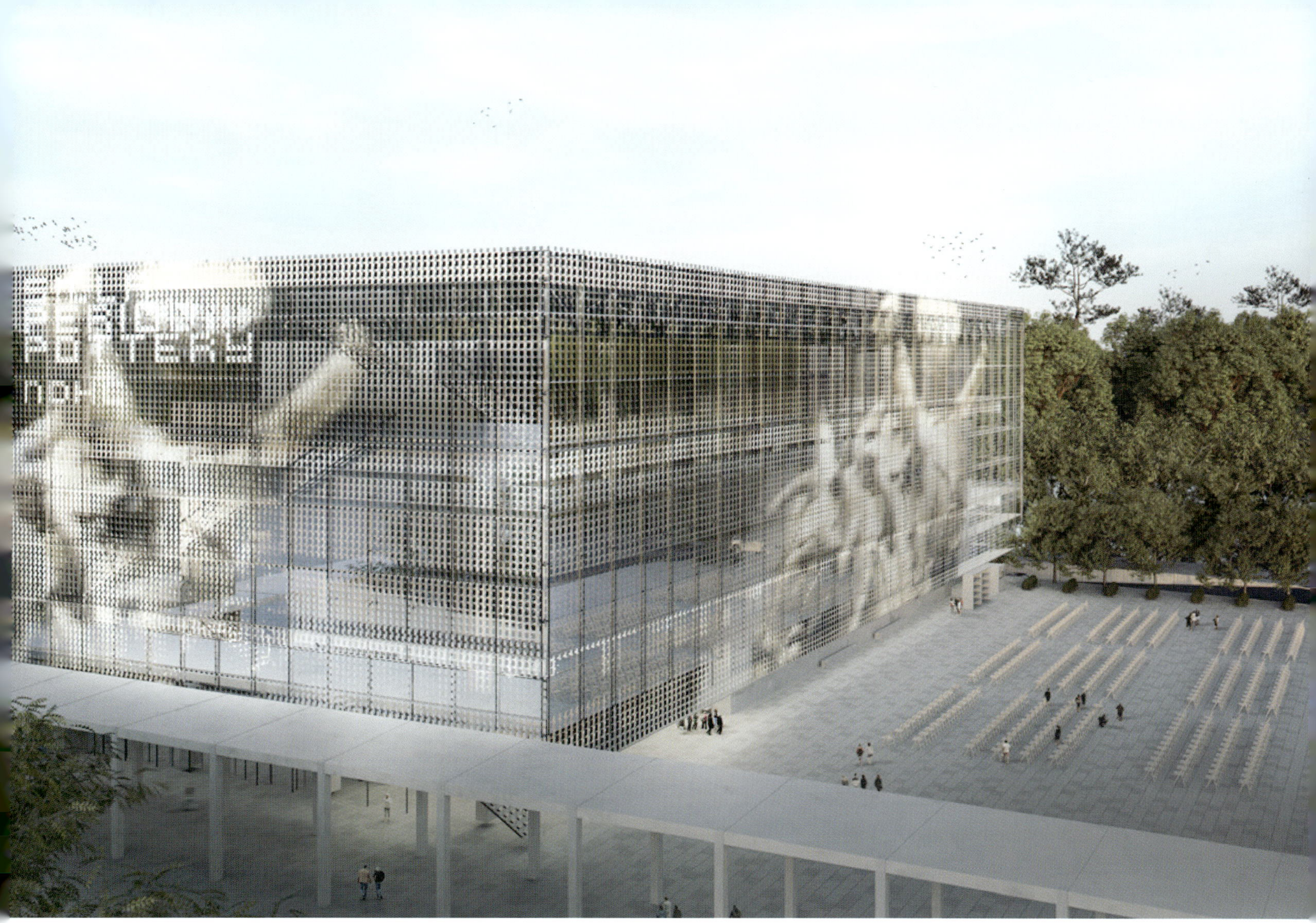

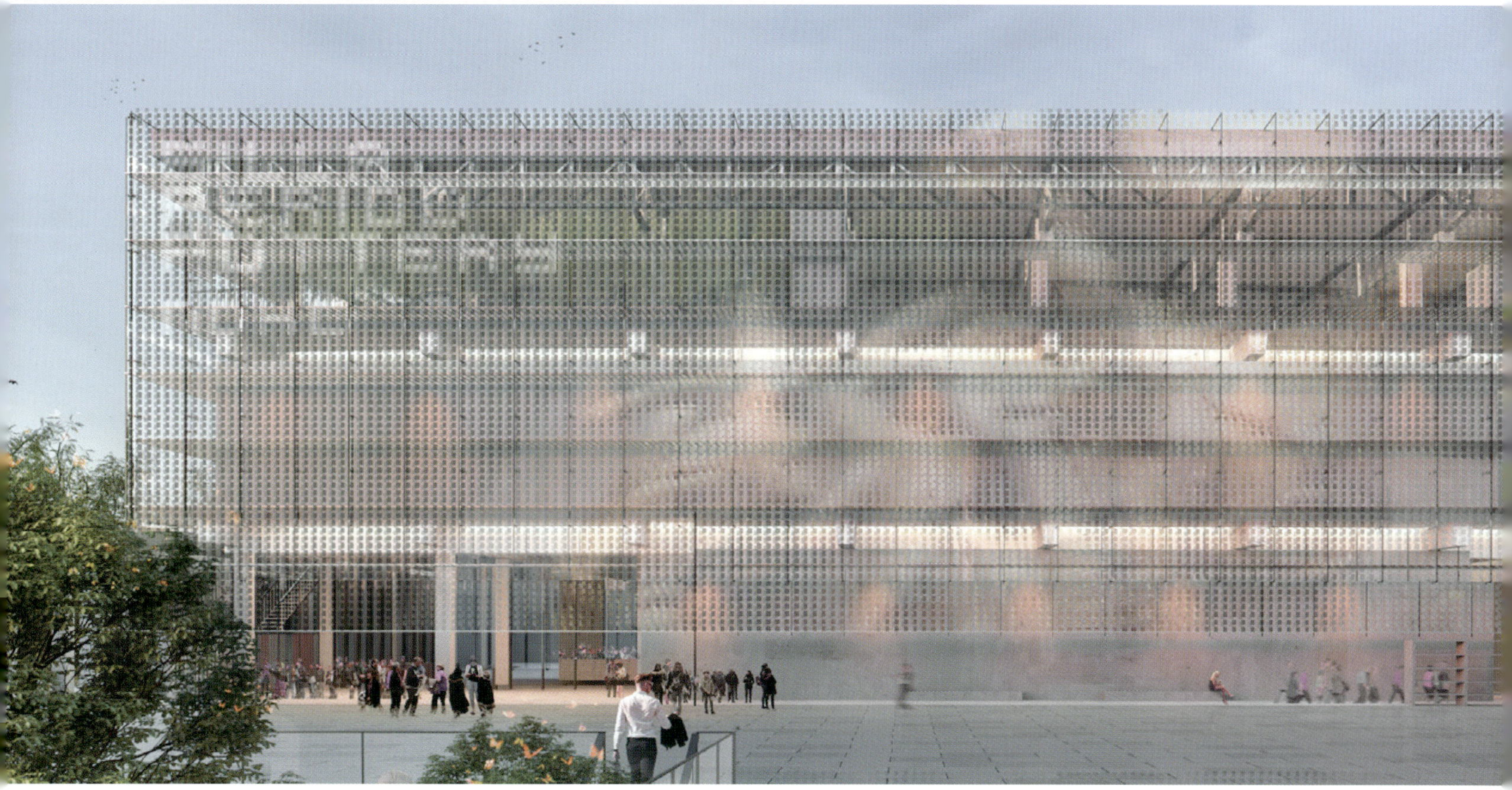

Winner _ 당선작

National Digital Heritage Center >>

▎Section Detail

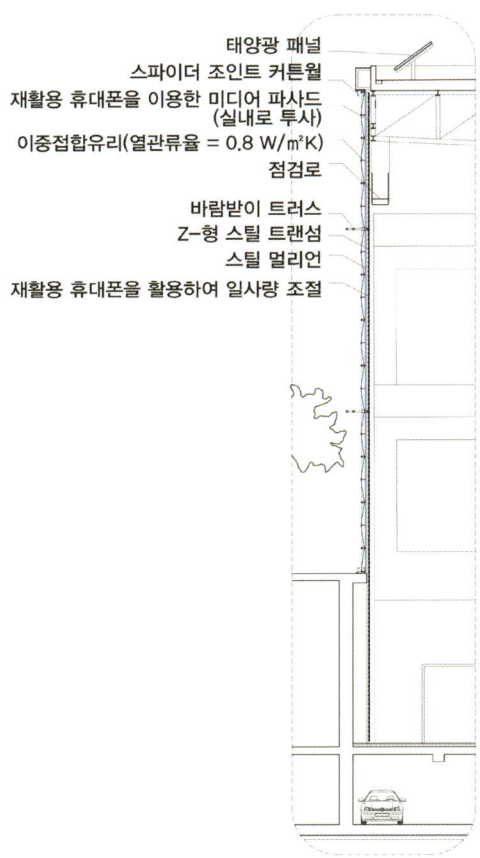

태양광 패널
스파이더 조인트 커튼월
재활용 휴대폰을 이용한 미디어 파사드
(실내로 투사)
이중접합유리(열관류율 = 0.8 W/㎡K)
점검로
바람받이 트러스
Z-형 스틸 트랜섬
스틸 멀리언
재활용 휴대폰을 활용하여 일사량 조절

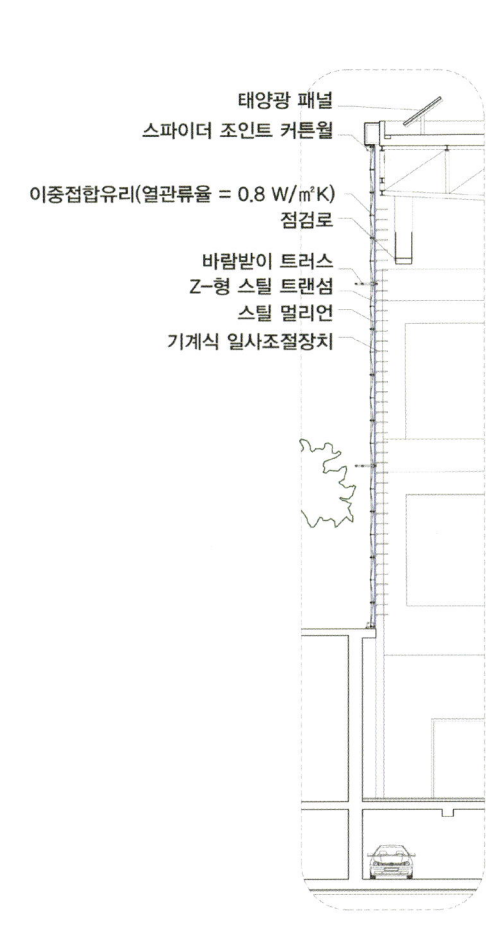

태양광 패널
스파이더 조인트 커튼월
이중접합유리(열관류율 = 0.8 W/㎡K)
점검로
바람받이 트러스
Z-형 스틸 트랜섬
스틸 멀리언
기계식 일사조절장치

Winner _ 당선작

1. Parking
2. Lobby
3. Exhibition center
4. Experience / Education

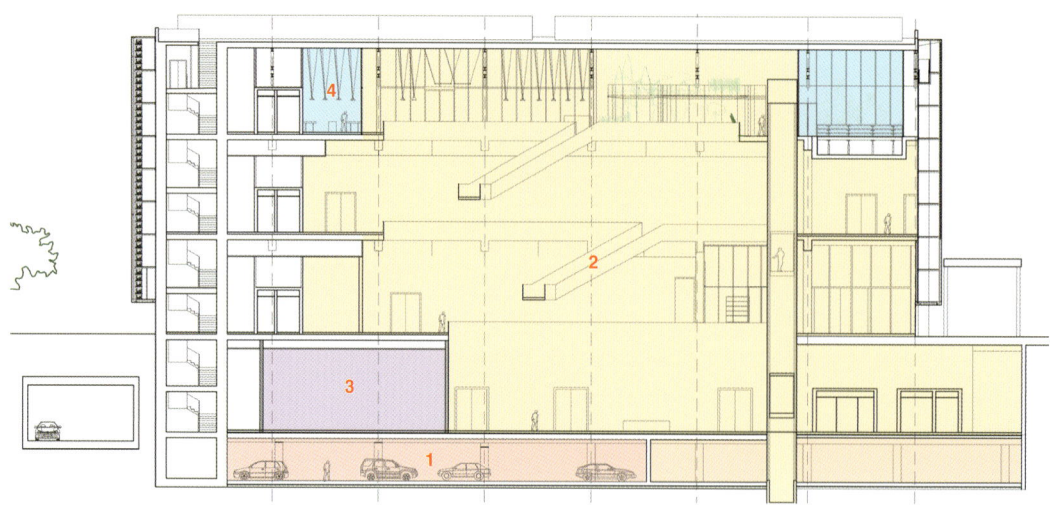

Longitudinal section

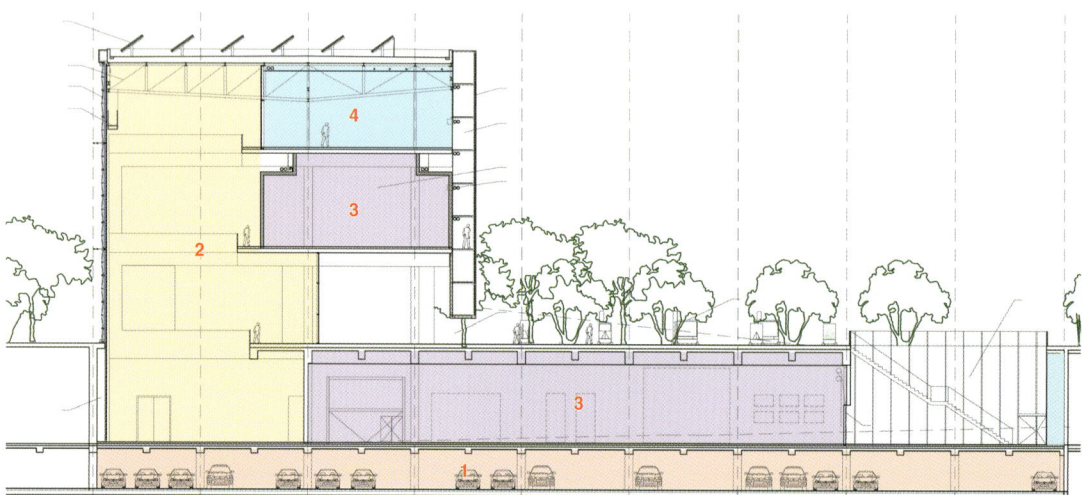

Cross section

National Digital Heritage Center >>

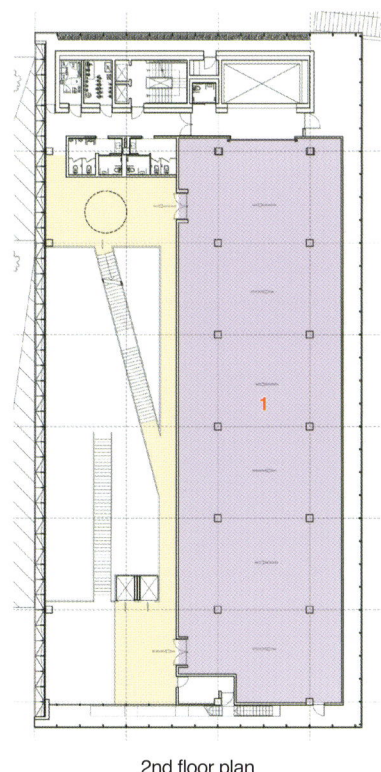

2nd floor plan

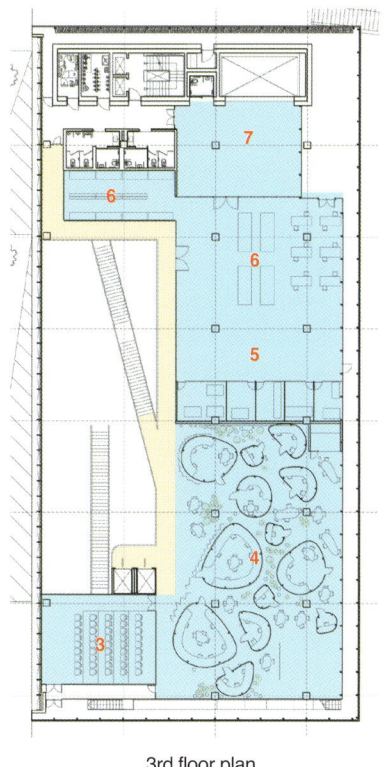

3rd floor plan

1. Lobby
2. Exhibition center
3. Sangsang hall
4. Creative studio
5. Experience / Education
6. Workshop
7. Education preparation room

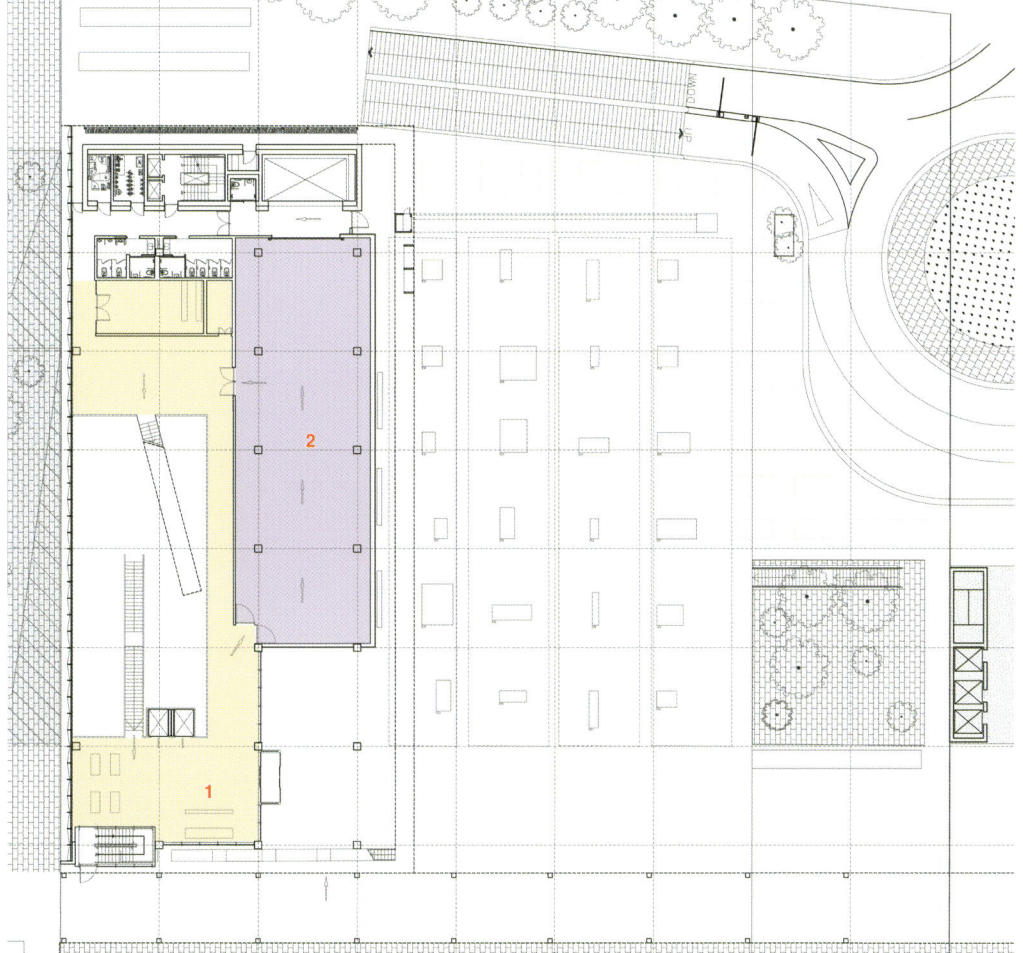

1st floor plan

125

Winner _ 당선작

National Jeongdong Theater
국립 정동극장

AUM&LEE ARCHITECTS&ASSOCIATES / Gwanpyo Yi, Min Lee
㈜엄&이종합건축사사무소 / 이관표, 이민

Jeong-dong contains the various memories of the city of Seoul. In the past, it was located on the outskirts of Hanyang, where foreign missions and religious facilities and schools for foreigners were built. During the period of the Korean Empire, it has a special history when Deoksugung in Jeongdong was used as the main palace instead of Gyeongbokgung. However, the current Jeong-dong does not reveal these historical memories, and it is only perceived as a promenade due to the lack of public space.

We propose Jeongdong Theater, which reveals the history of Jeong-dong and raises the value of Jeongdong-gil. The historical memory of the street was evoked by showing Jungmyeongjeon, which was not seen in Jeongdong-gil, and an Urban Living Room was planned where anyone could stay while communicating with Jeongdong-gil. People will gather in the space to share their daily lives, tell their stories, and make Jeongdong Theater a more enjoyable space. The new Jeongdong Theater will be a stage that vitalizes everyday life and enriches the story of Jeongdong-gil.

정동은 서울이라는 도시의 다양한 기억을 담은 장소이다. 과거에는 한양 외곽에 위치하여 외국 공관과 외국인들을 위한 종교시설과 학교 등이 들어섰으며, 대한제국 시기에는 경복궁 대신 정동의 덕수궁이 정궁으로 이용된 특수한 역사를 간직한 곳이다. 그러나 현재의 정동은 이러한 역사의 기억을 잘 드러내지 못하고 있으며, 또한 공공공간이 부족하여 사람들에게 산책 정도로만 인식되고 있다.

우리는 이러한 정동의 역사를 드러내고 정동길의 가치를 높이는 정동극장을 제안한다. 정동길에서 보이지 않았던 중명전을 드러내어 가로의 역사적 기억을 환기시키고, 정동길과 소통하며 누구나 머무를 수 있는 공공공간(Urban Living Room)을 계획하였다. 사람들은 이러한 공간에 모여 일상을 공유하고 자신의 이야기를 하며 정동극장을 더욱 즐거운 공간으로 만들어 갈 것이다. 새로운 정동극장은 일상에 활력을 더해주며 정동길의 이야기를 더욱 풍부하게 만드는 무대가 될 것이다.

Location : 8-11, Jeong-dong, Jung-gu, Seoul, Korea I **Function** : Cultural & Assembly facility I **Site area** : 1,634m I **Bldg. area** : 971m I **Total floor area** : 6,752m I **Stories** : B4, 3FL I **Structure** : Reinforced concrete, Steel frame I **Finish** : Clay brick, Curtain wall

위치 : 서울시 중구 정동 8-11 I **용도** : 문화 및 집회시설 I **규모** : 지하4층, 지상3층 I **구조** : 철근콘크리트, 철골 I **마감** : 점토벽돌, 커튼월 I **설계팀** : 성욱, 김종현, 이석영, 전원표, 김경종, 김민조, 이충희, 천서영, 김정환

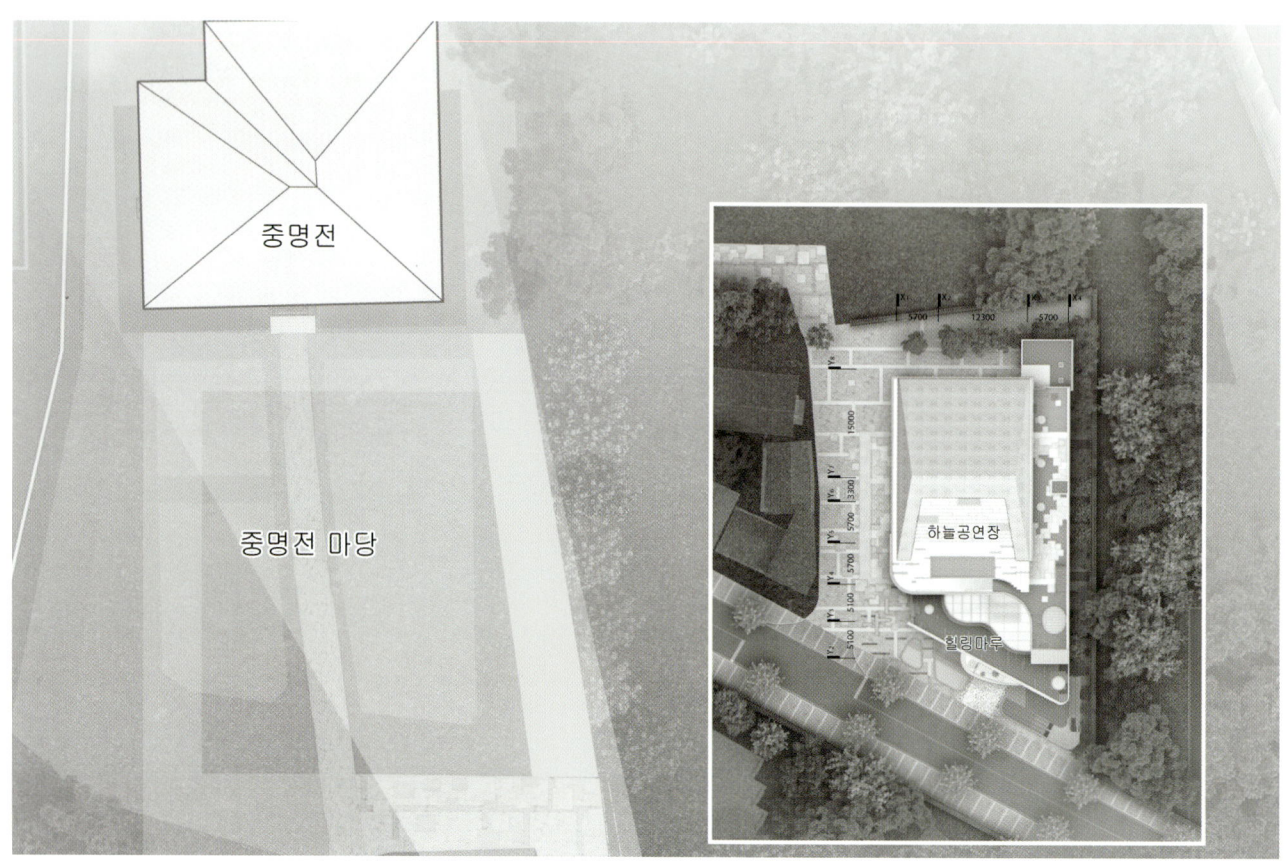

Site plan

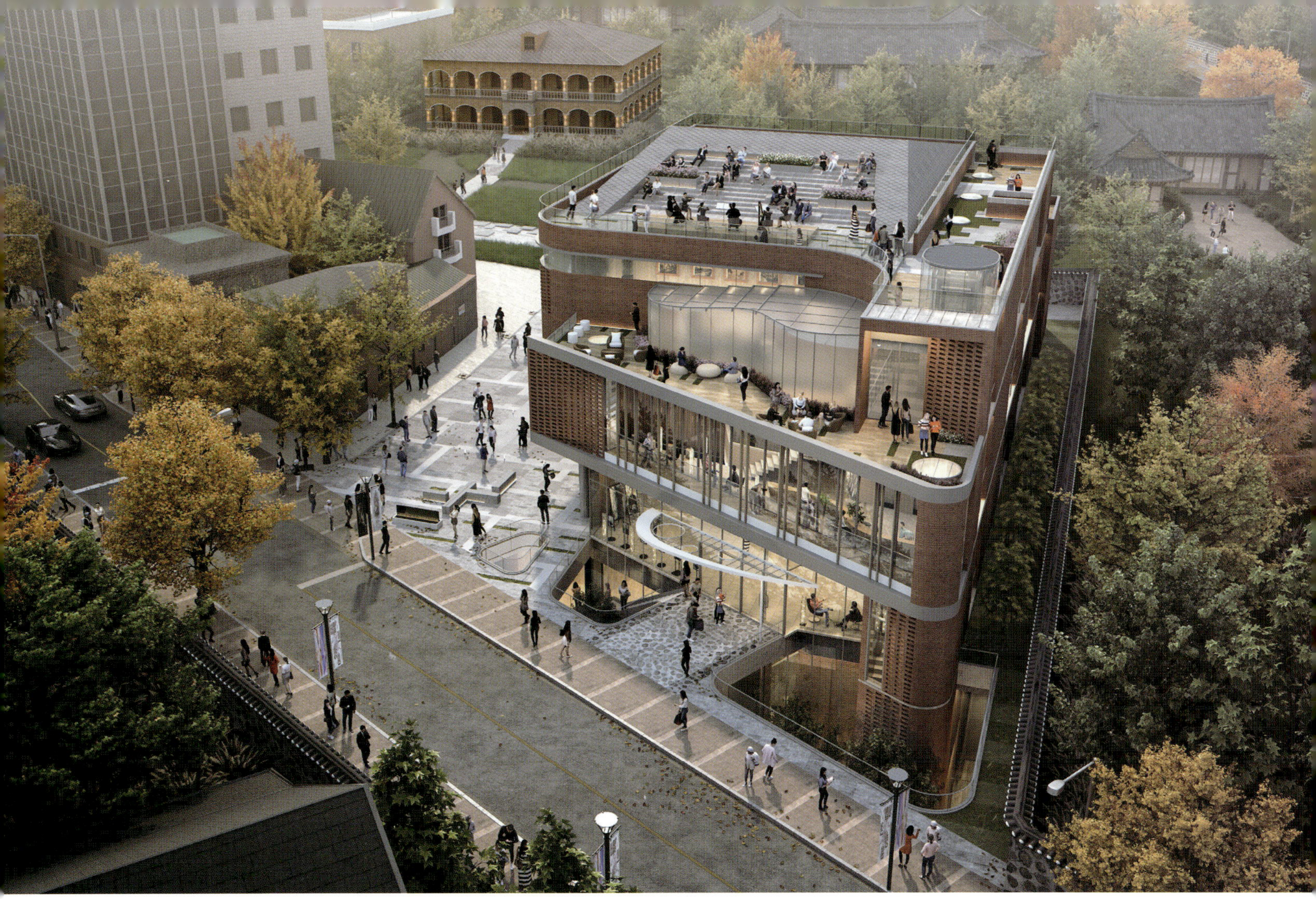

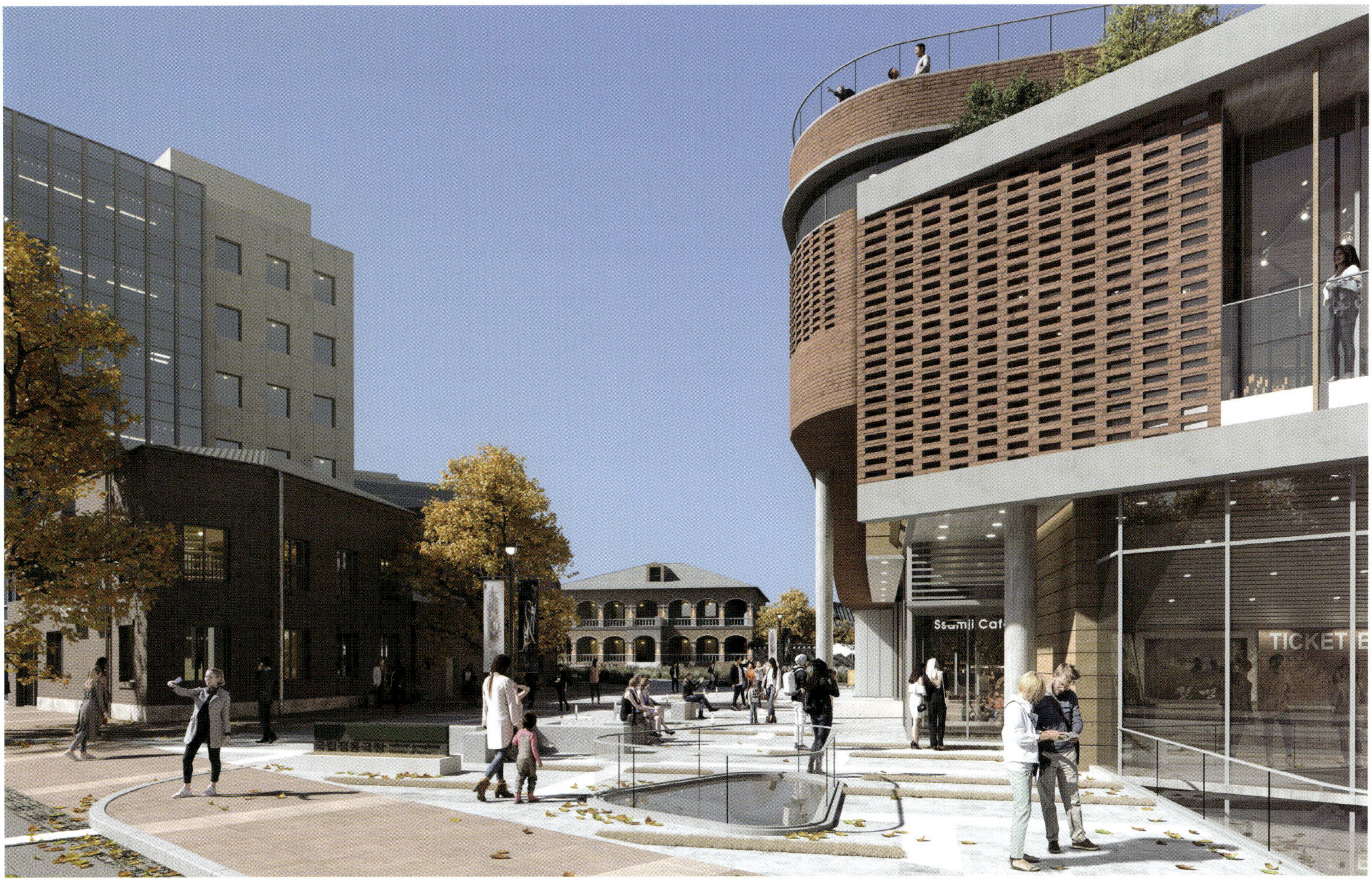
TICKET

Winner _ 당선작

▌Design Concept Ⅰ

- Theater as a stage in the city

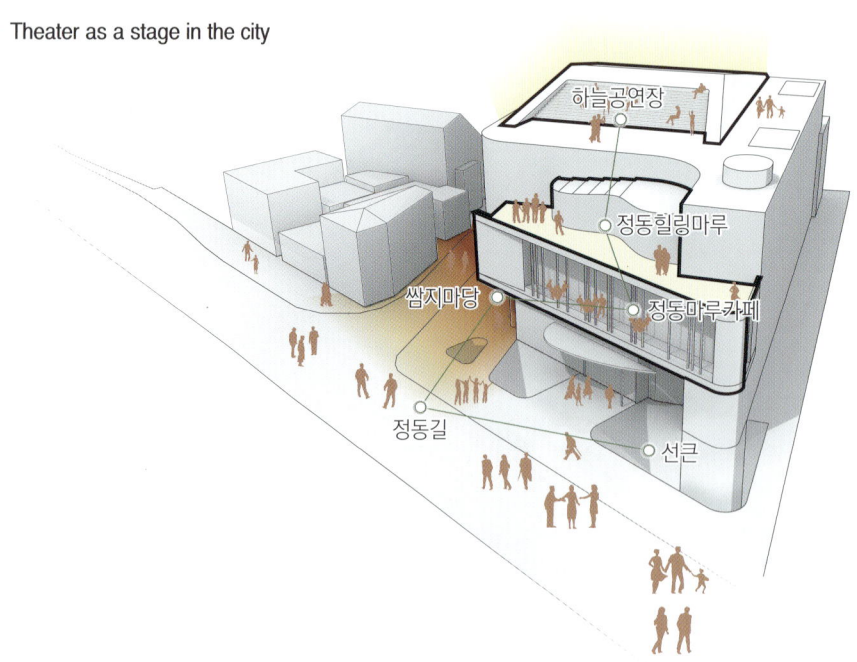

- 화려한 현대적 고층빌딩과 안쪽에 새로이 짓는 정동극장은 함께 무대의 배경이 된다.
 사람들은 마당에 모여 일상을 공유하고 스스로 무대에 올라 자신들의 이야기를 한다.
 정동길을 거닐던 사람들은 이들과 교감하며 정동 이야기의 관객이 된다.
 자연스럽게 정동극장 마당과 정동마루 카페로 들어와 무대와 하나가 되어 이야기를 만들어 간다.

National Jeongdong Theater >>

Design Concept Ⅱ

- Ssamzie Madang, the living room of the city

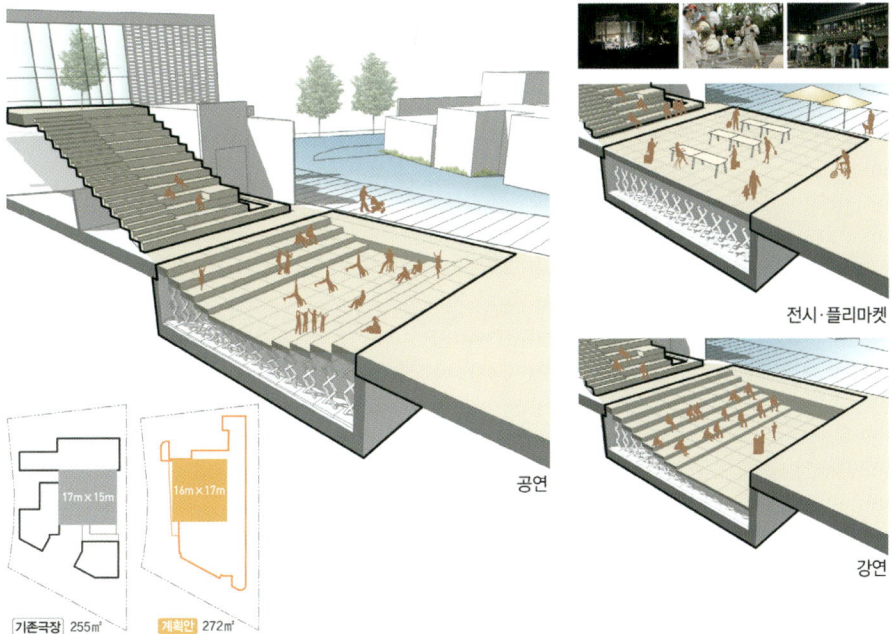

기존극장 255㎡   계획안 272㎡

공연
전시·플리마켓
강연

— 정동은 공공공간이 많지 않은 편이며 기존 정동극장의 쌈지마당(중정)은 정동의 몇 없는 쉼터였다. 우리는 그것과 물리적 크기가 유사한 마당을 계획해 기능을 충족하면서 다양한 행위를 할 수 있는 공간을 제안한다. 가변식 스테이지를 통해 많은 행사가 가능하고, 개방형 슬라이딩 도어를 활용하여 외부로 확장이 가능하다.

Winner _ 당선작

Section Plan

- Jeongdong theater with various stories

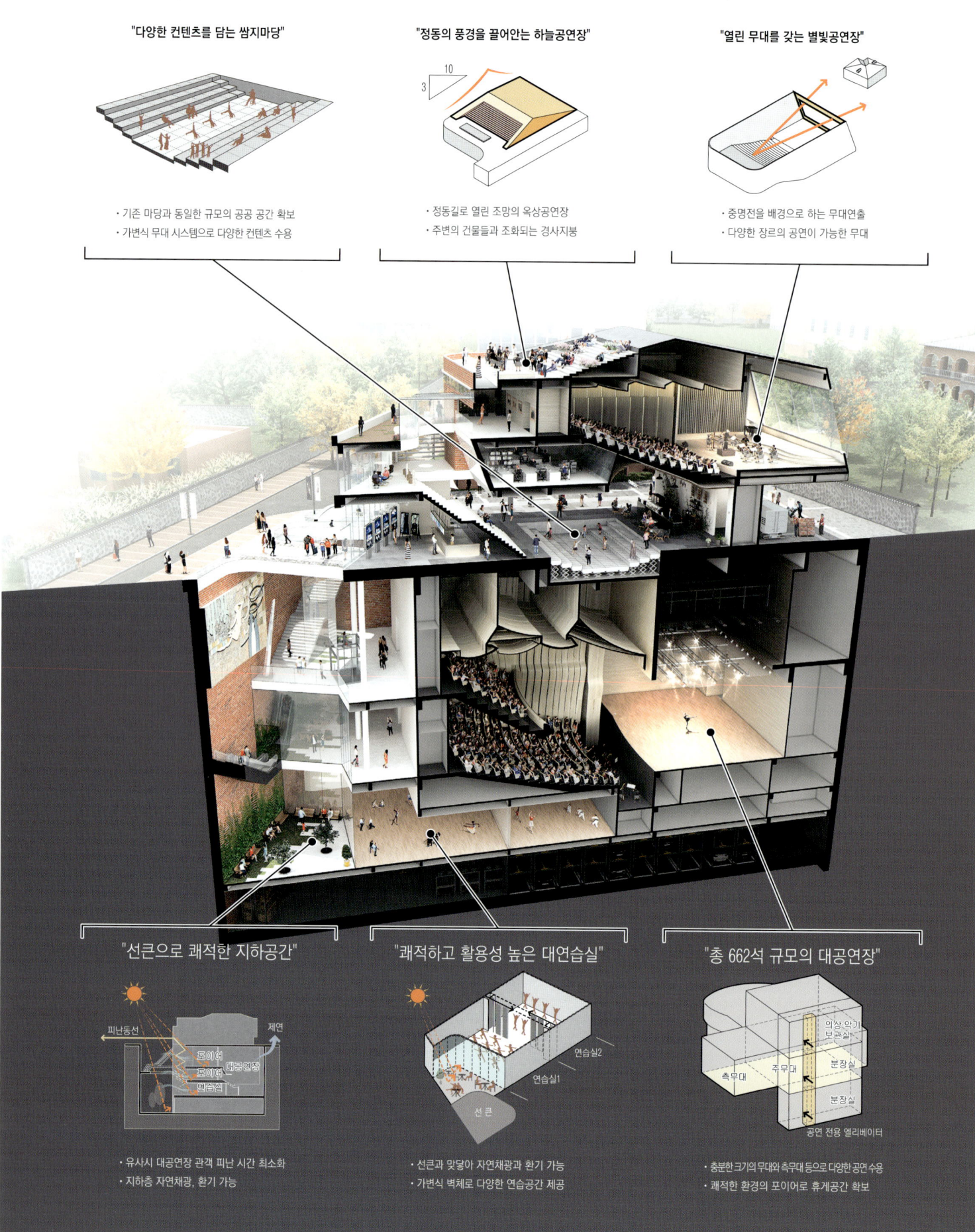

National Jeongdong Theater >>

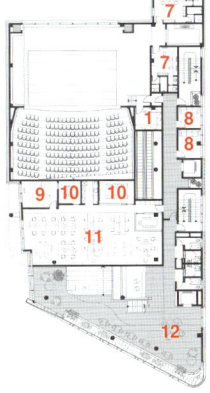
2nd floor plan

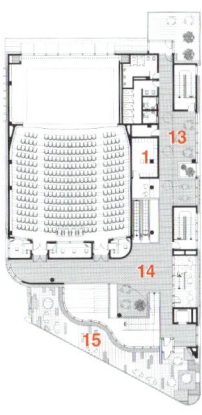
3rd floor plan

1. Storage
2. Machine room
3. Lounge
4. Loading deck
5. Yard
6. Lobby
7. Dressing room
8. Waiting room
9. Director's office
10. Conference room
11. Office
12. Cafe
13. Lounge
14. Foyer
15. Maru

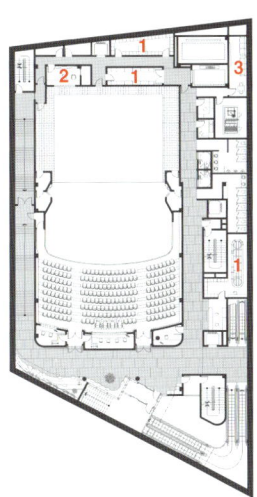
B3 floor plan

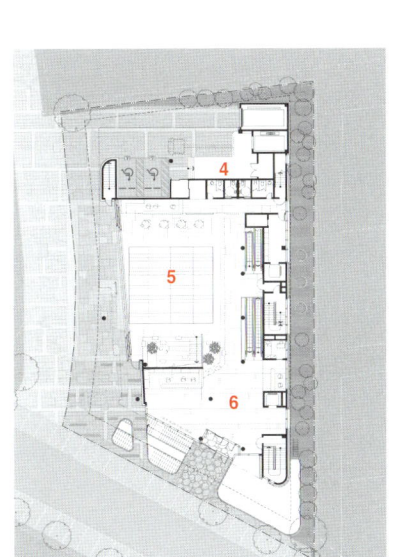
1st floor plan

2nd Prize _ 2등작

National Jeongdong Theater
국립 정동극장

SSK ARCHITEKTEN / Sooseok Kim
SSK 건축사사무소 / 김수석

Compared to other downtown areas, this area has cultural infrastructure, such as cultural assets, historical places, public institutions, historical schools, and art galleries, and Jeongdong Theater is located in the center. Despite the abundance of cultural resources, Jeongdong Church strongly preserves its symbolic identity and is widely known, but the identity of the space itself is hidden by strong surrounding monuments. Jeongdong Theater is surrounded by Deoksugung-gil and Jeongdong-gil and half of one side is open to the road, so the isolation is deepening. Moreover, the facade of the building facing the road is separated by a temporary wall to create a closed space composition, so openness is a challenge. Today when openness to citizens is emphasized, this closed space structure is the fundamental cause of weakening the original image of Jeongdong Theater as an open theater. Accordingly, the design plan was to face the physical limitations of the existing Jeongdong Church and suggest alternatives as a newly opened theater. Reducing the physical distance with citizens based on strong communication with surrounding contexts was the starting point of the design.

타 도심지에 비해 해당 지역은 문화재, 역사적 장소, 공공기관, 유서 깊은 학교, 미술관 등 문화적 인프라가 산재해 있는 지역으로 정동극장은 그 중심에 위치해있다. 이러한 풍부한 문화자원에 비해 정동교회는 그 상징적 정체성은 강하게 보존하며 널리 알려져 있으나, 공간 자체의 정체성은 주변의 강한 모뉴먼트들에 의해 가려진 상태다. 정동극장은 덕수궁길, 정동길에 둘러싸여 한 면 반 도로에 개방된 상태로 고립도가 심화되어 있는 상태다. 더욱이 도로에 면한 건물의 파사드가 가벽으로 분리되어 다소 폐쇄적인 공간구성을 취함으로써 개방적인 면에서 다소 취약점이 발견된다. 과거와 달리 시민에 대한 개방성이 강조되는 시점에서 이러한 폐쇄적인 공간구조는 열린 극장이라는 정동극장 본연의 색채를 약화시키는 근본적인 원인으로 판단된다. 이에 따라 본 설계안은 이러한 기존의 정동교회의 물리적인 한계를 직시하고 새롭게 열려있는 극장으로서 대안을 제시하고자 했다. 주변 컨텍스트들과의 강한 교감을 바탕으로 시민들과의 물리적 간격을 줄이는 것이 본 설계안의 시작점이 되었다.

Location : 43, Jeongdong-gil, Jung-gu, Seoul, Korea I **Function** : Theater I **Site area** : 1,673m I **Bldg. area** : 1,014m I **Total floor area** : 6,787m I **Stories** B4, 2FL I **Structure** : Steel framed reinforced concrete, Reinforced concrete, Precast concrete I **Finish** : Aluminium curtain wall, Ceramic panel, Exposed concrete

위치 : 서울시 중구 정동길 43 I 용도 : 극장 I 규모 : 지하4층, 지상2층 I 구조 : 철골철근콘크리트, 철근콘크리트, 프리캐스트콘크리트 I 마감 : 알루미늄커튼월, 세라믹패널, 노출콘크리트

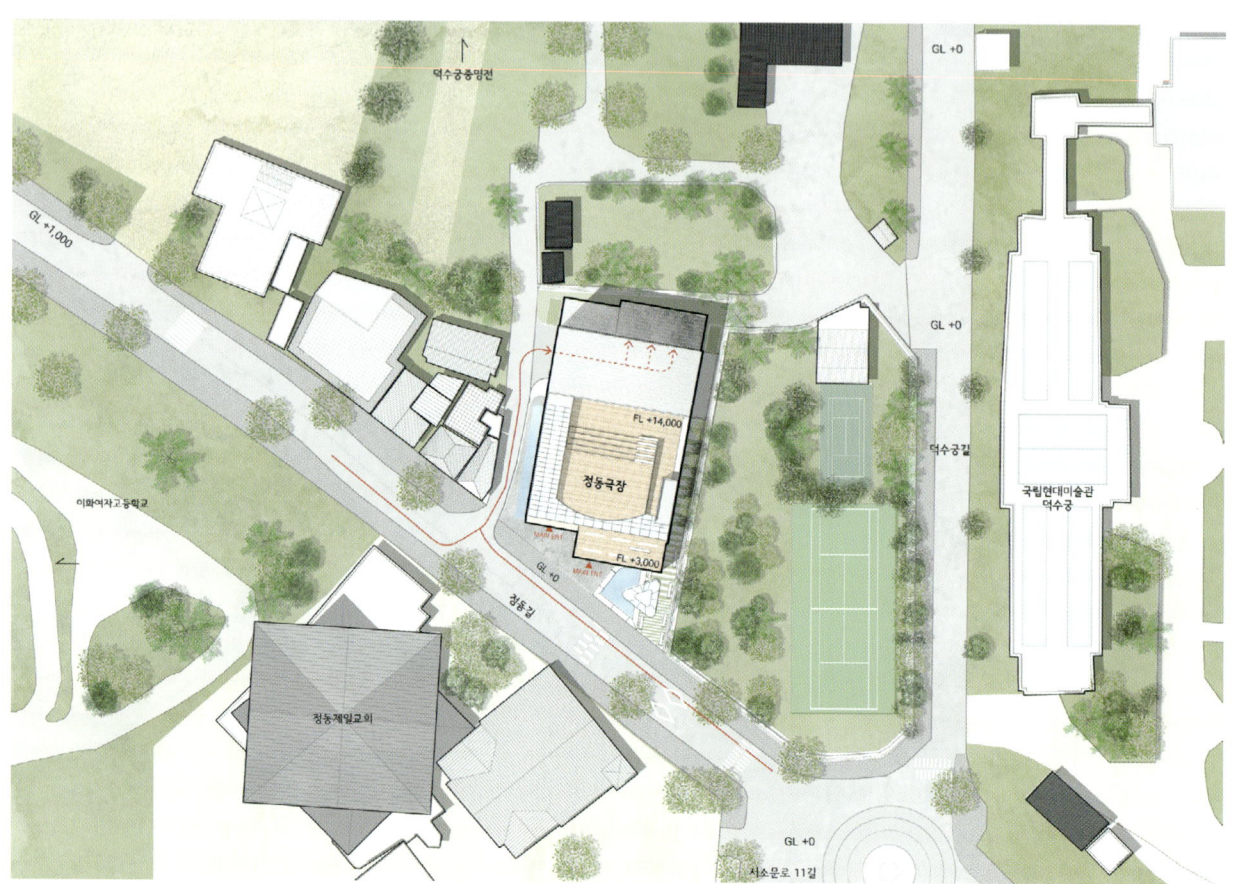

Site plan

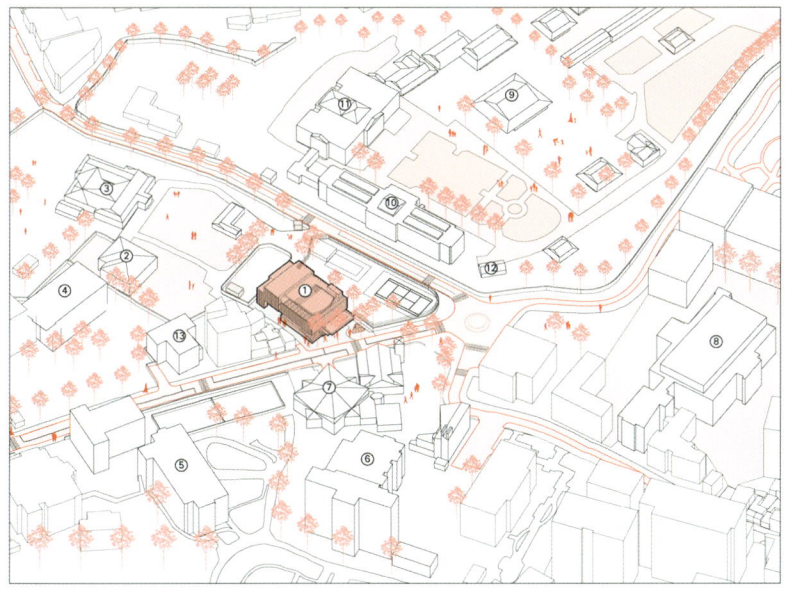

Site Analysis

- Path accessibility

- Urban void
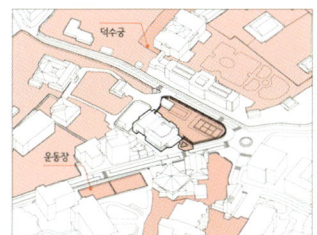

- Monument relationship
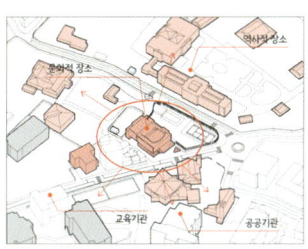

- Visiting and staying population

2nd Prize _ 2등작

■ Design Concept

- Volume optimization

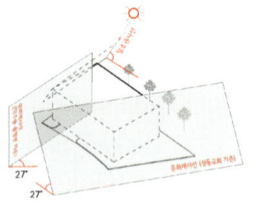

- 문화재 사선 제한 구역으로 둘러싸여 있으며 북측 경계는 일조권사선제한의 제약을 받는다.

- Lifting up mass

- 저층부 건물을 들어 올림(Lift Up)으로써 적극적인 개방성을 취했다.

- Deck level visual openness

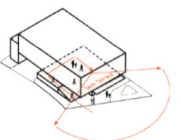

- 1층의 진입광장-계단-데크 테라스를 통해 저층부에서의 개방성을 강조했다.

- Reflection of surroundings

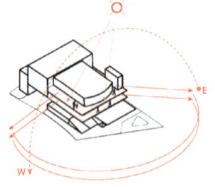

- 상층부에서 복합레벨을 통해 적극적으로 일조량을 유입할 수 있도록 했다.

- Adaptive facade

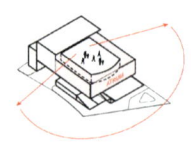

- 커튼월 상하부에 In-let 시스템을 도입하여 Sun Space로 기능하도록 의도했다.

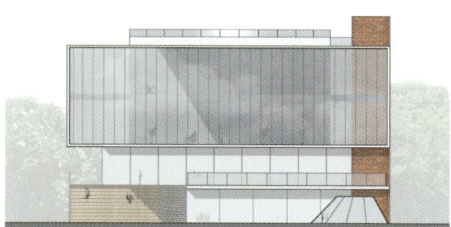

Front elevation

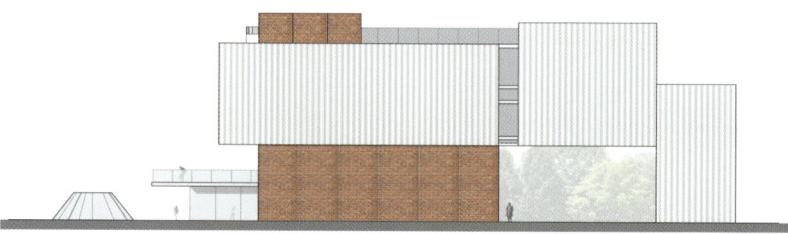

Right elevation

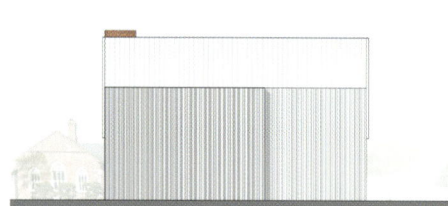

Rear elevation

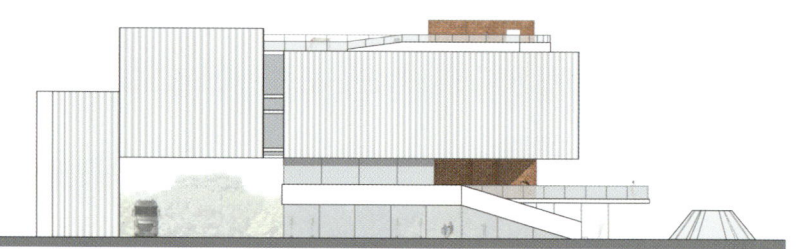

Left elevation

National Jeongdong Theater >>

■ Featured Space Description

- Atrium lounge

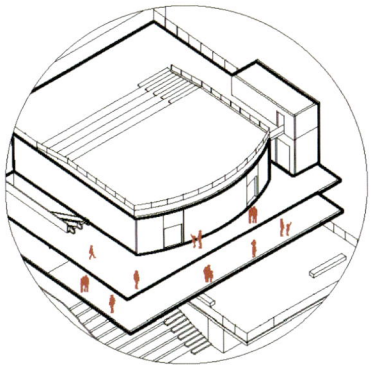

– 상부객석으로 접근가능한 Dress Circle과 하부객석 및 무대로 접근 가능한 Stall Lounge 레벨을 통합하는 아트리움을 통해 쾌적한 공연관람 환경 유지

- Roof terrace / Outdoor stage

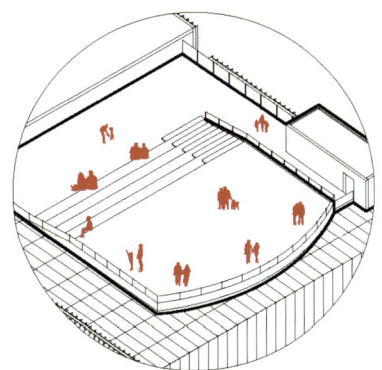

– 정동 일대 조망이 가능한 루프가든과 옥상에 조성된 소규모 야외 공연이 가능한 공연장을 통해 다양한 문화활동이 가능

- Deck terrace

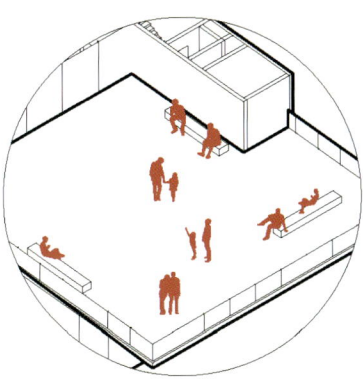

– 지상레벨에서 옥외계단을 통해 수월하게 접근 가능한 데크테라스를 통해 정동 일대를 아이레벨에서 조망하며 필로티 공간을 통해 다양한 외부활동 가능

- Stage area

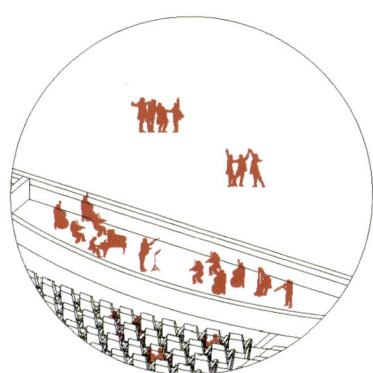

– 무대-오케스트라PIT-객석으로 이어지는 과정에서 객석에서의 최적의 뷰를 확보하고 PIT와 무대의 최적의 레벨 형성으로 관객은 물론 퍼포머들에게 최적의 공연을 위한 무대공간 제공

- Spiral atrium

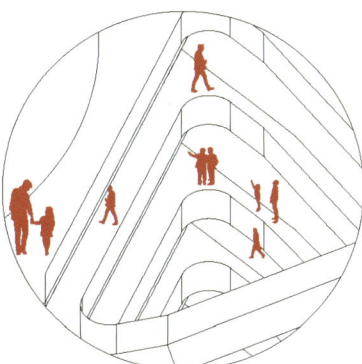

– 지하층 전층에서 지상레벨까지 직통으로 연결되는 스파이럴 계단과 이를 덮는 아트리움 공간을 통해 지하층 바닥에서도 밝고 쾌적한 관람환경을 조성하고 비상시 지상레벨로 즉각적인 대피가 가능

- Double height lobby

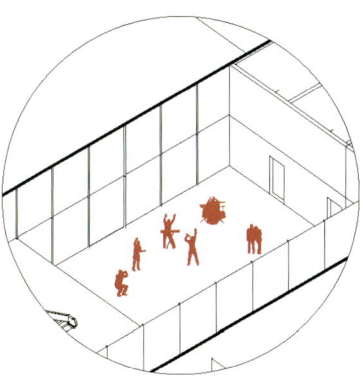

– 상부 소공연장 매스 하부에 형성된 로비는 Mezzanine 층으로 활용되는 Deck-Terrace와의 레벨차를 활용해 높은 층고의 개방감으로 방문객을 맞이함과 동시에 공연 전후 전이공간으로 기능

2nd Prize _ 2등작

1. Roof garden / Stage
2. Dress circle
3. Stall lounge
4. Auditorium
5. Stage
6. Deck
7. Lobby
8. Parking
9. Office
10. Academy room
11. Practice room
12. Dressing room
13. Waiting room
14. Machine room
15. Orchestra pit
16. Stage basement
17. Control room
18. Corridor

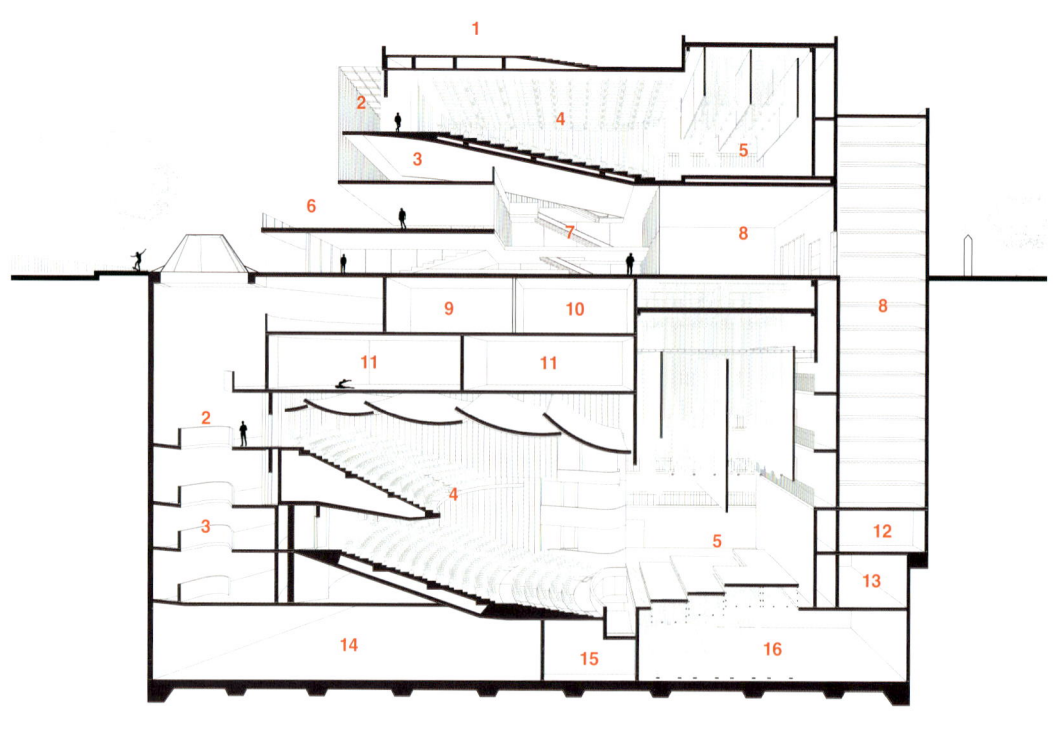

Section I

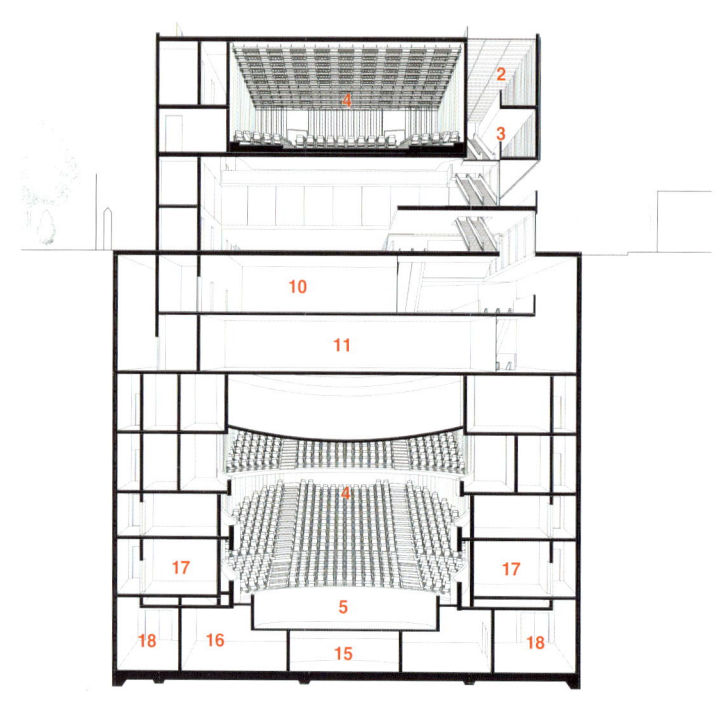

Section II

National Jeongdong Theater >>

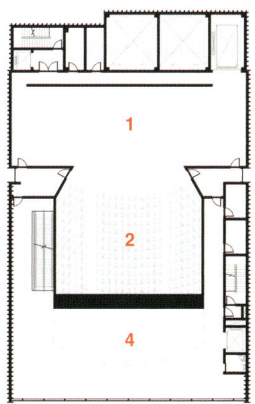

1. Stage
2. Auditorium
3. Control room
4. Stall lounge
5. Dress circle
6. Hall
7. Academy room
8. Office
9. Representative room
10. Meeting room
11. Cafe
12. Lobby
13. Entrance square
14. Atrium

2nd floor plan

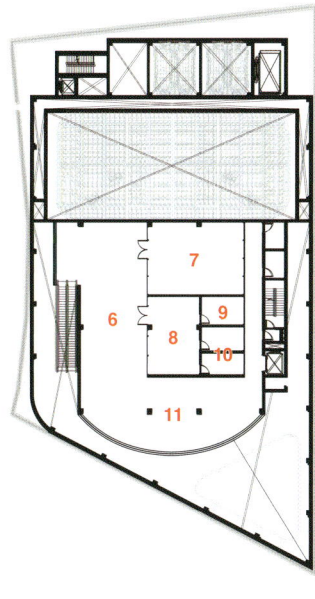

B1 floor plan

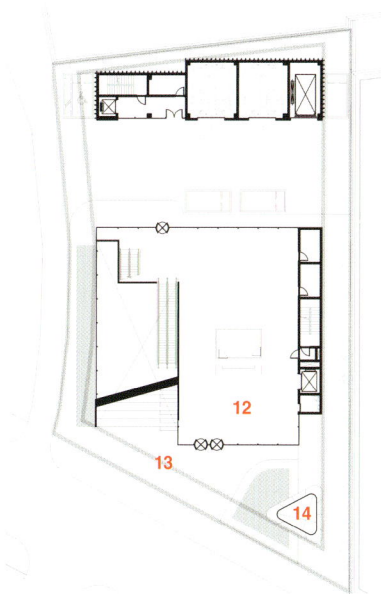

1st floor plan

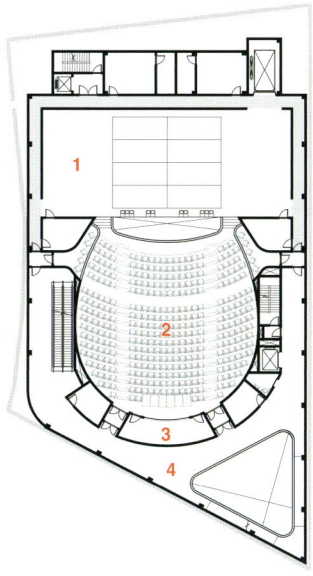

B3 floor plan

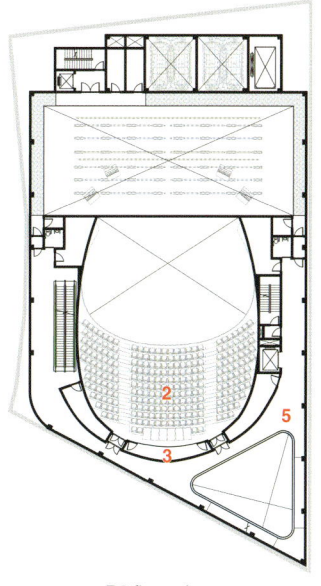

B2 floor plan

Winner _ 당선작

House for Film & Media in Stuttgart
슈투트가르트 영화 & 미디어 하우스

DMAA(Delugan Meissl Associated Architects)
DMAA(델루간 마이슬 아키텍츠)

The new Haus für Film und Medien in Stuttgart has an open, inviting and communicative appearance. The key role of the facade is to combine maximum transparency with an ability to adapt in line with the building's wide range of interior settings. The grille-like solar protection elements that extend from the facade enable it to fulfil this role by providing shade, while also opening up broad views into and out of the building. This transparency, together with the ability of the facade to act as a multi-media projection surface, allows the building to communicate with both the city as a whole and its immediate surroundings.

Inside the building, the investigation of film and media takes place along a clearly defined path lined with versatile surfaces that offer visitors and users the opportunity to engage in a wide range of activities.

At street level, the generous entrance stair of the Haus für Film und Medien merges with the public realm, creating a setting – based on the red carpet – for visitors to enter the building and rise to the foyer that is very consciously located on the 'Bel Étage'.

The open restaurant space over which the building extends, protecting it from the rain, can also be understood as a further transitional zone between the building and the outside that invites passers-by to rest awhile in the open air.

In order to optimise the transparency between inside and outside in the lower parts of the building, the two cinemas are placed as windowless volumes on the uppermost levels. The multifunctional spaces are located directly above the entrance level, which allows the educational, administrative, gastronomic, exhibition and waiting areas to be organised between these along a circulation route with a variety of spatial sequences. The visual relationship with the exterior and, in particular, the panoramic view of the city of Stuttgart plays a key role along the entire length of this route.

Location : Leonhardsplatz 70182, Stuttgart, Germany I **Function** : Cultural facility I **Site area** : 1,128㎡ I **Total floor area** : 9,500㎡ I **Stories** : B2, 7FL I **Client** : Landeshauptstadt Stuttgart Referat Wirtschaft, Finanzen und Beteiligungen I **Project manager** : Sebastian Brunke, Bernd Heger I **Design team** : Marinke Boehm-Kneidinger, Magdalena Czech, Dusan Sekulic / Tom Peter-Hindelang, Gregor Hilpert

위치 : 독일 슈투트가르트 I 용도 : 문화시설 I 규모 : 지하2층, 지상7층

슈투트가르트의 새로운 영화미디어센터는 개방적이고 맞이하며 소통이 가능한 외관을 가지고 있다. 파사드의 핵심 역할은 건물의 다양한 내부 환경에 따라 적응하면서 투명성을 극대화하는 것이다. 파사드에서 확장된 그릴 모양의 태양열 차단 요소는 차양 역할을 할 뿐만 아니라 건물 안팎으로 탁 트인 전망을 제공한다. 이 투명성을 통해 파사드는 멀티미디어 투영 표면으로 작용하고 건물은 도시 전체 및 주변 환경과 소통할 수 있다.

건물 내부에서는 방문자와 사용자가 다양한 활동에 참여할 수 있도록 다목적 표면이 줄지어 있는 명확하게 정의된 경로를 따라 영화 및 미디어 연구가 이루어진다.

도로 레벨에서는 이 센터의 넓은 입구 계단은 공공 영역과 합쳐져 방문자가 건물에 들어가 매우 의식적으로 '벨레 타주'에 있는 로비로 올라갈 수 있는 레드 카펫이 깔린 무대를 만든다.

건물이 확장되어 비로부터 건물을 보호하는 열린 식당 공간은 행인들이 야외에서 잠시 쉴 수 있도록 건물과 외부 사이의 전이 공간이 될 수 있다.

건물 하부의 내부와 외부 사이 투명성을 최적화하기 위해 두 영화관을 최상층에 창문 없는 볼륨으로 배치했다. 다기능 공간은 입구 레벨 바로 위에 위치하여 교육, 행정, 식사, 전시 및 대기 공간이 다양한 공간 시퀀스로 동선을 따라 사이 공간에 구성될 수 있다. 외부와의 시각적 관계, 특히 슈투트가르트 시의 탁 트인 전망은 이 경로의 전체를 따라 중요한 역할을 한다.

Winner _ 당선작

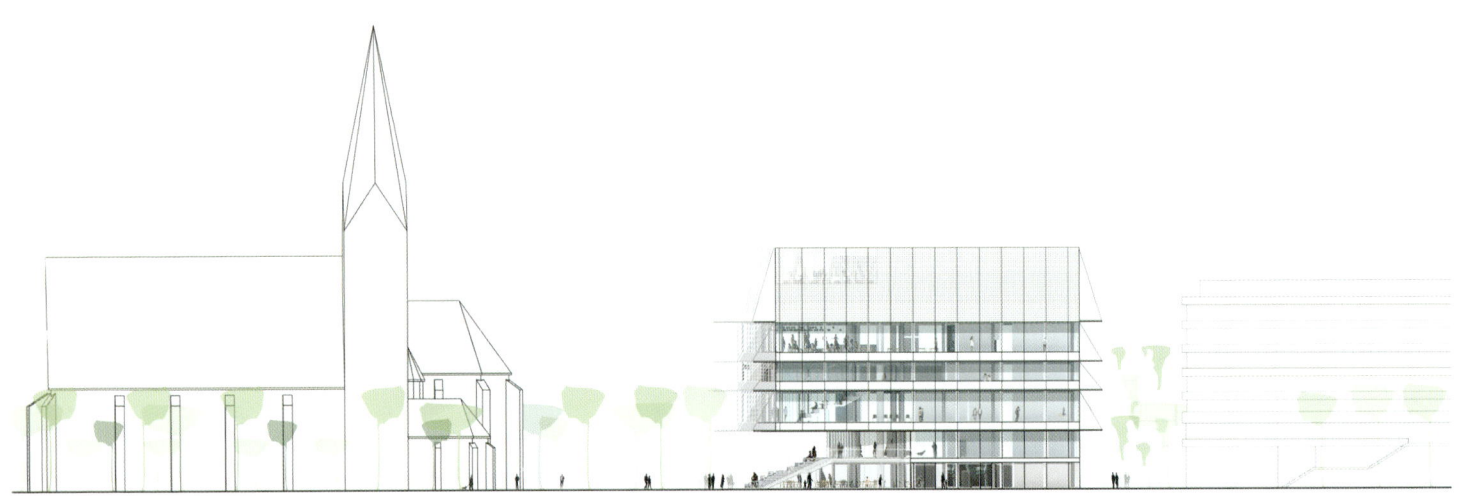

Elevation

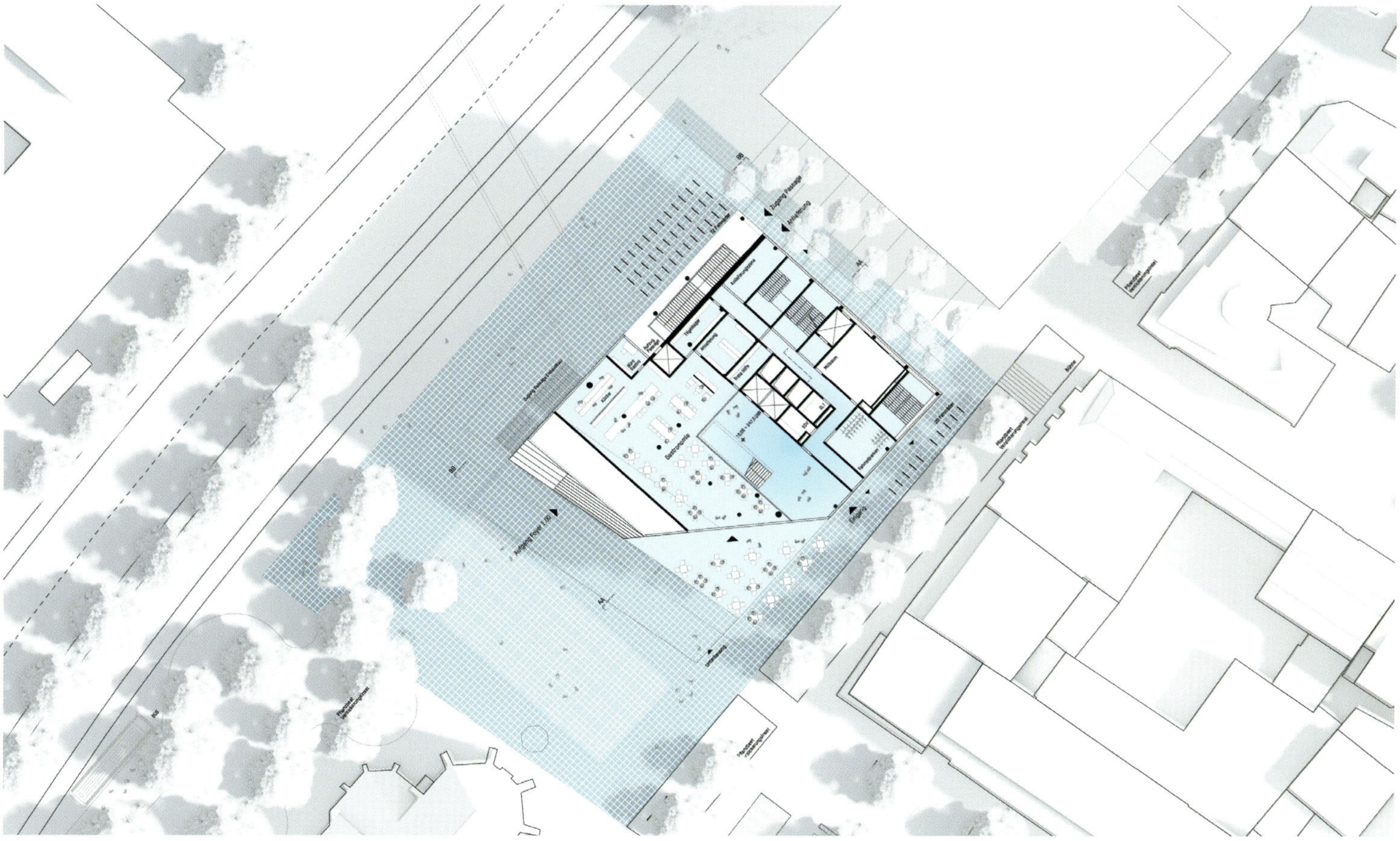

House for Film & Media in Stuttgart >>

Floor plan

Winner _ 당선작

Montreal Holocaust Museum
몬트리올 홀로코스트 뮤지엄

KPMB Architects + Daoust Lestage Lizotte Stecker Architecture
KPMB 아키텍츠 + 다우스 레스테이지 리조트 스텍커 아키텍처

Based on the pillars of memory, education, and community, the striking new building will contain multiple exhibition spaces, classrooms, an auditorium, a memorial garden, and a dedicated survivor testimony room. Construction on the new museum will begin in the fall of 2023.
The MHM is moving from its current location in response to growing demand for its educational programs about the Holocaust, genocide, and human rights. Facing a rise of racism, antisemitism, and discrimination, the new MHM will have a broader impact in galvanizing communities throughout Quebec and Canada to fight all forms of hatred and persecution.

기억, 교육 및 커뮤니티 필라를 기반으로 하는 이 인상적인 새 건물에는 여러 전시 공간, 교실, 강당, 기념 정원 및 헌신한 생존자 증언실이 포함될 것이다. 박물관 신축공사는 2023년 가을에 시작된다.
MHM은 홀로코스트, 대량 학살, 인권에 대한 교육 프로그램에 대한 수요 증가에 대응하여 현재 위치에서 이전했다. 인종 차별주의, 반유대주의 및 차별의 증가에 직면하여 새로운 MHM은 모든 형태의 증오와 박해에 맞서 싸우기 위해 퀘벡과 캐나다 전역의 커뮤니티에 활력을 불어넣는 데 광범위한 영향을 미칠 것이다.

Location : Montreal, Canada | **Function** : Cultural facility | **Total floor area** : 4,336㎡ | **Stories** : B1, 3FL

위치 : 캐나다 몬트리올 | **용도** : 문화시설 | **연면적** : 4,336㎡ | **규모** : 지하1층, 지상3층

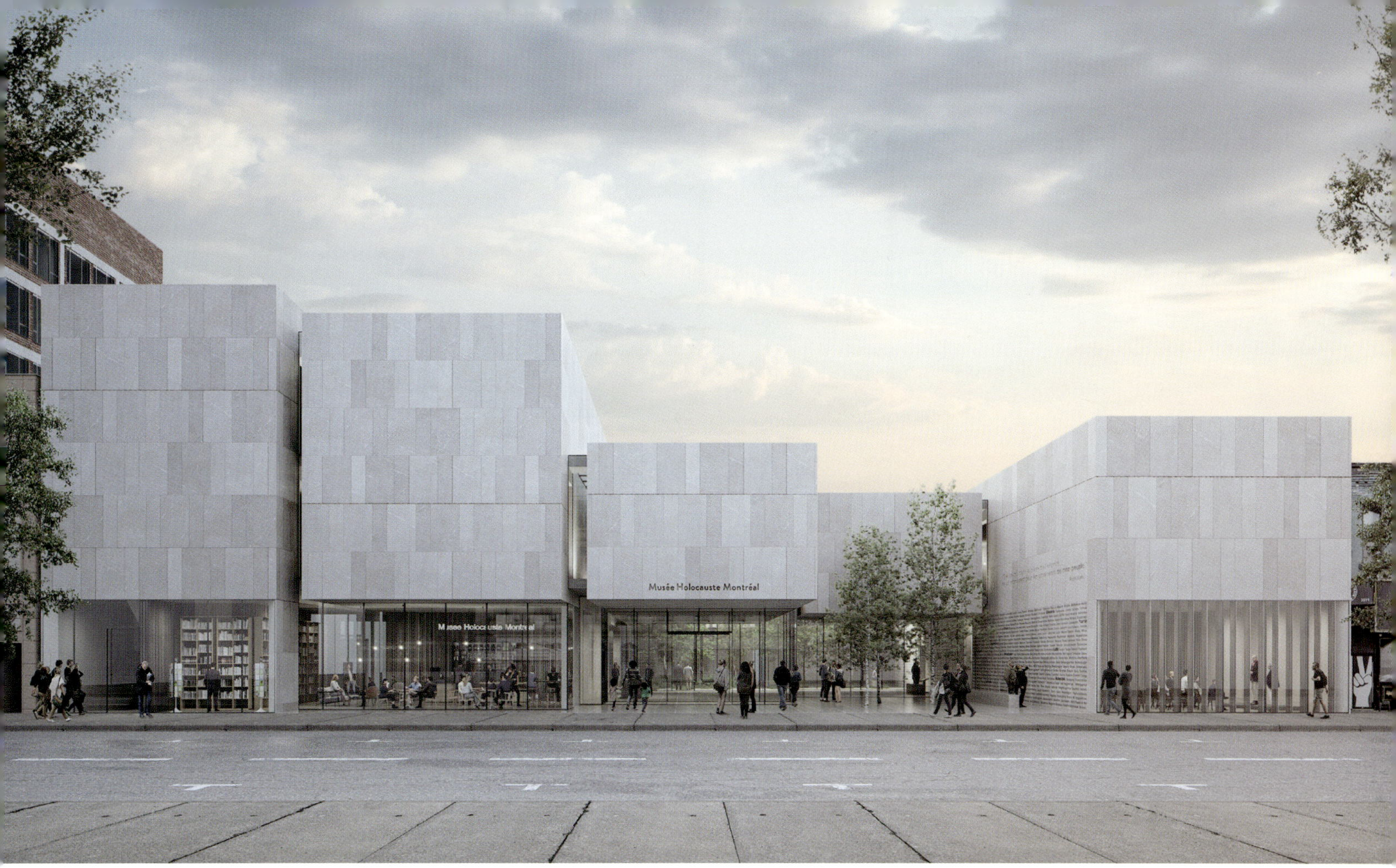
Musée Holocauste Montréal

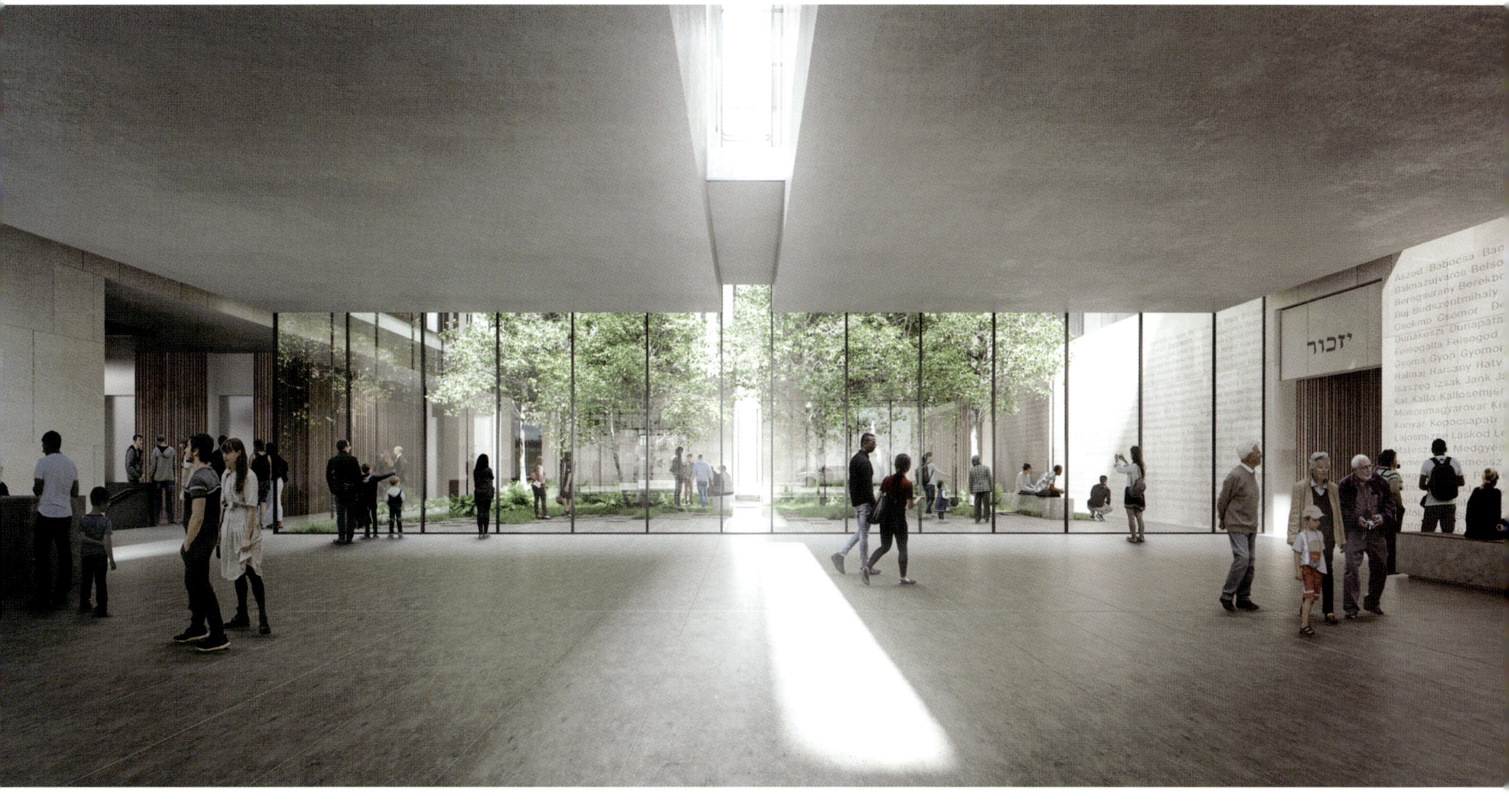
יזכור

Winner _ 당선작

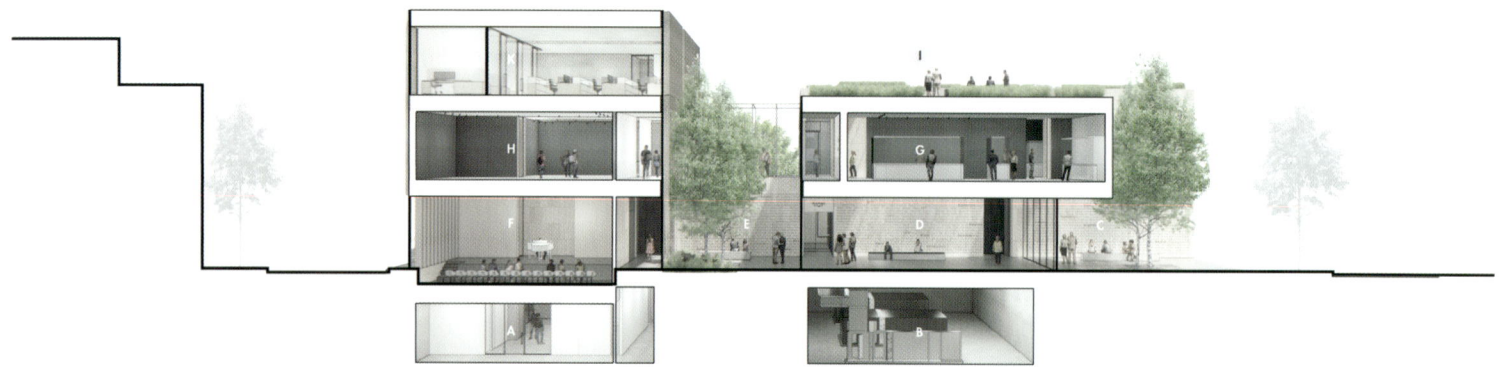

Section I

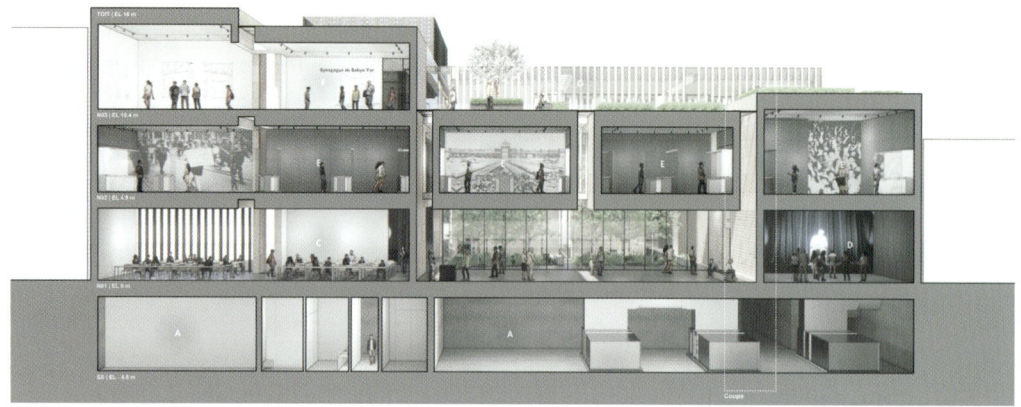

Section II

Montreal Holocaust Museum >>

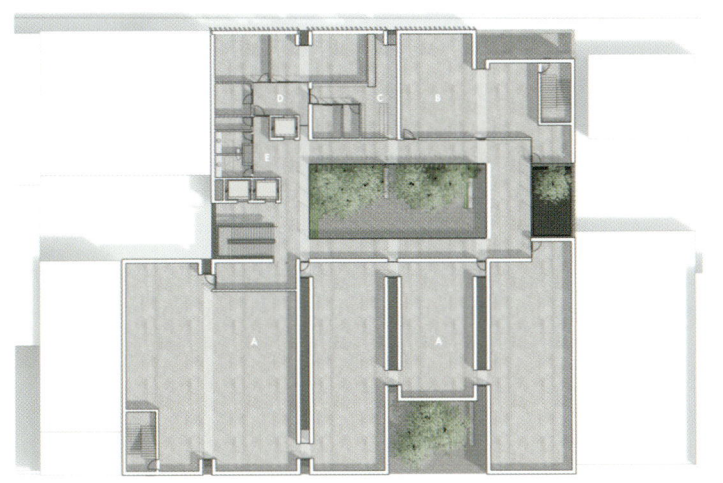

2nd floor plan

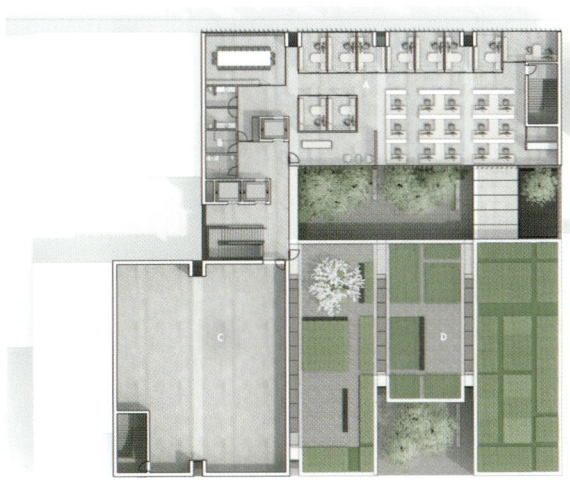

3rd floor plan

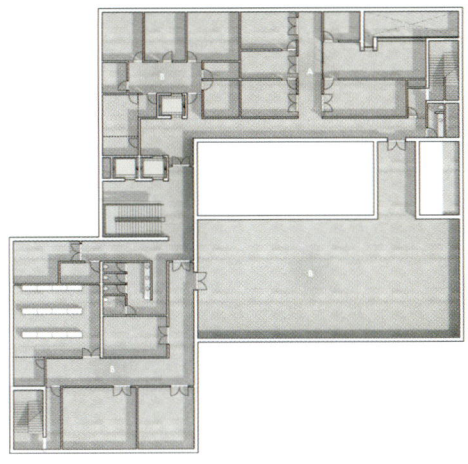

B1 floor plan

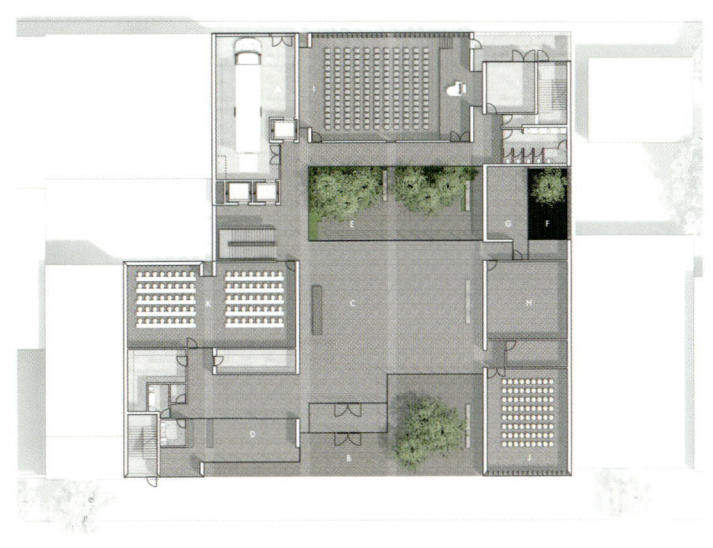

Ground floor plan

Winner _ 당선작

Turku Music Centre
투르쿠 음악 센터

PES-Architects / Tuomas Silvennoinen
PES-아키텍츠 / 투오마스 실벤노이넨

The new Turku Music Centre will form an inspiring new link in the chain of vibrant cultural asites on the banks of the Aura River in central Turku, Finland. In addition to providing a new home with world-class acoustics for the Turku Philharmonic Orchestra, the Music Centre will be an accessible and inviting cultural venue that encourages encounters, communication and new forms of interaction between diverse user groups throughout the day.

The new building seeks to integrate respectfully into the important cultural and historical context at Independence Square, flanked by the City Theatre and Wäinö Aaltonen Museum, with a major government office building at the back of the site. Instead of creating a bold, disruptive icon, the architecture seeks dialogue with the existing built environment, including material references such as the use of copper facade cladding. At the same time, the building has a distinctive, clearly contemporary character of its own, with soft, curving forms that appear to be shaped by the flow of the river.

The public spaces, lobbies, foyers and cafe all face the street and river. The vertical openings in the copper-clad timber and glass facade allow varying views of the river from within and make the building open and inviting from outside.

The rounded form of the building, with the main concert hall situated diagonally, allows the main entrance to be placed at the corner of the site, opening both towards the City Theatre and the riverfront. This creates a maximally open space on the small plot and retains lines of sight from the existing buildings towards the river.

The urban square formed between the Music Centre and City Theatre serves as a shared space for events and activities, inspiring artistic collaboration. There is also a direct indoor connection between the two buildings.

The 1,300-seat main concert hall takes the form of a modified shoebox, with curved balconies that wrap around the stalls and stage to create an immersive space where the audience and orchestra are enveloped by music. The acoustic design principles are directly derived from the acclaimed shoebox concert halls of the 19th century, providing for excellent lateral reflections and a rich reverberation. These key aspects of the model are retained, while a stronger visual connection is created to the musicians on stage. This results in a unique design customised to the ambitions of the project: the acoustic excellence of the shoebox and visual intimacy reconciled.

Location : Turku, Finland I **Function** : Cultural facility I **Stories** : 4FL I **Client** : City of Turku I **Design team** : Marko Kivistö (Laidun-design Oy), Martin Lukasczyk (project manager), Willem Barendregt, Martin Genet, Hannes Halttunen, Anniina Ikäheimo, Mikko Karppanen, Kai Lindvall, Satu Mattila, Toivo Moustgaard, Yan Peng, Jarkko Salminen, Yizhou Zhao

위치 : 핀란드 투르쿠 | **용도** : 문화시설 | **규모** : 지상4층

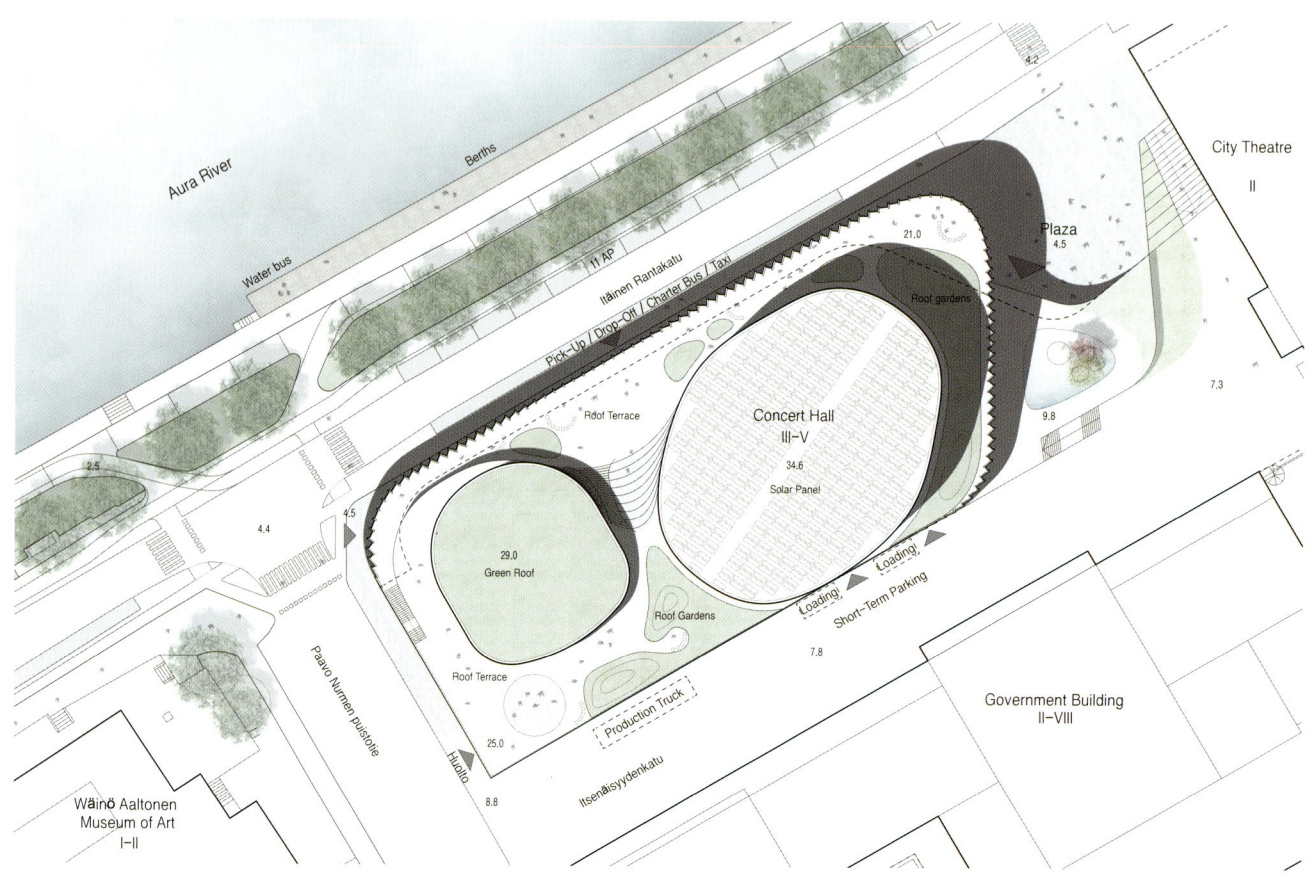

Site plan

Design Concept

VIEWS
Good views to the Aurajoki river have been preserved. Some building functions have been pushed underground to enable a plaza.

ROTATION
Concert hall rotation gives a more spacious grand foyer.

SHAPE
With its free formed shape the grand foyer inner wall resembles distantly Aurajoki riverbed from an era before the city.

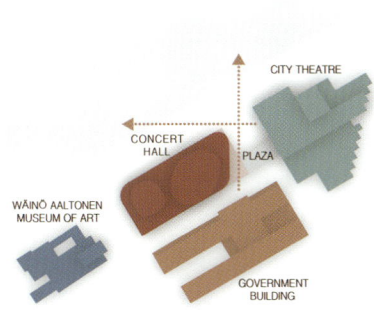

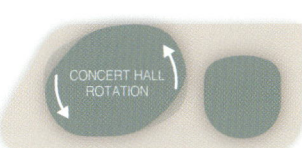

PUBLIC SPACE
Concert hall will be connected to the theatre with an inner connection and a plaza.

AUDIENCE
People will flow inside the building naturally. The whole building is easily accessible.

DIVISION
The grand foyer inner wall cuts the building in two parts dividing public areas from private areas.

North-east elevation

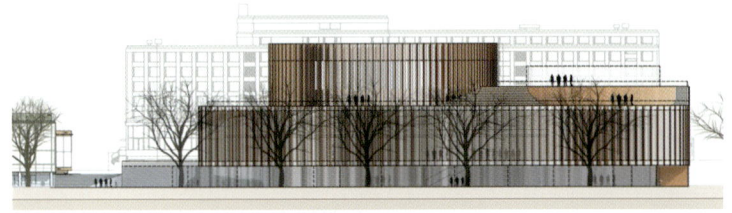

North-west elevation

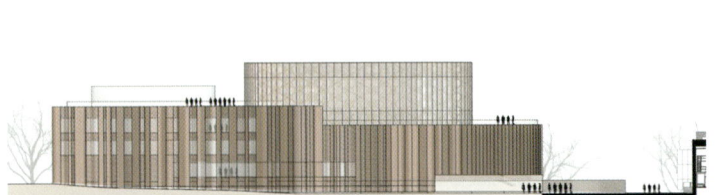

South-east elevation

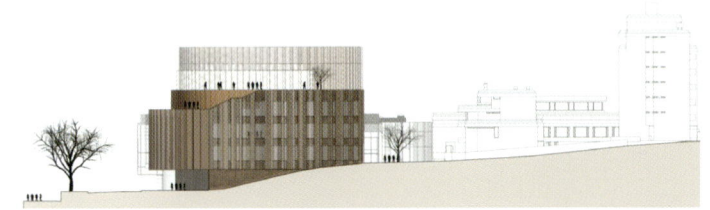

South-west elevation

새로운 투르쿠 음악 센터는 핀란드 투르쿠 중심부의 아우라 강둑에 있는 활기찬 문화 유적지 사슬에서 고무적인 새로운 연결 고리를 형성할 것이다. 투르쿠 필하모닉 오케스트라를 위한 최고의 음향 시설을 갖춘 새로운 보금자리를 제공할 뿐만 아니라 하루 종일 다양한 사용자 그룹 간의 만남, 소통, 새로운 교류를 장려하는 접근 가능하고 매력적인 문화 공간이 될 것이다.

새 건물은 시립극장과 봐이네 알토넨 미술관 옆에 있는 독립 광장의 중요한 문화, 역사적 맥락에 정중하게 스며들고자 하며, 부지 후면에는 주요 관공서 건물이 있다. 대담하고 파격적인 상징성을 만들기 보다는 구리 외피를 사용하는 등 재료를 참고하면서 기존 건축 환경과의 대화를 추구한다. 동시에, 건물은 강물의 흐름으로 형성된 것처럼 보이는 부드러운 곡선 형태로 독특하고 현대적인 특성을 갖는다.

공공 공간, 로비, 카페는 모두 거리와 강을 마주한다. 구리로 덮인 목재와 유리 파사드의 수직 개구부를 통해 내부에서는 강의 다양한 풍경을 조망할 수 있고 외부에서는 건물이 개방적이고 매력적으로 보인다.

건물은 둥근 형태이고 메인 콘서트홀이 비스듬하게 놓여 있으며 주출입구는 대지의 모서리에 배치되어 시립극장과 강변으로 열려 있다. 따라서 작은 필지에서도 공간이 최대한 열리고 기존 건물에서의 시선이 강을 향하게 된다.

음악 센터와 시립극장 사이에 형성된 도시광장은 행사와 활동을 위한 공유 공간 역할을 하며 예술적 협업을 고취한다. 또한, 두 건물 사이에 직접적인 내부 연결이 있다. 1,300석 규모의 메인 콘서트홀은 구두상자 모습과 같으며, 곡선형 발코니는 1등석과 무대를 감싸고 있어 관객과 오케스트라를 음악으로 감싸는 몰입형 공간이 만들어진다. 음향 설계 원칙은 19세기의 찬사를 받은 구두상자 형태의 콘서트홀에서 직접 파생되어 우수한 측음 반향과 풍부한 잔향을 제공한다. 이러한 사항들이 유지되면서 동시에 무대 위의 음악가들로 강력한 시각적 연결이 만들어진다. 그 결과, 구두상자 콘서트홀의 음향적 우수성과 시각적 교감이 조화를 이룬다는 프로젝트의 포부에 맞게 독특한 디자인이 완성되었다.

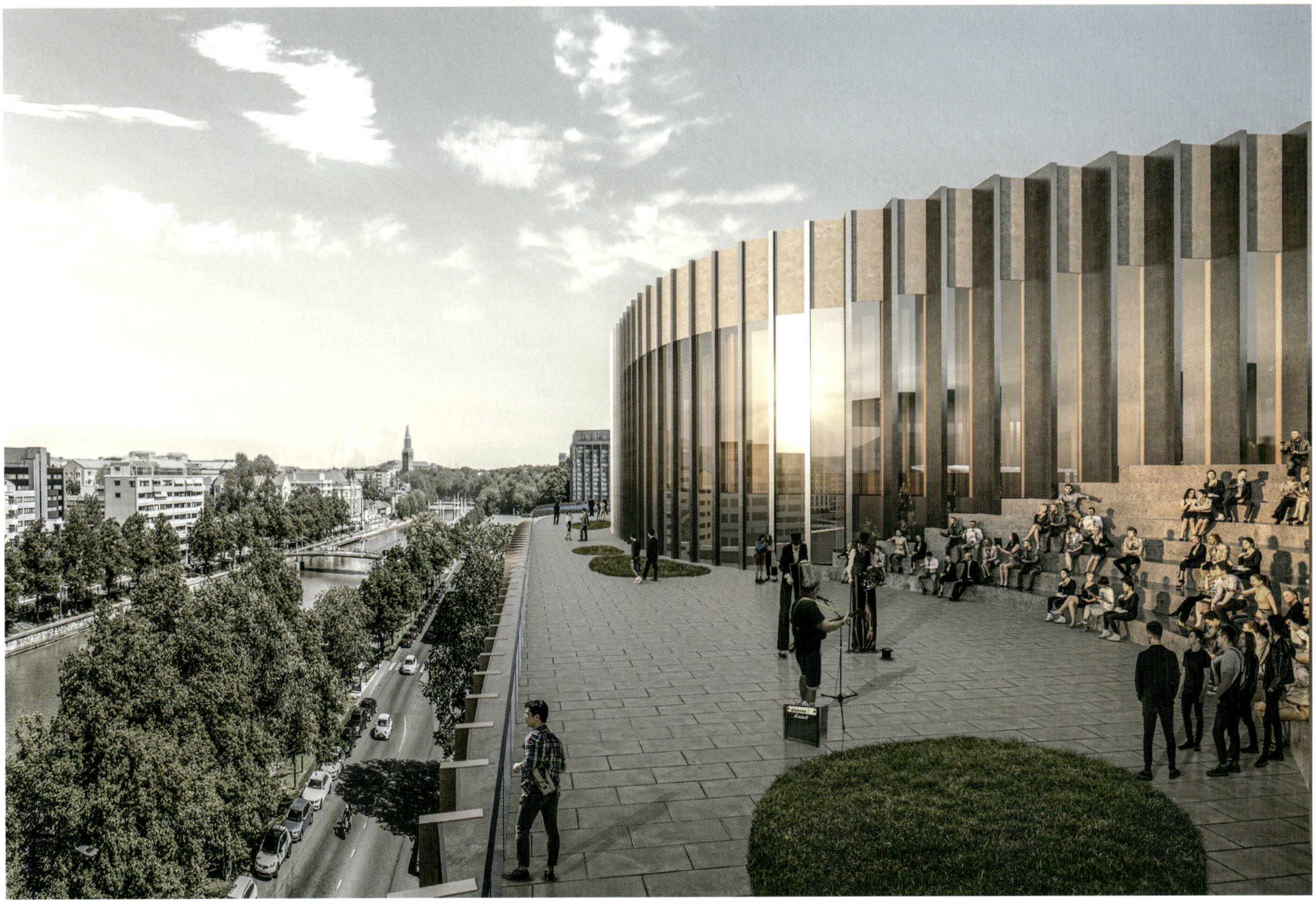

Axonometric

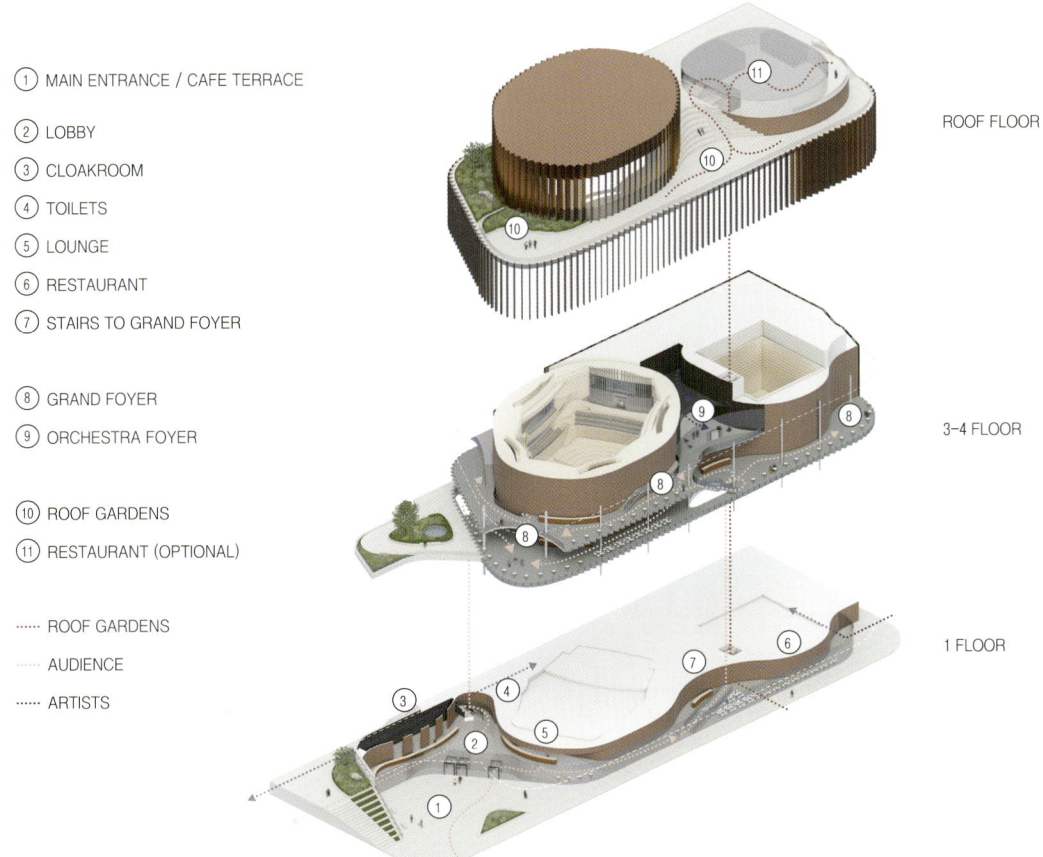

1. MAIN ENTRANCE / CAFE TERRACE
2. LOBBY
3. CLOAKROOM
4. TOILETS
5. LOUNGE
6. RESTAURANT
7. STAIRS TO GRAND FOYER

8. GRAND FOYER
9. ORCHESTRA FOYER

10. ROOF GARDENS
11. RESTAURANT (OPTIONAL)

...... ROOF GARDENS
...... AUDIENCE
...... ARTISTS

ROOF FLOOR

3-4 FLOOR

1 FLOOR

Winner _ 당선작

Turku Music Centre

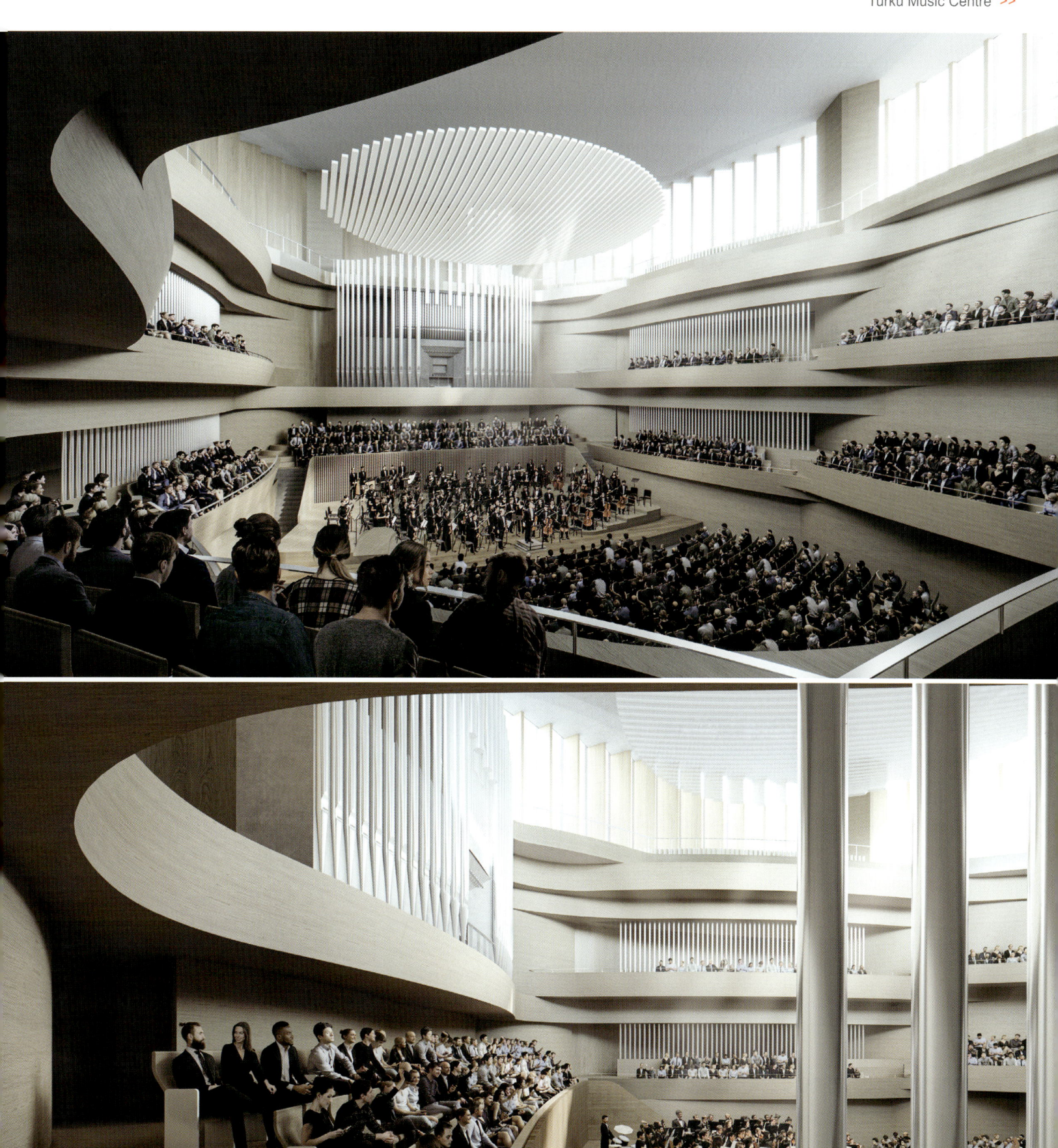

1. Public
2. Personnel
3. Concert hall
4. Technical
5. Artists
6. Services
7. Main facade

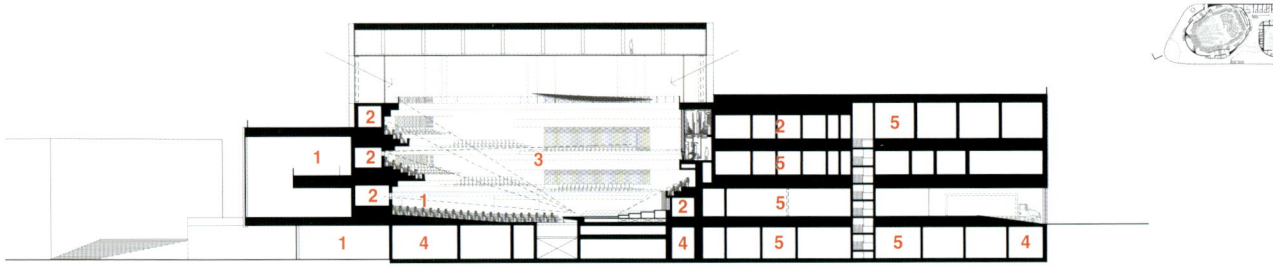

Section I

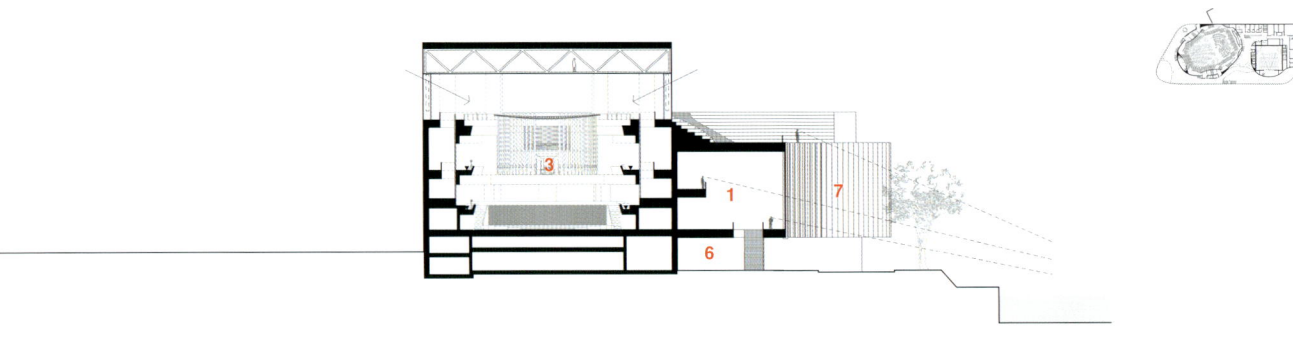

Section II

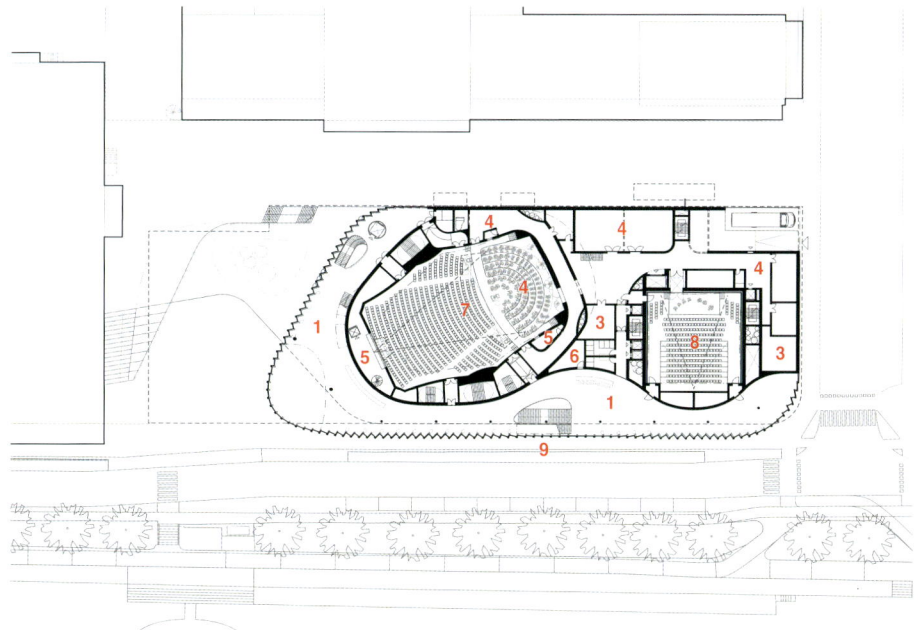

1. Public
2. Services
3. Storage
4. Artists
5. Personnel
6. Technical
7. Concert hall
8. Multipurpose hall
9. Main facade

2nd floor plan

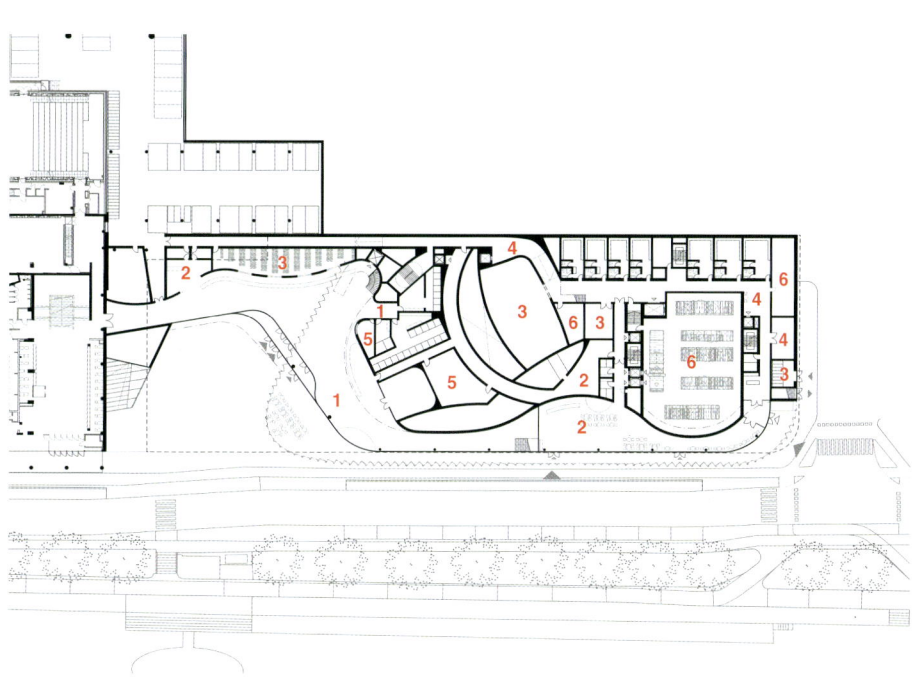

1st floor plan

Winner _ 당선작

Wuxi Art Museum
우시 미술관

Ennead Architects
엔네아드 아키텍츠

As a proposed new center of art and culture in the region, the Wuxi Museum and Art Park builds on the cues of the Chinese garden tradition whose legacy has long been part of the Wuxi-Suzhou-Shanghai area.
Set within Shangxianhe Wetland Park, the museum's site design and architectural form closely connect to the natural context and integrate the museum experience with the natural environment.
The architecture is conceived as a metaphorical Taihu Scholar Stone: a contemplative and intricate spatial structure that simultaneously invites one's spirit in but also sits quietly as a meditative object amidst the broader natural context.
The building becomes a set piece in a larger overall composition, highlighting views in and out of the museum through its cracks and crevices that are seen as a series of subtractions and removals from a larger massing, similar to the erosion of spirit stones by the natural forces of wind and water. As a result, the garden metaphor inspires not only a formal proposition but an experiential one as well, providing an ever-transforming journey of art and nature through a carefully composed choreography that unfolds something new with each successive step.

이 지역의 새로운 예술문화 중심지로 제안된 우시 미술관은 우시-쑤저우-상하이 지역에서 오랫동안 전해 내려오는 중국 정원 전통의 단서를 기반으로 건설된다.
상시안허 습지공원 내에 위치한 미술관의 부지 설계와 건축 양식은 자연 맥락과 밀접하게 연결되어 있으며 미술관 경험을 자연 환경과 통합한다.
이 건축물은 타이후 스콜라 스톤으로 비유된다. 사색적이고 복잡한 공간 구조는 영혼을 일깨우면서도 넓은 자연 맥락 속에서 사색의 오브제로 조용히 앉아 있다.
건물은 일반적인 전체 구성을 이루며, 바람과 물의 자연적인 힘에 의한 영혼석의 침식처럼 큰 매스에서 빼고 제거한 것 같은 틈을 통해 미술관 안팎의 전망을 강조한다. 따라서 정원이라는 표현은 형식적인 제안뿐만 아니라 경험적인 제안에도 영감을 주며, 연속적인 단계마다 새로운 것을 신중히 구성하여 예술과 자연의 끊임없이 변화하는 여정을 보여준다.

Location : Wuxi, China | **Function** : Cultural facility | **Total floor area** : 30,000㎡ | **Design team** : Thomas Wong, Kevin McClurkan, Grace Chen, Brian Masuda, Devin Murray, Edward Chang, Yan Ding, Geoffrey Hughes, Weiwei Kuang, Oliver Li, Yimika Osunsanya, Lanxi Sun | **Landscape architect** : West 8 | **Renderings** : ennead architects LLP
위치 : 중국 우시 | 용도 : 문화시설

Wuxi Art Museum >>

Site Components

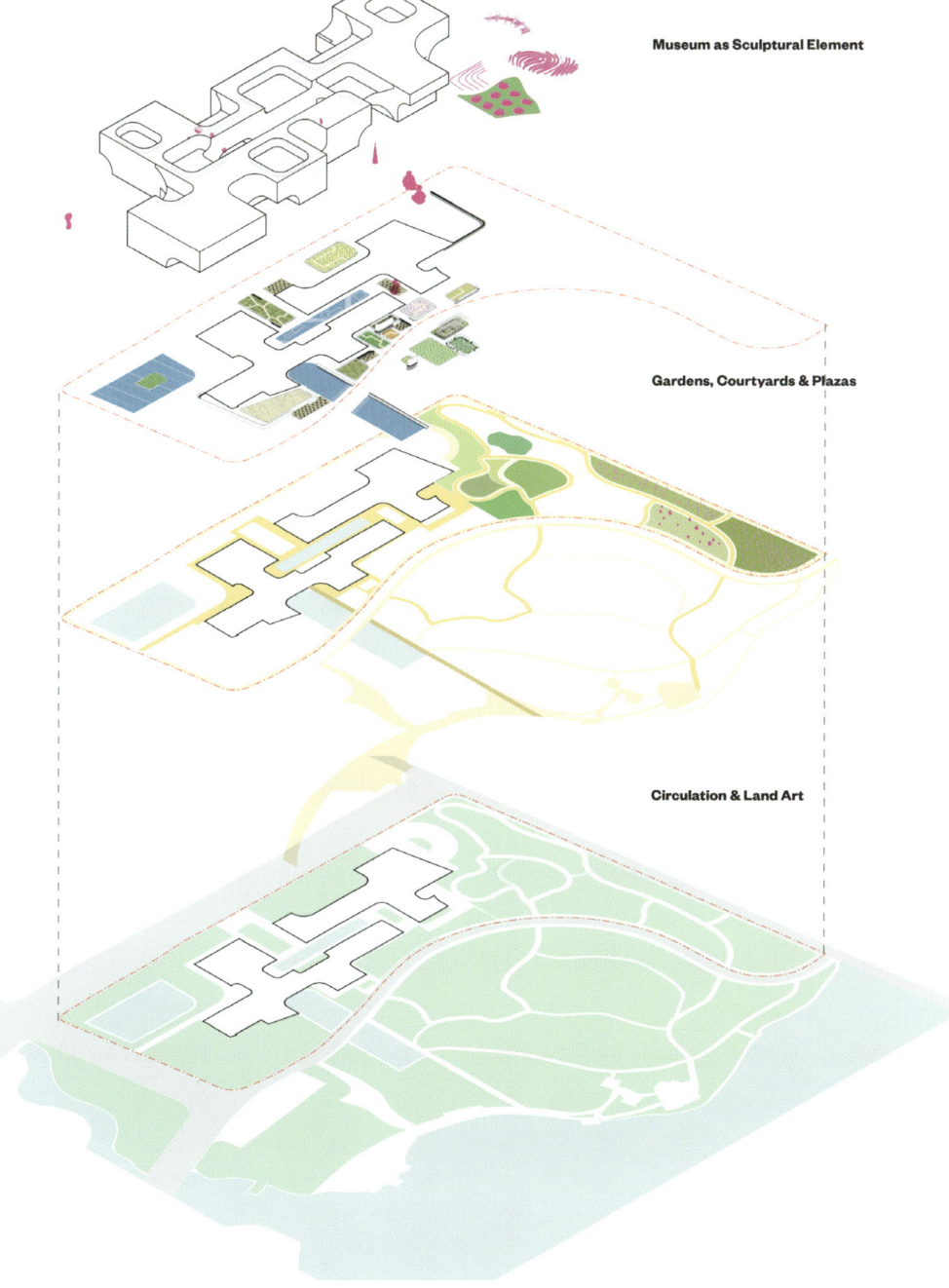

Museum as Sculptural Element

Gardens, Courtyards & Plazas

Circulation & Land Art

Blue-Green Softscape & Wetland Approach

Winner _ 당선작

Wuxi Art Museum >>

■ Section Perspective

1. Central court
2. Cafe
3. Lecture hall
4. Grand exhibition
5. Open exhibition
6. Shaded terrace
7. The orchard

Winner _ 당선작

Philharmonic Hall
필하모닉 홀

BIG / Bjarke Ingels, Brian Yang
BIG / 비야케 잉겔스, 브리안 양

Located at the intersection between the Vltava riverfront and Prague's Cultural Mile, the Vltava Philharmonic Hall will connect the Old Town's traditional cultural scene with the Holešovice neighborhood's modern art scene and become a new civic heart for Prague and the surrounding public realm.

An essential public building for the Holešovice district and a new focal point for Prague, the new Vltava Philharmonic extends horizontally and vertically in all directions to create key urban connections and form a recognizable landmark for surrounding communities near and far. Ascending from the Vltava River to the skyline of Prague, a series of grand public plazas will become a new symbol of inclusionary architecture, welcoming the multitude of Prague's vibrant urban life to flow across, through, under and over the new concert hall.

A contemporary extension of Prague's dramatic urban topography, the Vltava Philharmonic Hall is composed as a cascade of outdoor destinations from the waterfront on the river to the city's iconic skyline. By raising and lowering the corners of the building at multiple touch points, the public spaces connect and allow activities to spill in and out of the building on every side: towards the river, the square, the street, and the alley. Visitors are drawn in from all forms of arrival, with carefully chosen programs inviting them to explore the music venues inside or climb the elegant, arced roofs of the Hall.

Beyond being a major cultural destination for Prague, the building is crafted to maximize its potential to host external uses and special events. The venues are carefully designed to maximize flexibility for a range of uses - from the boldness of contemporary music styles to theater performances and digital exhibitions. At the buildings summit, an elegant hyperbolic structure spans over the Vltava Hall and forms the iconic ceiling of restaurant and event ballroom framing views of the historic city center of Prague and becoming a major destination for social gatherings of all kinds.

Location : Prague, Czech Republic I **Function** : Cultural facility I **Site area** : 49,715m² I **Client** : Prague Institute of Planning and Development I **Project leader** : Shane Dalke I **Project manager** : Luca Nicoletti I **Design team** : Sarkis Sarkisyan, Giulia Orlando, Matthew Oravec, Giulia Vanni, Jan Magasanik, Jeremias Sas Iros, Jonathan Chester, Khaled Magdy Zaki Ahmed Elfeky, Sorcha Burke, Clara Elma Margareta Karlsson, Mads Primdahl Rokkjær, Ondrej Slunecko, Tania-Cristina Farcas, Polina Galantseva, Yanis Amasri Sierra, Paula Madrid I **Landscape** : Giulia Frittoli, Eleanor Gibson, Jialin Liang

위치 : 체코 프라하 I 용도 : 문화시설

Location

Vltava Philharmonic Site
Convent of St. Agnes
Rudolfinum
Museum Kampa
National Theatre
Žofín Palace
Mánes Gallery

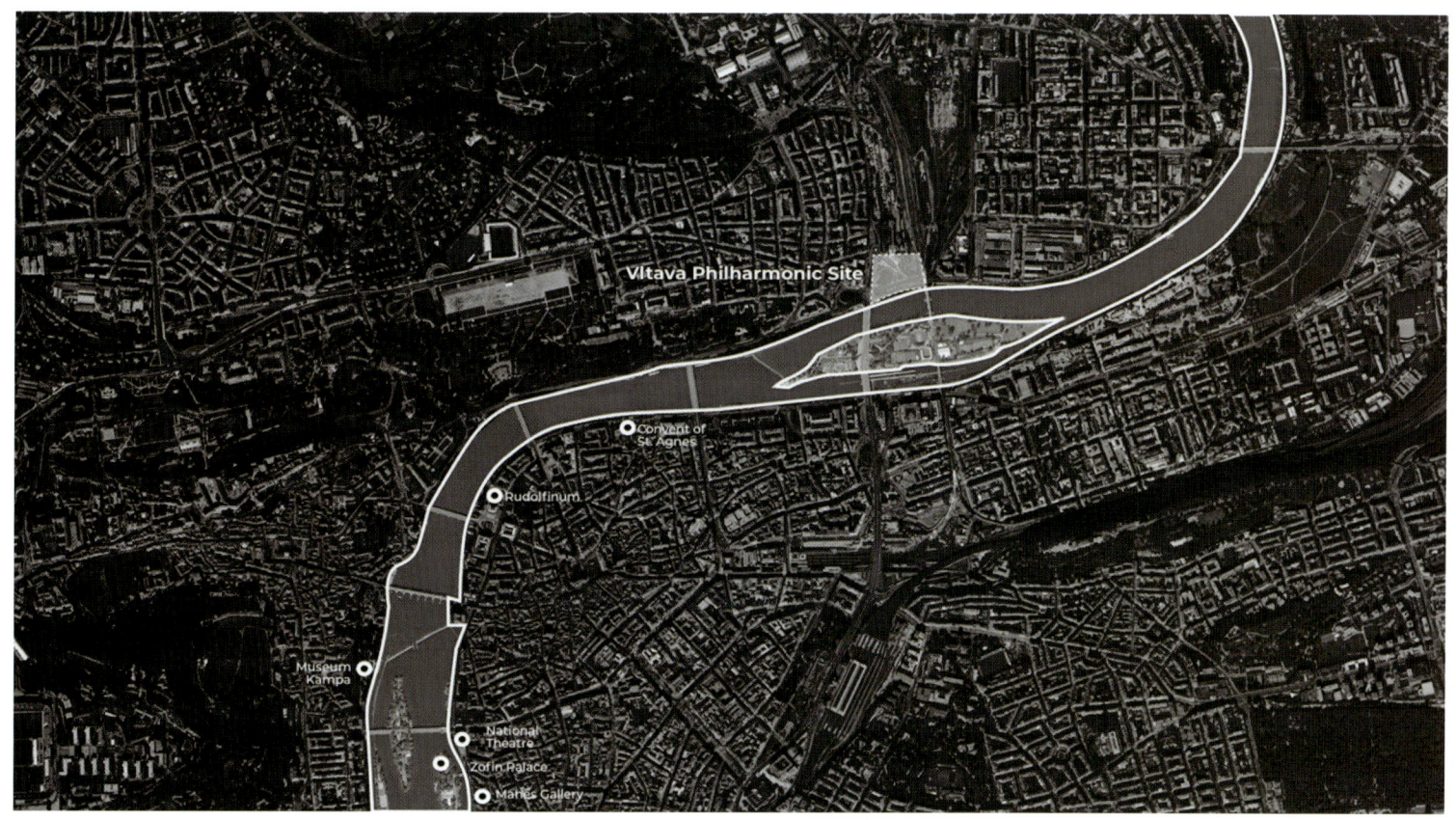

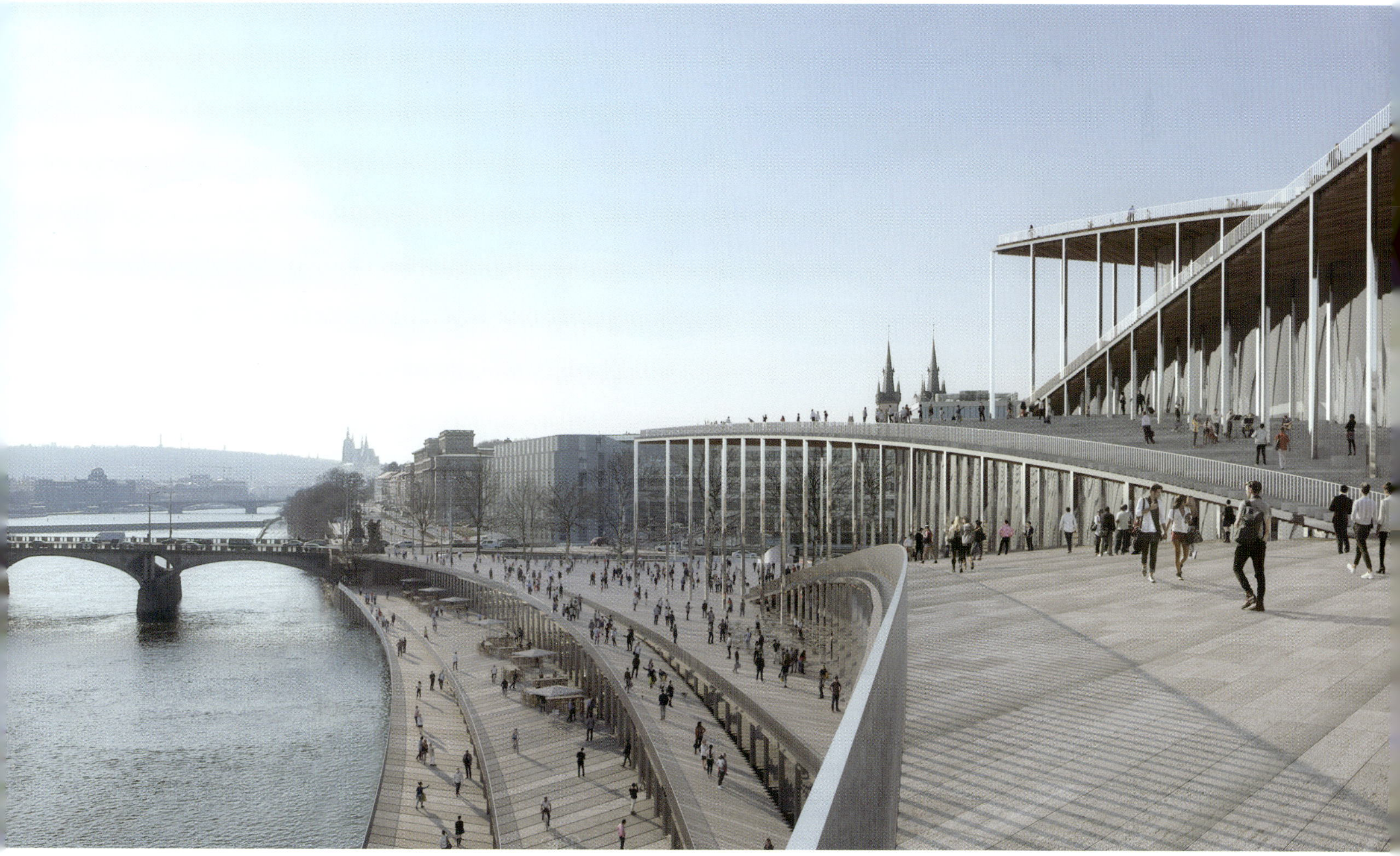

Winner _ 당선작

Model

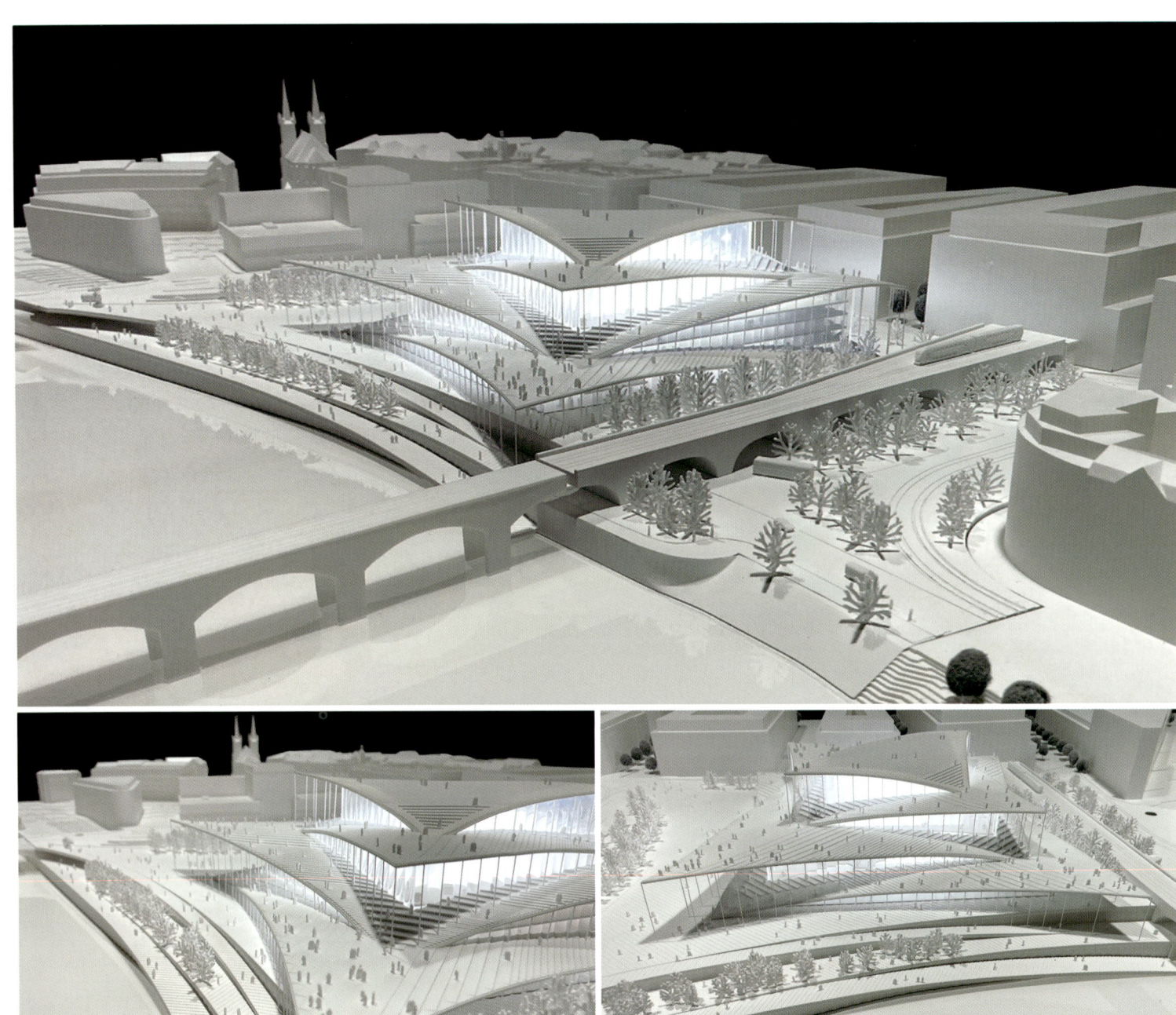

블타바 강변과 프라하 컬처럴 마일의 교차점에 위치한 블타바 필하모닉 홀은 구시가지의 전통 문화 모습과 홀레쇼비체 마을의 현대 미술 모습을 연결하고 프라하와 주변 공공 영역의 새로운 중심지가 될 것이다.

홀레쇼비체 지구의 필수적인 공공 건물이자 프라하의 새로운 중심지인 블타바 필하모닉 홀은 동서남북 사방에서 수평적, 수직적으로 확장되어 주요 도시 연결점을 생성하고 가깝고 먼 주변 지역사회에서 인식 가능한 랜드마크를 형성한다. 블타바 강에서 프라하의 스카이라인까지 이어지는 일련의 대형 공공 광장은 통합 건축의 새로운 상징이 될 것이며, 새로운 콘서트 홀 전체에 프라하의 활기찬 도시 생활이 펼쳐질 것이다.

프라하의 극적인 도시 지형을 현대적으로 확장한 블타바 필하모닉 홀은 강변에서 상징적인 도시 스카이라인까지 계단식 야외 장소로 구성된다. 여러 접점에서 건물의 모서리를 올리거나 내림으로써 공공 공간이 연결되고 강, 광장, 거리, 골목 등 모든 측면에서 건물 안팎으로 활동이 활발히 이루어진다. 내부에서 음악 공연장을 탐색하거나 홀의 우아한 아치형 지붕을 오르도록 신중하게 선택된 프로그램을 통해 모든 형태의 도착점에서 방문객을 끌어들인다.

프라하의 주요 문화 장소가 되는 것 외에도 이 건물은 외부 사용과 특별 행사를 주최하기 위해 건설되었다. 웅장한 현대 음악부터 극장 공연과 디지털 전시회에 이르기까지 최대한 유연하게 다양한 용도로 활용되도록 신중하게 설계되었다. 건물 꼭대기의 세련된 쌍곡선 구조는 블타바 홀 전체를 아우르며 레스토랑의 상징적인 천장을 형성하고, 연회장은 역사적인 프라하 도심의 전망을 담아내고 모든 사교 모임을 위한 주요 장소가 된다.

Philharmonic Hall >>

Winner _ 당선작

Jinju Public Science Museum
진주 공립 전문과학관

VOID architects / Kusang Lee, Kiuck Zhang
㈜보이드아키텍트 건축사사무소 / 이규상, 장기욱

Jinju Public Science Museum 'Mysterious Cube'
As a central institution for science education, the public science museum has played an important role as a central institution of science education in discovering future jobs that can be applied by mythicizing technological development and enhancing practical capabilities to implement it. The unusual imagination cannot undermine the foundation of the Science Museum as a cultural exhibition space and educational space. The fantasy created by aerospace science sometimes makes us forget its academic practicality.

When I imagined that this public science museum would be newly created in Jinju, I tried to avoid the disconnected spatial experience with black boxes. I reconsidered the prototype of the science museum of 'Curious Room', which induces surprise and social communication through the complex overlap of the new.

A cube with a fluid facade using the moire cannot be seen from the outside. The infinite space is observed inside the cube when following the hidden entrance next to the railroad tracks with the history of Jinju. The interior of the facade combined with media art technology provides a foundation to overcome the physical limitations, and the floating interior volume with various heights becomes a frame of ideas that frees the exhibition planner from the limitations of the ground. Exhibition spaces and experience spaces where you can experience various exhibits from different angles will be scattered in one single space.

진주 공립 전문과학관 'Mysterious Cube'
기술 발전의 신화화를 통해 응용할 수 있는 미래 먹거리를 찾아내고, 이를 실행할 수 있는 실용적인 역량을 발전시키는데, 공립과학관은 과학교육의 중심 기관으로서 중요한 역할을 해 왔다. 아무리 도발적인 상상력을 발휘하더라도, 문화전시 공간이자 교육 공간으로서 과학관의 근간을 훼손할 수는 없다. 하지만 항공우주 과학이 보여주는 판타지는 때로 그 학문으로서의 실용성을 잊게 만들곤 한다.

이를 다루는 공립과학관이 진주에 새롭게 생겨나는 것을 상상해 보았을 때, 블랙박스들로 이뤄진 단절된 공간적 경험은 피하고자 하였다. 새로움의 복합적인 중첩을 통해 놀라움과 사회적 소통을 유발했던 '호기심의 방'이라는 과학관의 원형을 다시 고민하였다.

무아레 현상을 이용한 유동적인 파사드로 이뤄진 큐브는, 겉에서 볼 때는 그 실체를 알 수 없다. 진주의 역사를 담은 철길 옆의 감추어진 출입구를 따라 들어간 큐브의 내부에서는 무한한 공간이 관찰된다. 미디어아트 기술과 결합한 파사드의 내면은 물리적 제한을 넘을 수 있는 바탕을 제공하며, 다양한 높이를 가지고 부유하는 내부 볼륨은 전시 기획자를 지면의 한계로부터 벗어나게 하는 아이디어의 틀이 된다. 하나의 단일 공간 속에 다양한 전시물을 다각도에서 경험할 수 있는 전시 공간과 체험 공간이 흩뿌려져 있을 것이다.

Location : 446-1, Manggyeong-dong, Jinju-si, Gyeongsangnam-do, Korea I **Function** : Cultural & Assembly facility I **Site area** : 7,890m² I **Bldg. area** : 1,972m² I **Total floor area** : 6,000m² I **Stories** : B1, 4FL I **Structure** : Reinforced concrete, Steel frame I **Finish** : Laminated glass

위치 : 경상남도 진주시 망경동 446-1번지 외 5필지 I **용도** : 문화 및 집회시설 I **규모** : 지하1층, 지상4층 I **구조** : 철근콘크리트, 철골 I **마감** : 접합유리 I **설계팀** : 금태연, 남정훈

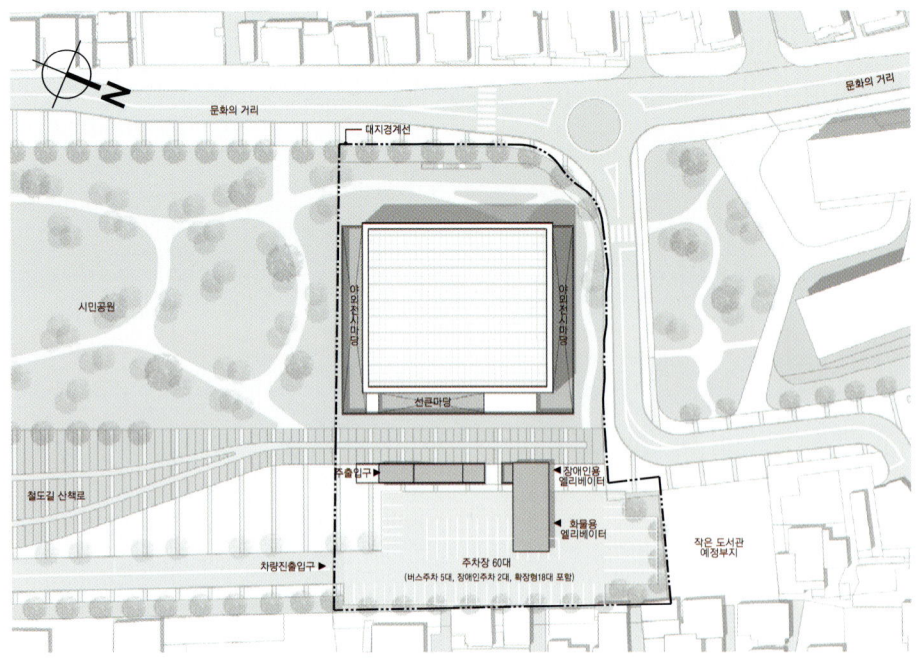

Site plan

- 진주의 역사를 담은 선로를 복원한다.
- 계획된 보행로와 차량동선을 반영할 수 있도록 건축영역을 설정한다.

공원과 박물관, 기찻길의 한켠에서 자신을 과시하지 않도록 높이를 조정하고 입구를 만들어낸다.

공립 과학관으로서 필요한 공공적 기능을 만족하고, 주변과의 소통을 위한 틈을 만들어낸다.

■ Design Concept

・Recycling & Adapt

・Consideration & Withdraw

・Function & Communication

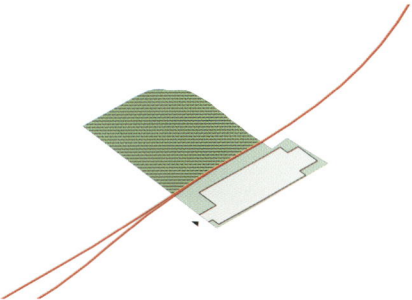

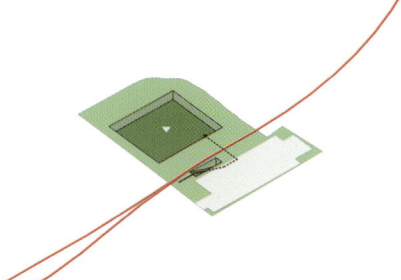

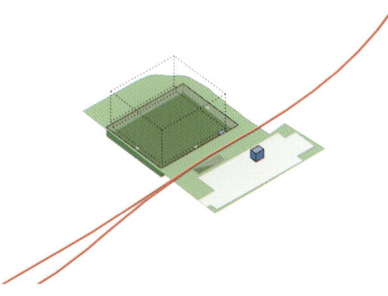

- 진주의 역사를 담은 선로를 복원한다.
- 계획된 보행로와 차량동선을 반영할 수 있도록 건축영역을 설정한다.

- 공원과 박물관, 기찻길의 한켠에서 자신을 과시하지 않도록 높이를 조정하고 입구를 만들어낸다.

- 공립 과학관으로서 필요한 공공적 기능을 만족하고, 주변과의 소통을 위한 틈을 만들어낸다.

1. 9.5mm 2 layers of gypsum board
2. Rigid urethane foam insulation
3. Low-E pair glass aluminum windows and doors
4. 12.8mm laminated glass (random vertical line etch print)
5. Outer junction: weather-resistant silicone joint
6. Inner junction: structural Silicone Joints
7. 19.6mm laminated glass
8. Holes for ventilation inside the facade (Φ10 @ 600)
9. 6mm high-strength transparent polycarbonate L-shaped plate
10. M12 high tension bolt
11. M12 set anchor
12. Anti-vibration pad inside polycarbonate plate
13. Transparent epoxy adhesive

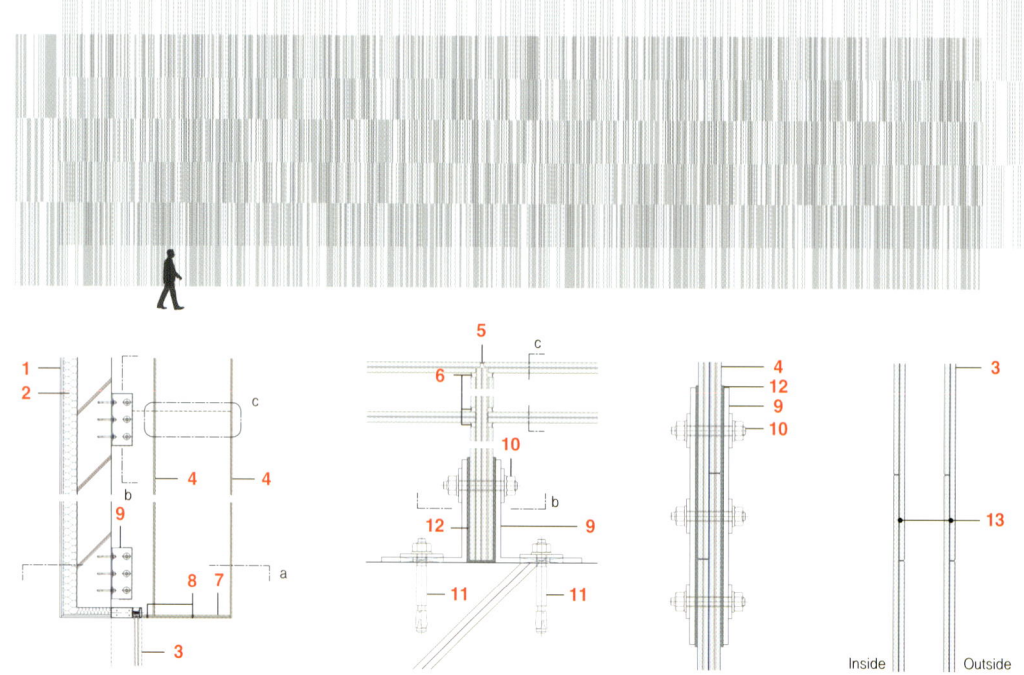

Moiré facade detail - Section

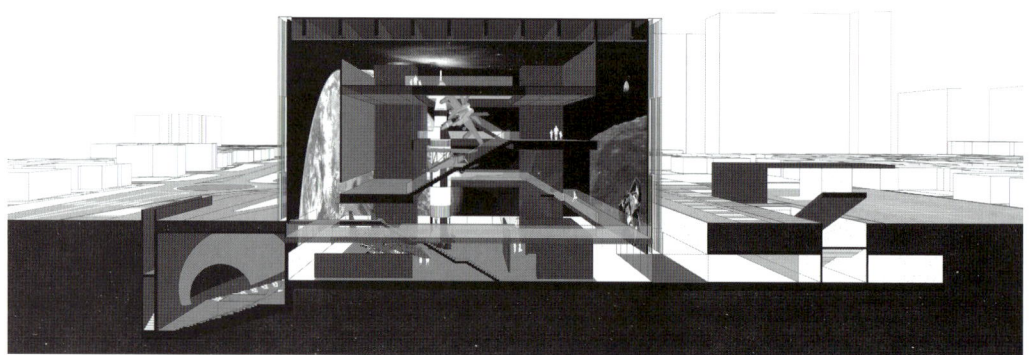

Section

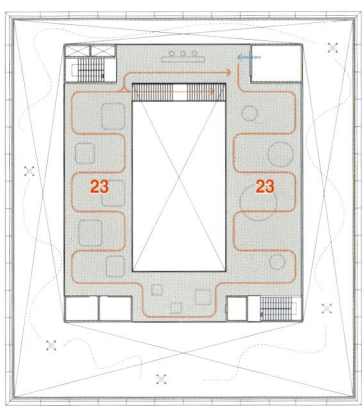
4th floor plan

1. Electrical room
2. Machine room
3. Emergency room
4. Planetarium
5. 4D experience facility
6. Smart studio
7. Multipurpose room
8. Storage
9. Wind break room
10. Education preparation room
11. Operation preparation room
12. Meeting room
13. Tea making room
14. Warehouse & Library
15. Astronomical observation experience
16. ISS exploration
17. Atmospheric media wall
18. UAM boarding experience
19. Drone control experience
20. Flight boarding experience
21. Flight simulator
22. Paper airplane experience
23. Special exhibition

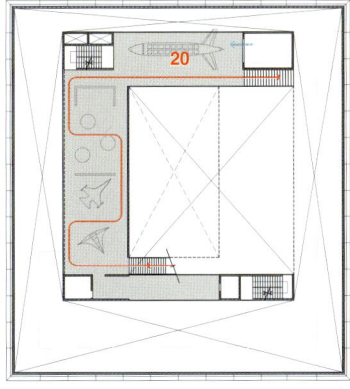
2nd floor plan

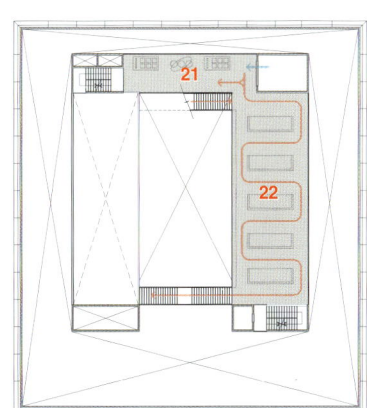
3rd floor plan

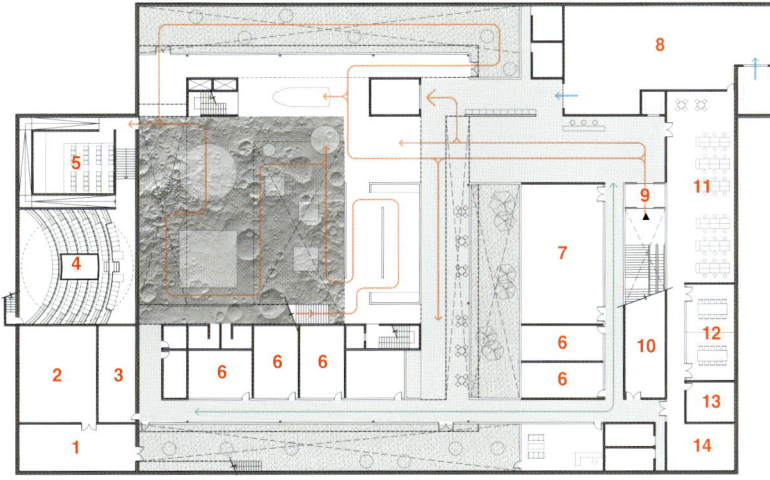
Ground floor plan

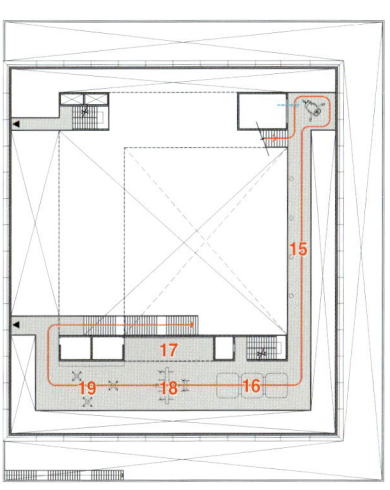
1st floor plan

Saskatoon New Central Library
새스커툰 뉴 센트럴 라이브러리

Formline + Chevalier Morales + Architecture49
폼라인 + 슈발리에 모랄레스 + 아키텍처49

Thoughtful urban and site design shape the 142,000 square foot library, which will play an important role in linking Saskatoon's green spaces to Remai Modern Gallery, street shops, and foot traffic along Second Avenue. Unlike other Canadian cities, Saskatoon has a lower park-to-person ratio. City-wide, there are 4.4 hectares of green space for every 1,000 residents, compared to 10 in Halifax and 42 in Calgary. This observation drives decisions on public space, building form, and cladding. 45% of the site will be landscaped public space at both the north and south ends. An oval plan geometry will allow the building to taper at the south end and shift to the west, defining a key public space for live reading, hanging out, and meeting for ceremony.

The geometry and plan strategies will be enforced in building sections along the triple-glazed perimeter windows along Second Avenue, which will step back with angled surfaces, evoking the iconic form, lightness, and luminosity of teepees. The transparent and translucent skin of the building will take advantage of and diffuse the crisp natural light of the prairies. Inspired by the Saskatchewan landscape and local architecture traditions, the library will also evoke the Métis log cabin with exposed glulam beams and a cross-laminated timber strategy that not only provides structure, but also psychic warmth.

The mass timber structure of CLT and glulam beams and columns will rise from a concrete plinth on the ground floor. Bearing on this and radiating out from the plinth are innovative glulam slab bands, set along the radii of the oval plan. The glulam bands will redefine wood construction at a large public scale. Angled glulam columns will run along the public elevation on Second Avenue, where the leaves of lapped glass panels are located, providing air intake for natural ventilation.

The architecture team has created major contemporary libraries throughout Quebec and brings a deep well of knowledge to decisions on the layout and interiors of Saskatoon's new central library. Clever public spaces will be arrayed throughout the building, from a children's theatre, to a community kitchen and a bright learning and sharing circle. The library will also house book and media collections, multimedia labs, and cafes. Accessible entrances, luminous staircases, floor openings, and intimate circulation spaces will intertwine, allowing visitors to move freely, find books, and meet friends. Saskatoon's new central library will showcase the best of library design and community-building for the Reconciliation Era and generations to come.

Location : Saskatoon, Canada I **Function** : Cultural facility I **Total floor area** : 13,192㎡ I **Stories** : B1, 4FL
위치 : 캐나다 새스커툰 I 용도 : 문화시설 I 규모 : 지하1층, 지상4층

Urban plan

167

Winner _ 당선작

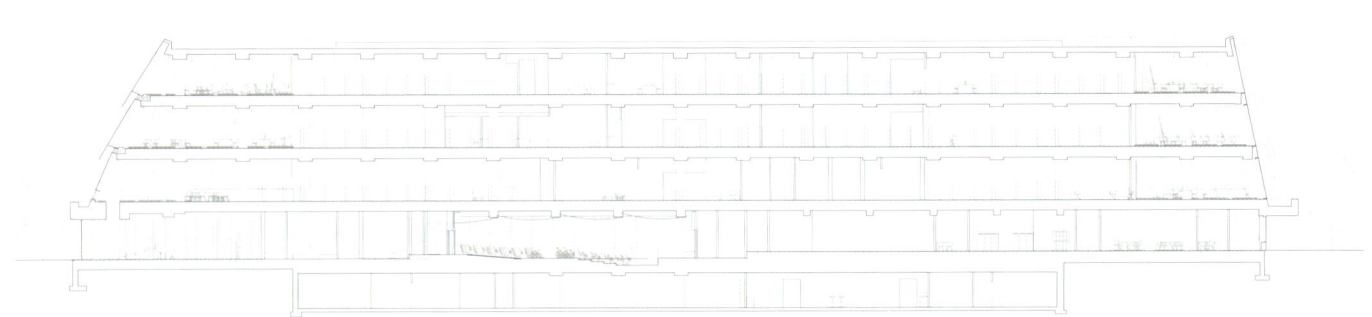

Section I

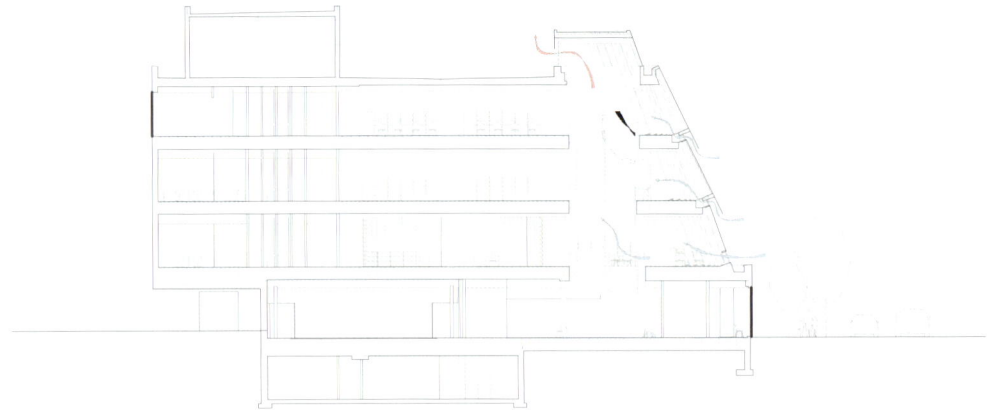

Section II

사려 깊은 도시 및 대지 디자인은 새스커툰의 녹지 공간을 레마이 모던 갤러리, 거리 상점 및 세컨드 에비뉴를 따라 유동인구로 연결하는 데 중요한 역할을 할 142,000평방피트의 도서관을 형성한다. 다른 캐나다 도시와 달리 새스커툰은 공원 대 사람 비율이 낮다. 도시 전체에서 1,000명당 녹지 면적은 4.4헥타르로 핼리팩스의 10헥타르, 캘거리의 42헥타르와 비교된다. 이 관찰은 공공 장소, 건물 형태 및 클래딩에 대한 결정을 유도한다. 부지의 45%는 북쪽과 남쪽 끝 모두에서 조경된 공공공간이 될 것이다. 타원형 평면 기하학은 건물이 남쪽 끝에서 가늘어지고 서쪽으로 이동할 수 있게 하여 읽고, 어울리고 만남을 위한 주요 공공공간을 정의한다.

기하학 및 계획 전략은 2번가를 따라 3중 유리로 된 주변 창을 따라 건물 단면에 적용되며, 각진 표면으로 뒤로 물러나서 천막의 상징적 형태, 가벼움 및 광도를 불러일으킨다. 건물의 투명하고 반투명한 피부는 대초원의 선명한 자연광을 활용하고 확산한다. 서스캐처원의 풍경과 지역 건축 전통에서 영감을 받은 이 도서관은 노출된 집성재 빔과 구조뿐만 아니라 심리적 따뜻함을 제공하는 교차 적층 목재 전략으로 메티스 통나무집을 연상시킨다.

CLT와 집성재 빔 및 기둥의 매스팀버 구조는 1층의 콘크리트 플린트에서 올라간다. 이를 지지하고 주각에서 방사되는 것은 타원형 평면의 반경을 따라 설정된 혁신적인 집성판 슬래브 밴드이다. 집성재 밴드는 대규모 공공 규모에서 목재 건축을 재정의할 것이다. 각진 집성 기둥은 자연 환기를 위한 공기 흡입구를 제공하는 겹쳐진 유리 패널의 잎이 있는 세컨드 에비뉴의 공공 입면을 따라 이어진다.

건축 팀은 퀘벡 전역에 주요 현대 도서관을 만들고 새스커툰의 새로운 중앙 도서관의 배치와 인테리어에 대한 결정에 깊이 있는 지식을 제공한다. 영리한 공공공간은 어린이 극장에서 커뮤니티 키친, 밝은 학습 및 공유 서클에 이르기까지 건물 전체에 배치된다. 도서관에는 책과 미디어 컬렉션, 멀티미디어 연구실, 카페도 있다. 접근 가능한 입구, 빛나는 계단, 바닥 개구부 및 친밀한 순환 공간이 얽혀 방문객이 자유롭게 이동하고 책을 찾고 친구를 만날 수 있다. 새스커툰의 새로운 중앙 도서관은 화해 시대와 다음 세대를 위한 최고의 도서관 디자인과 커뮤니티 구축을 선보일 것이다.

Winner _ 당선작

Saskatoon New Central Library >>

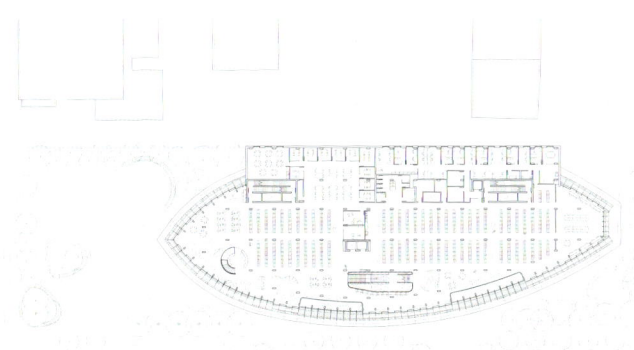

3rd floor plan

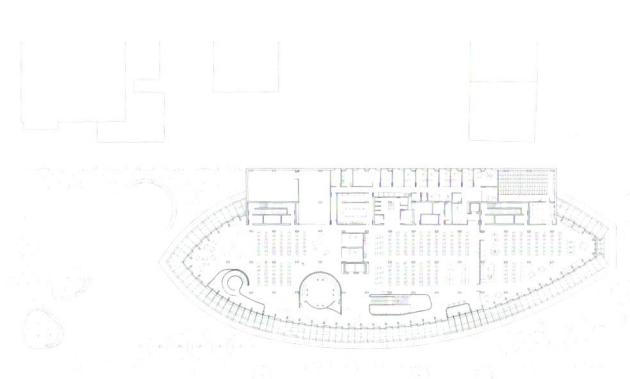

4th floor plan

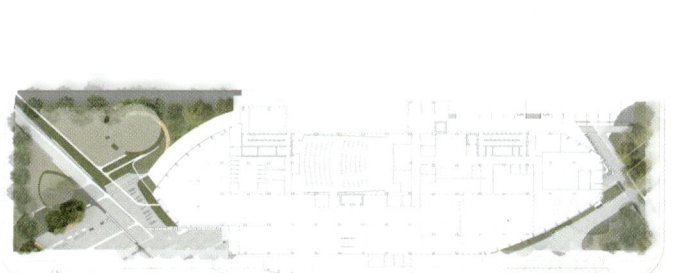

1st floor plan

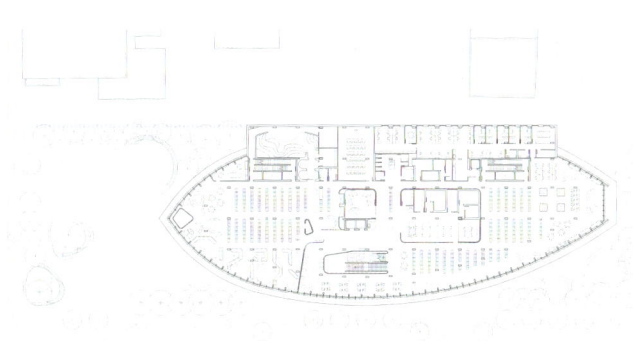

2nd floor plan

171

Winner _ 당선작

Terezín Ghetto Museum
테레진 게토 박물관

Steven Holl Architects / Steven Holl + SKUPINA / Marcela Steinbachová
스티븐홀 아키텍츠 / 스티븐홀 + 스쿠피나 / 마르셀라 스타인바코바

Terezín, founded in 1780 as a military fortress, served as a Jewish Ghetto during World War II where an estimated 33,000 people died. The existing Terezín Ghetto Museum honors the individuals whose lives were taken during this tragic moment in history, and the new design will be a memorial of hope and light.

Tower of Light, a new addition, is a contemplative space to experience spectral light phenomena from daylight refracted into a spectrum of colors—the colors of humanity. At night, it glows as a beacon through the darkness. The Tower of Light recalls 'Moon Landscape,' a drawing made by Petr Ginz. Born in Prague on February 1, 1928, Ginz was deported to the Terezín concentration camp where he made this imaginative drawing of a view of the earth from the moon. In 1944, Ginz was deported to Auschwitz and gassed to death at the age of sixteen.

The Tower of Light is a hopeful new presence in the center of Terezín, gently ascending above the surrounding buildings toward the sky. The design also includes a renovation of the existing museum and exhibition design, new parks and green spaces surrounding the site, updated parking, and a new information center.

1780년에 군사 요새로 세워진 테레진은 약 33,000명의 사망자가 있었던 제2차 세계대전 동안 유대인의 거주 지역이었다. 기존의 테레진 게토 박물관은 역사상 가장 비극적인 순간에 목숨을 잃은 사람들을 기리며, 새로운 디자인은 희망과 빛의 기념비가 될 것이다.

'빛의 타워'는 인류의 색상 스펙트럼으로 굴절된 일광의 스펙트럼 빛 현상을 경험할 수 있는 사색의 공간이다. 밤에는 어둠 속의 등불처럼 빛난다. 빛의 타워는 페트르 긴츠가 그린 '달 풍경'을 떠올리게 한다. 1928년 2월 1일 프라하에서 태어난 긴츠는 테레진 강제 수용소로 이송되어 달에서 본 지구의 모습을 상상하며 그렸다. 1944년 긴츠는 아우슈비츠로 이송되었고 16살 때 가스로 사망했다.

빛의 타워는 테레진 중심에서 새로운 희망이 되어 하늘을 향해 주변 건물 위로 완만하게 올라간다. 이 디자인에는 기존 박물관 및 전시 디자인의 리모델링, 부지 주변의 새로운 공원 및 녹지 공간, 개선된 주차장, 새로운 정보센터도 포함된다.

Location : Terezín, Usti, Czech Republic I **Function** : Cultural facility I **Client** : Terezín Memorial I **Associate architects** : Veronika Tichá, Jan Mojka, Marie Harigelová, Michael Haddy, Talya Polat, Michal Nohejl Obrazek.org (visualizations)

위치 : 체코 우스티 테레진 I **용도** : 문화시설

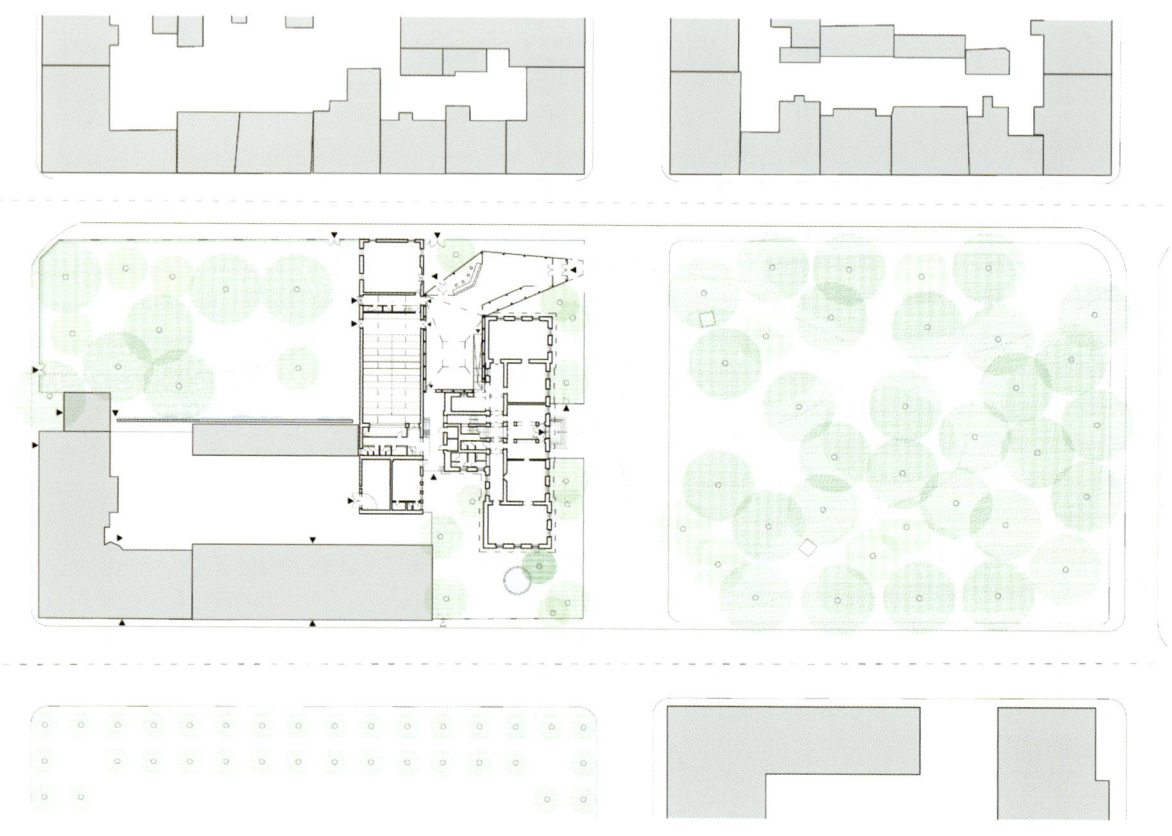

Site plan

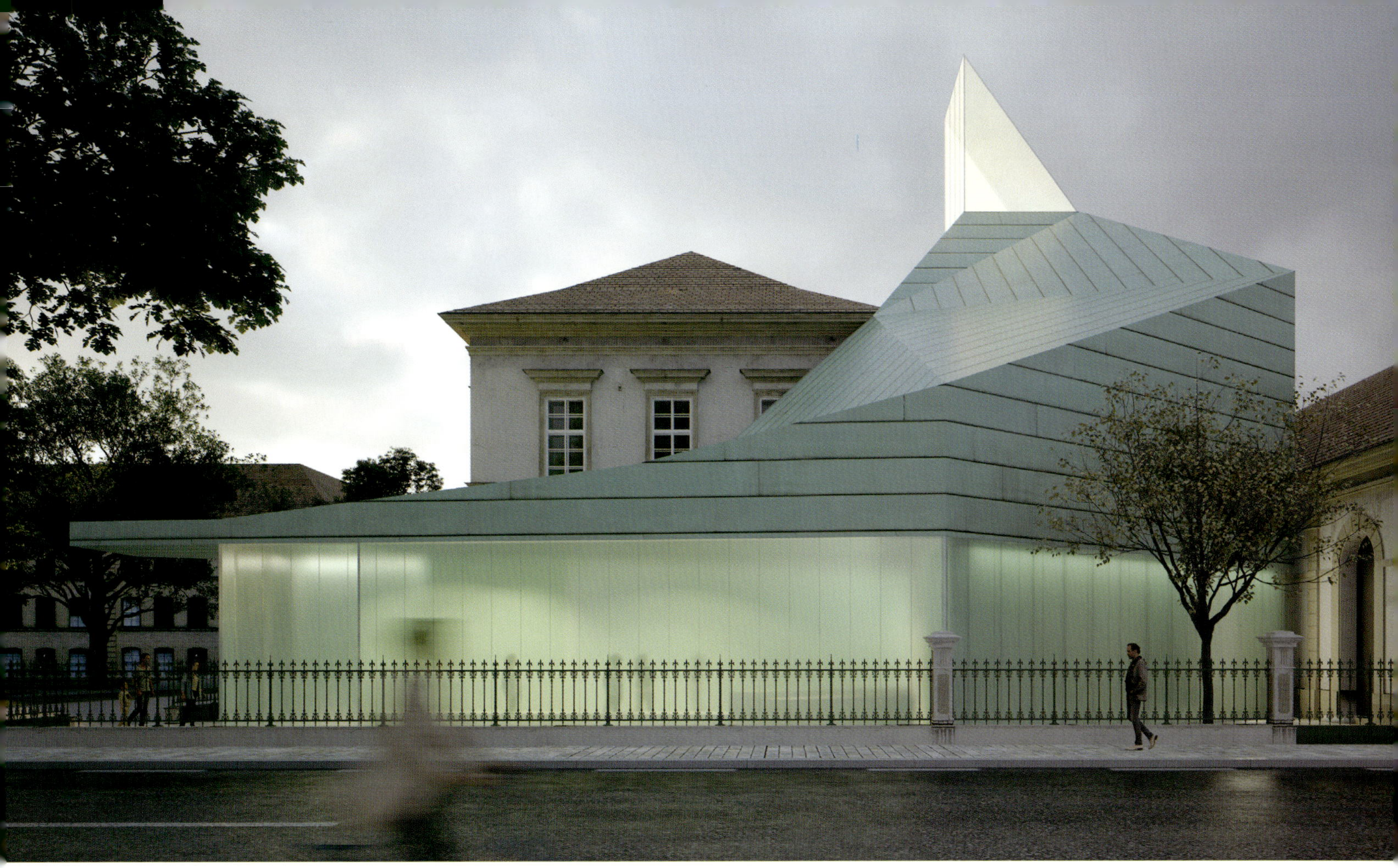

TOWER of LIGHT
EXPANSION of TEREZIN GHETTO MUSEUM
PRISMATIC GLASS FACES SOUTH
EXISTING BUILDINGS
VERDEGRIS COPPER, OR Rhine zink
NO MULLION CHANNEL GLASS
2 8 22
S.H.
ELEVATOR GOES TO ATTIC LEVEL
GLASS ELEV
GLASS STAIR
WHITE PLASTER
PRISMATIC LIGHT
WHITE PLASTER
STONE FLOOR & WALLS
WC
WC
BENCH & COATS
CHANNEL GLASS
TOWER of LIGHT/TEREZIN
2 8 22
NIGHT: TOWER OF LIGHT
TEREZIN 2/9/22 S.H.
NIGHT: TOWER of Light
TEREZIN
GLOWING IN DARK
2 9 22
S.H.

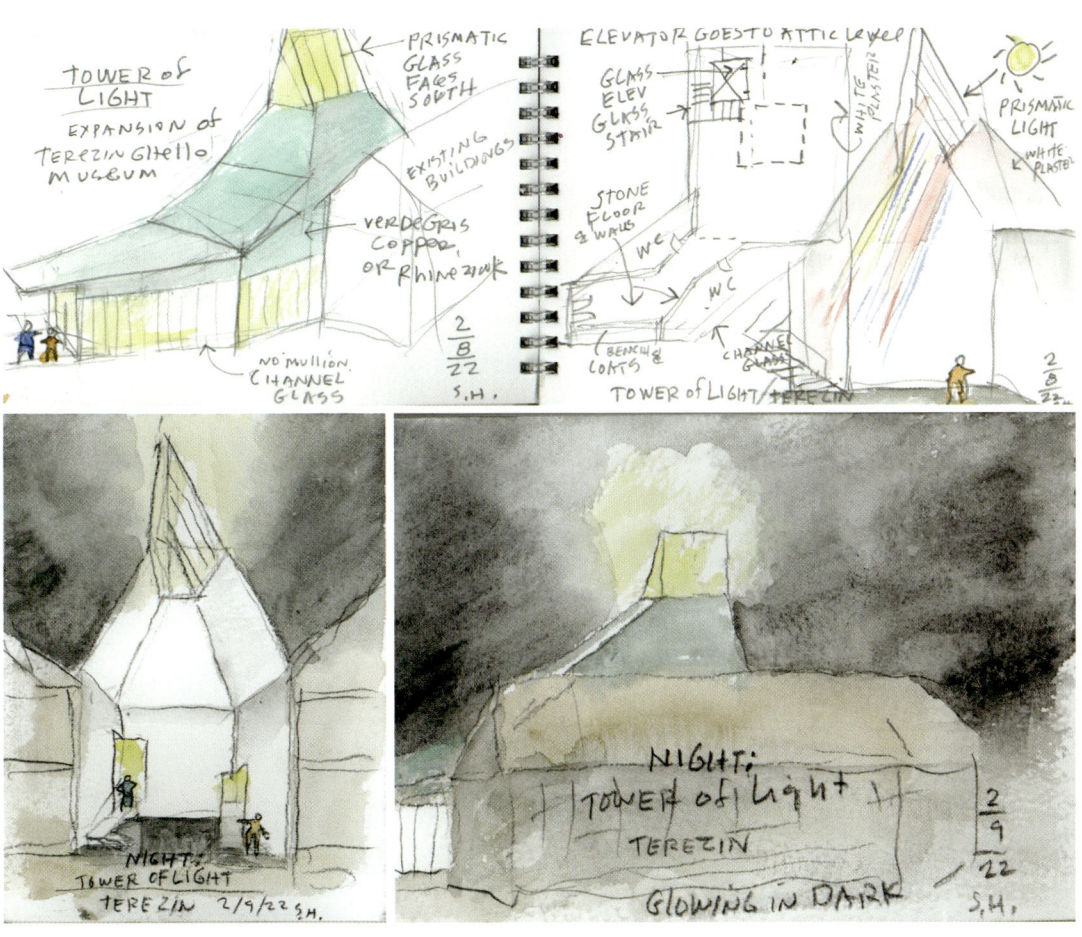

Sketch

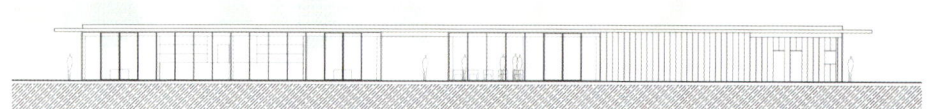
Elevation I

Elevation II

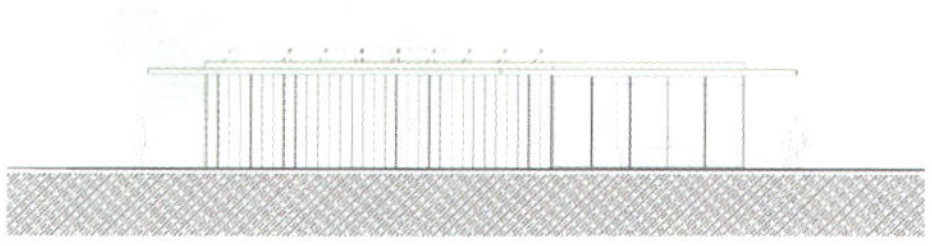

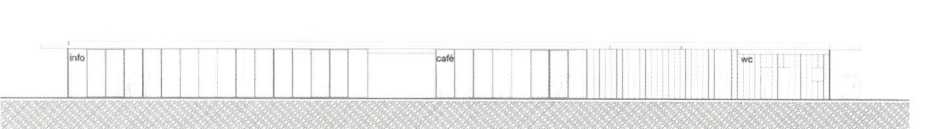
Elevation III

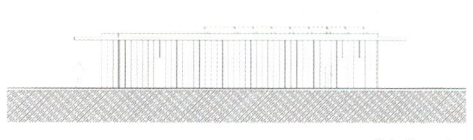
Elevation IV

Terezín Ghetto Museum >>

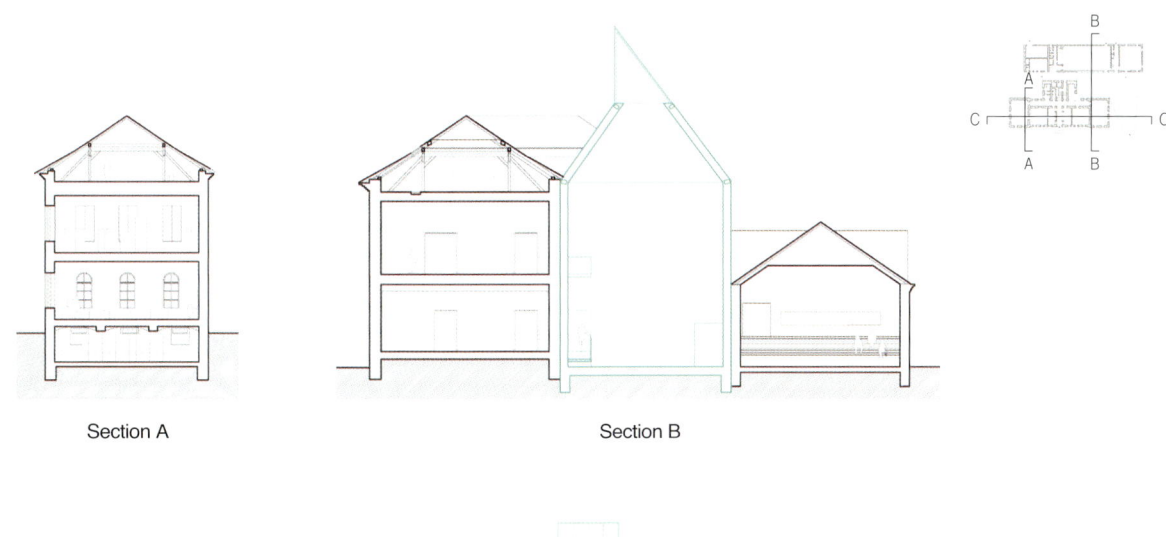

Section A

Section B

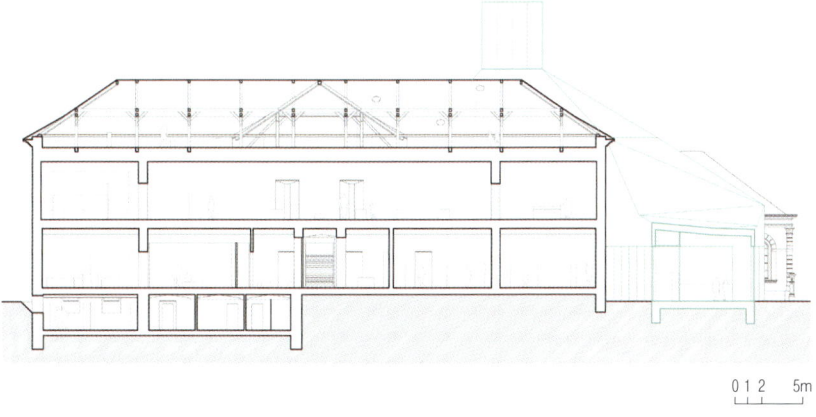

0 1 2 5m

Section C

175

Winner _ 당선작

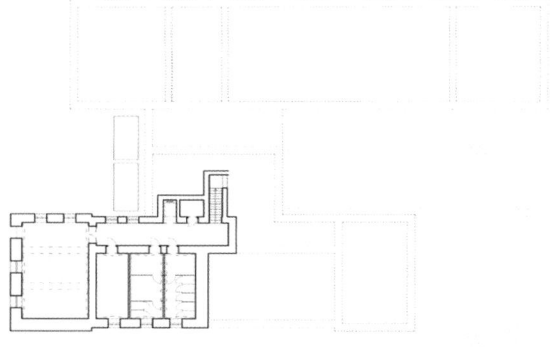

4th floor plan

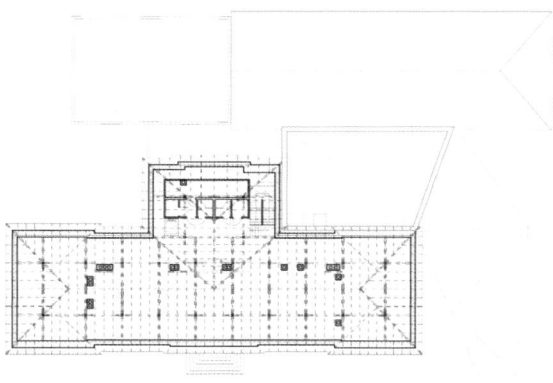

3rd floor plan

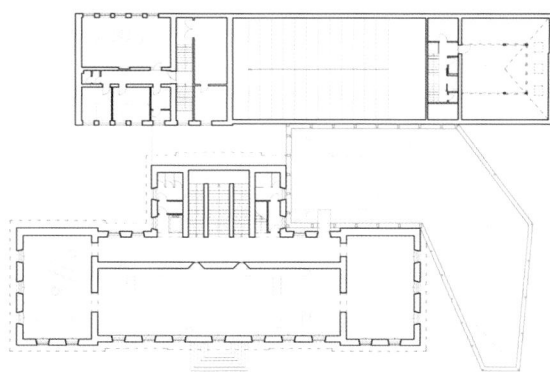

2nd floor plan

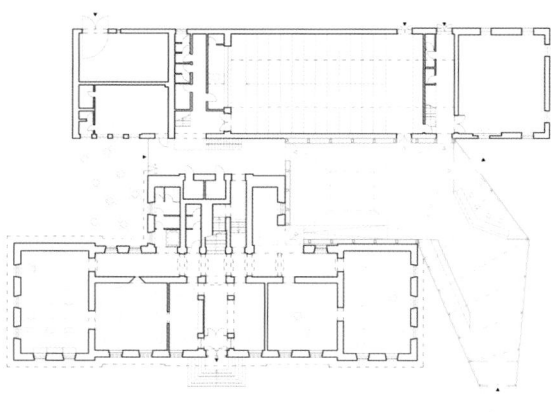

1st floor plan

National Intangible Heritage Center Miryang Branch

국립무형유산원 밀양 분원

SHINHAN ARCHITECTS & ENGINEERS / Inho Jeong, Sanghoon Kim
신한종합건축사사무소 / 정인호, 김상훈

Gayeondang_A big festival of intangible heritage

Ga_A beautiful streetscape in harmony with Arirang Street
Yeon_A big festival of intangible heritage
Dang_A new yard facing the Miryang River

Sammun-dong, Miryang-si, Gyeongsangnam-do is surrounded by Miryang River, so it looks like an island. Miryang, the hometown of Yeongnamru, one of the three major pavilions of the Joseon Dynasty, and Miryang Arirang, the representative folk song of Gyeongsang Province, is a city with outstanding cultural, artistic and humanistic characteristics. The Miryang Branch of the National Intangible Heritage Center will be built in Sammun-dong, Miryang-si with beautiful natural scenery and various local events, and its value is endless. With the slogan of conveying tomorrow with yesterday, it will be another trace of tradition that will allow us to recognize the value of intangible heritage and pass it on to the next generation through transmission.

佳宴堂 (가연당) '흥겨운 무형유산의 대잔치'

佳 아름다울 가 : 아리랑거리와 조화되는 아름다운 가로경관
宴 잔치 연 : 흥겨운 무형유산의 대잔치
堂 마당 당 : 밀양강을 바라보는 새로운 마당

경상남도 밀양시 삼문동은 밀양강의 강줄기가 감싸고돌아 마치 섬과 같은 모습을 하고 있는 장소이다. 조선시대 3대 누각 영남루와 경상도 대표 민요 밀양 아리랑의 고향인 밀양시는 문화적, 예술적, 인문학적 특징이 두드러지는 도시이다. 수려한 자연경관과 다양한 볼거리의 지역 행사가 열리는 밀양시 삼문동에 지어질 국립무형유산원 밀양 분원의 가치는 무궁무진하다. 어제를 담아 내일을 전하겠다는 슬로건과 맞물려 무형유산의 가치를 알아보고 전수를 통한 계승으로 이어져 후대에까지 아름다운 문화를 전파할 수 있는 또 하나의 전통의 발자취가 될 것이다.

Location : 271, Sammun-dong, Miryang-si, Gyeongsangnam-do, Korea I **Function** : Cultural & Assmbly facility I **Site area** : 7,000m² I **Bldg. area** : 4,093m² I **Total floor area** : 10,973m² I **Stories** : B1, 3FL I **Structure** : Reinforced concrete I **Finish** : Long brick, White concrete, Low-E pair glass, Wood louver

위치 : 경상남도 밀양시 삼문동 271일원 I **용도** : 문화 및 집회시설 I **규모** : 지하1층, 지상3층 I **구조** : 철근콘크리트 I **마감** : 롱브릭, 화이트콘크리트, 로이복층유리, 목재루버 I **설계팀** : 최은철, 민주홍, 허연식, 김예원, 정구환, 이지웅, 박정원, 박예랑, 안주연, 임연수

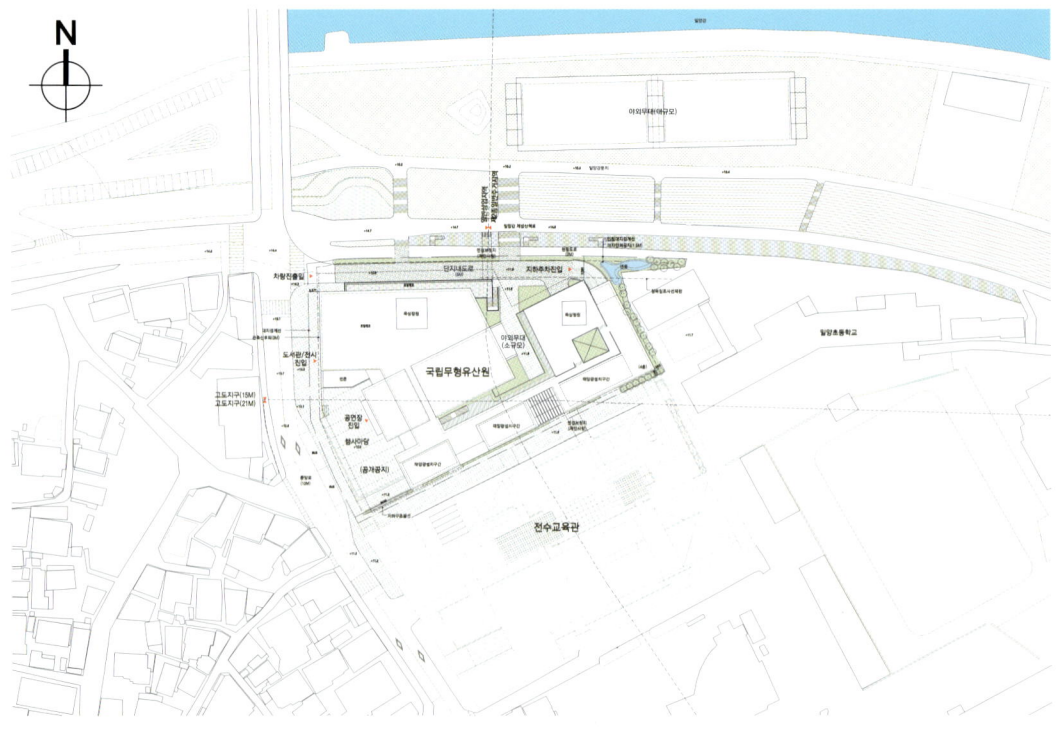

Site plan

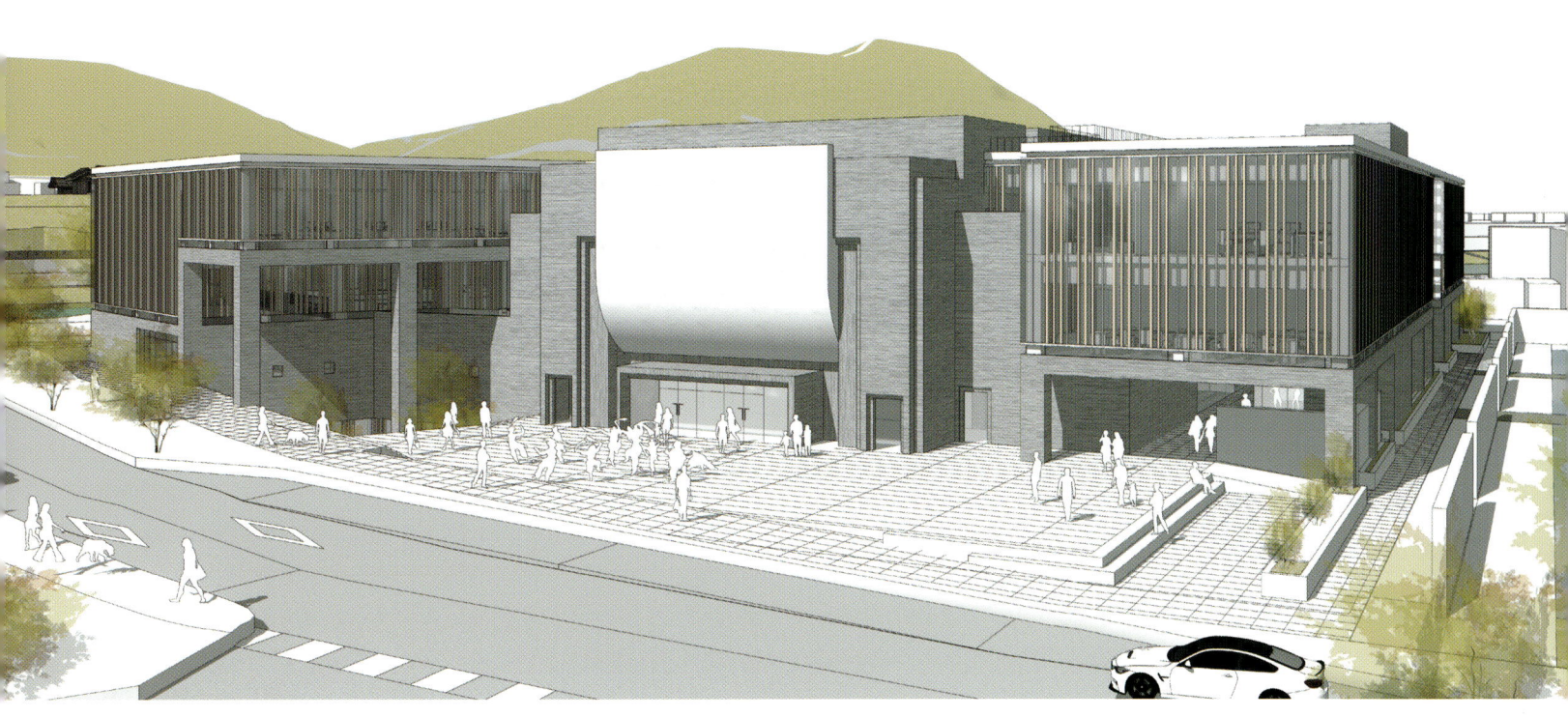

▌Site Concept

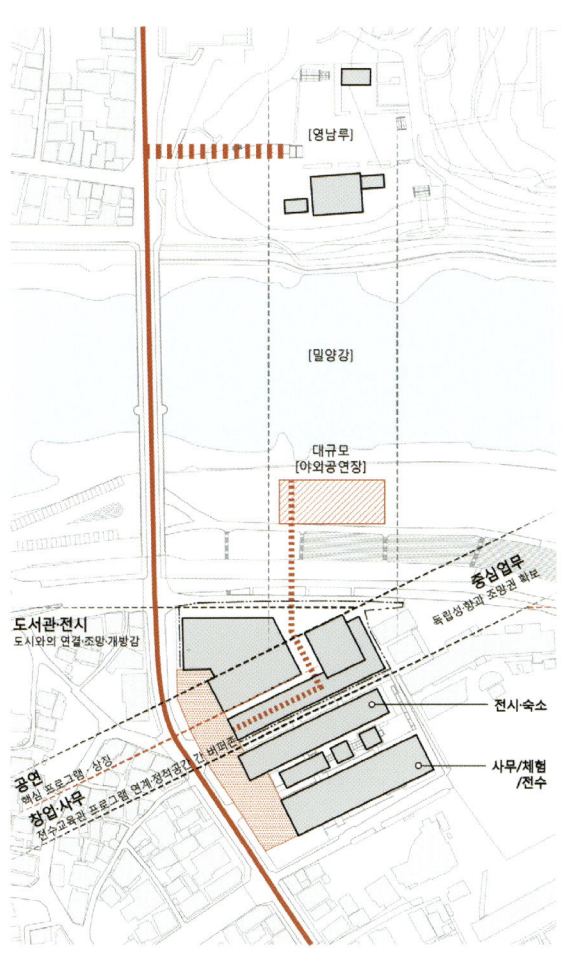

▌Circulation Plan

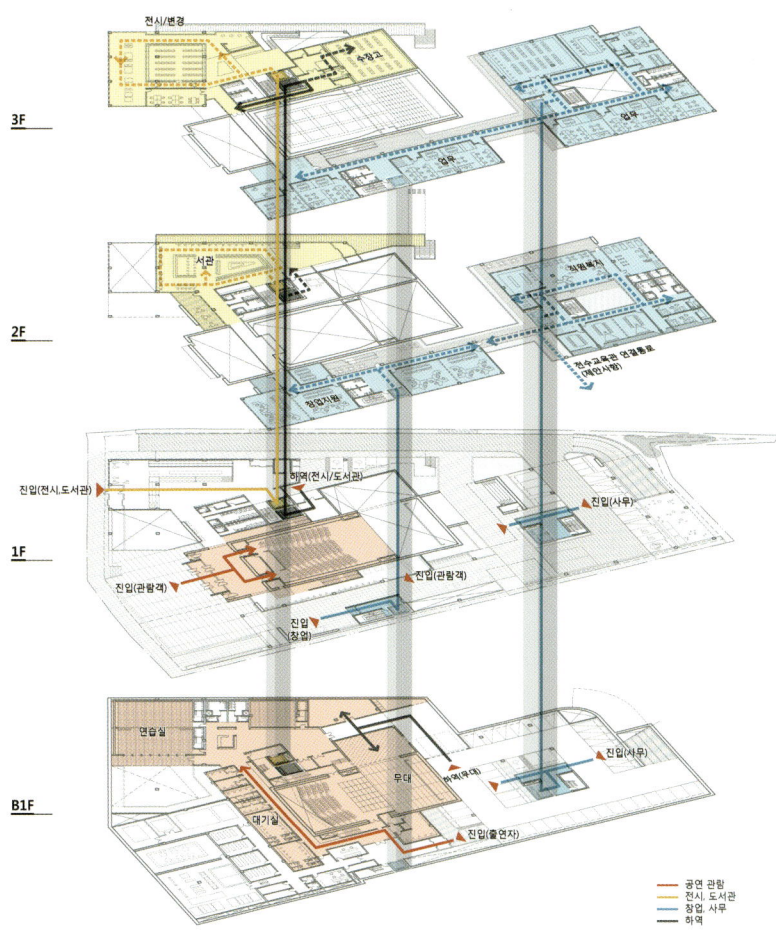

Winner _ 당선작

■ Design Direction

- Activation of Arirang street

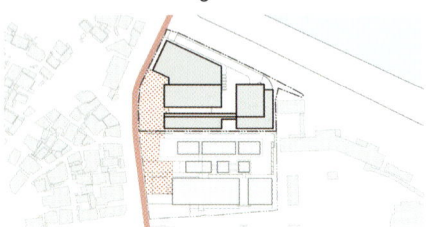

- 주진입부에 넓은 행사마당 조성
- 전수교육관과 이어지는 오픈 스페이스 계획

- Link with Miryang riverside park

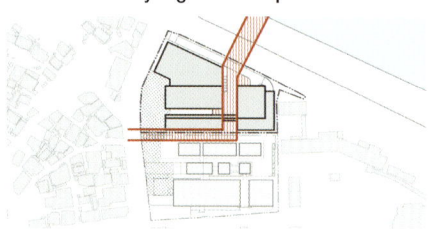

- 서측 주 진입구부터 밀양강둔턱까지 연결되는 외부공간 조성
- 수변 동선 유입 활성화 및 시설 연계성 향상

- View of Miryang river·Yeongnamru pavilion

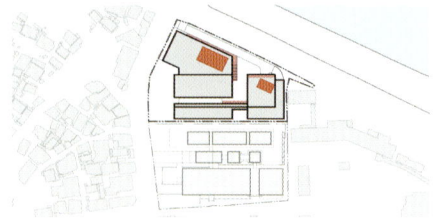

- 야외부대와 여러 구간으로 열린 외부공간 조성
- 서로 독립되면서 상호연광성을 위한 외부공간계획

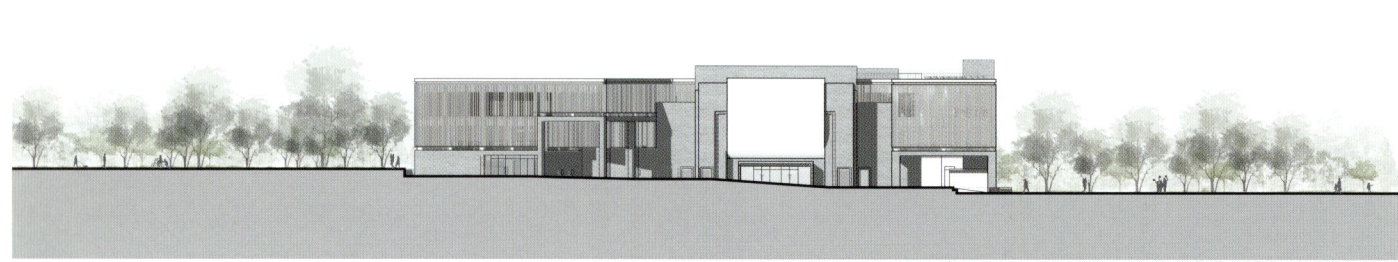

Front elevation

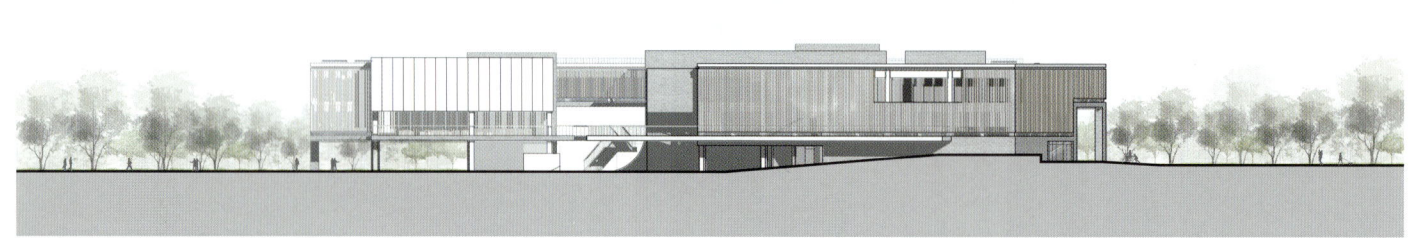

Left elevation

National Intangible Heritage Center Miryang Branch

Winner _ 당선작

■ Structure Plan

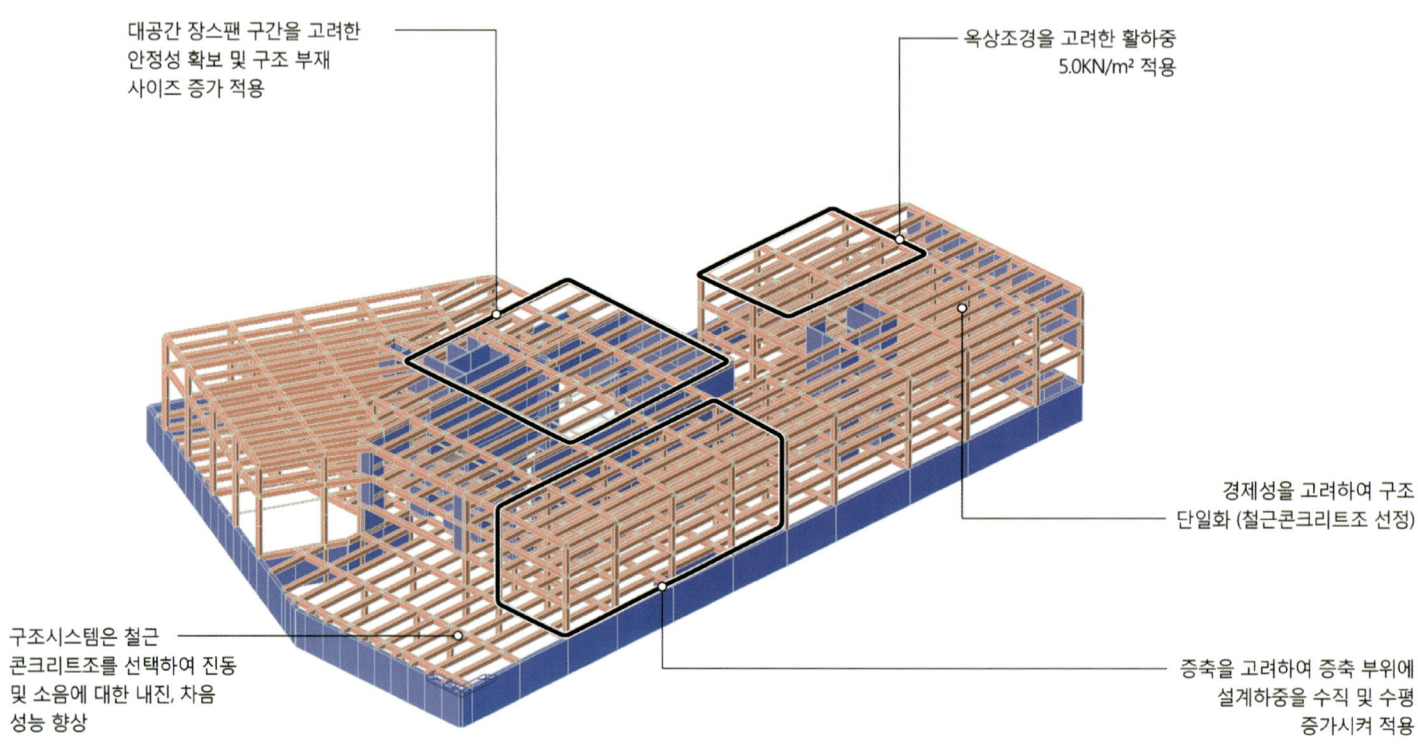

- 대공간 장스팬 구간을 고려한 안정성 확보 및 구조 부재 사이즈 증가 적용
- 옥상조경을 고려한 활하중 5.0KN/m² 적용
- 경제성을 고려하여 구조 단일화 (철근콘크리트조 선정)
- 구조시스템은 철근 콘크리트조를 선택하여 진동 및 소음에 대한 내진, 차음 성능 향상
- 증축을 고려하여 증축 부위에 설계하중을 수직 및 수평 증가시켜 적용

National Intangible Heritage Center Miryang Branch >>

Winner _ 당선작

1. Electrical room
2. Communications room
3. Cast waiting room
4. Concert hall
5. Parking
6. Lobby
7. Outdoor stage
8. Cafeteria
9. Staff welfare space
10. Document arrange room
11. Head office
12. Practice room
13. Machine room
14. Arirang hall
15. Small library
16. Office sharing
17. Exhibition area
18. Business room

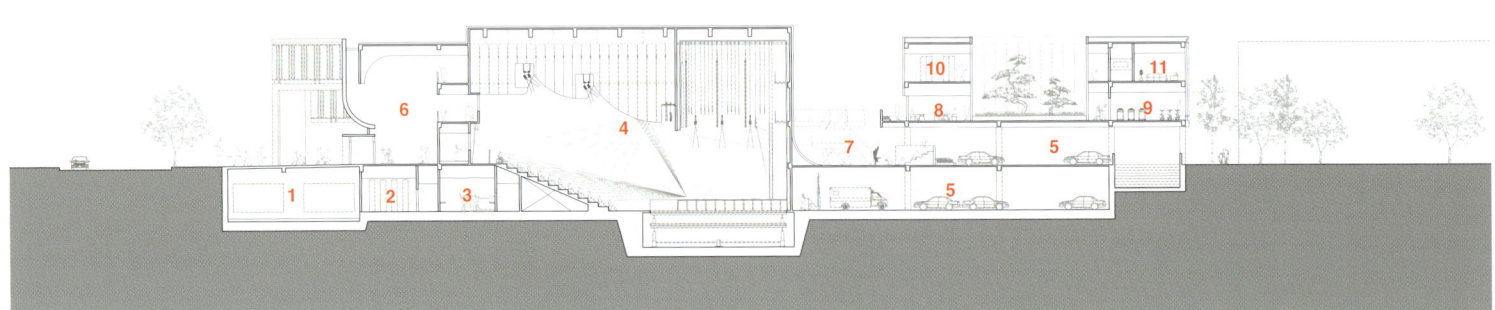

Cross section

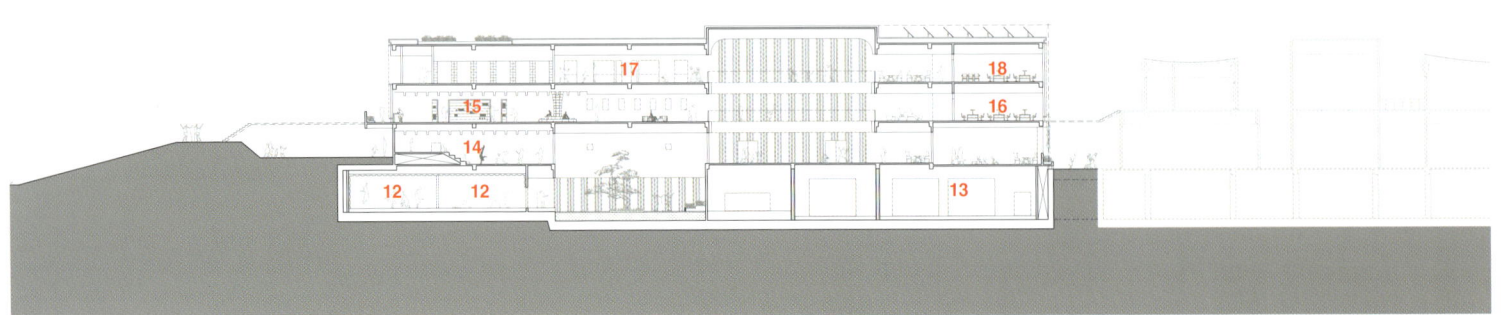

Longitudinal section

National Intangible Heritage Center Miryang Branch >>

3rd floor plan

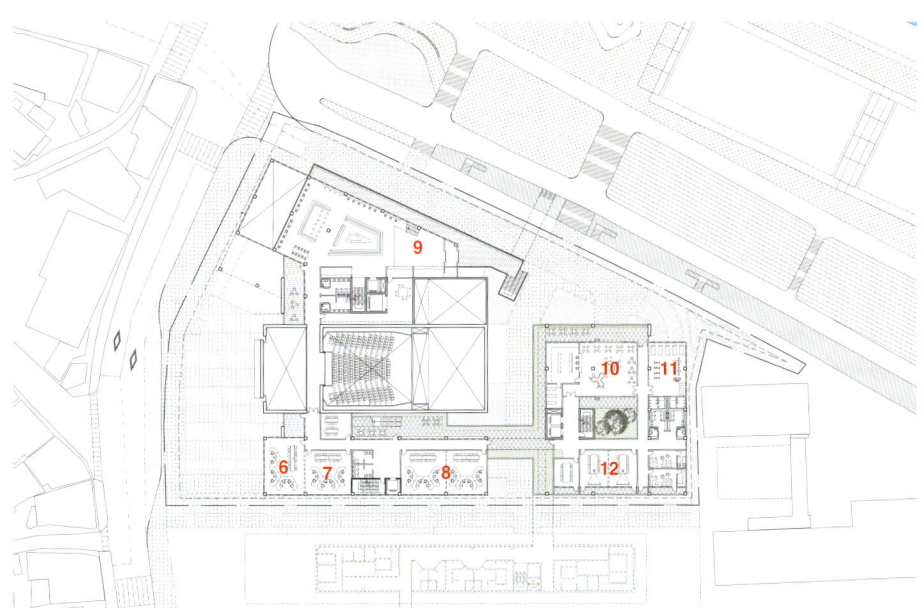

2nd floor plan

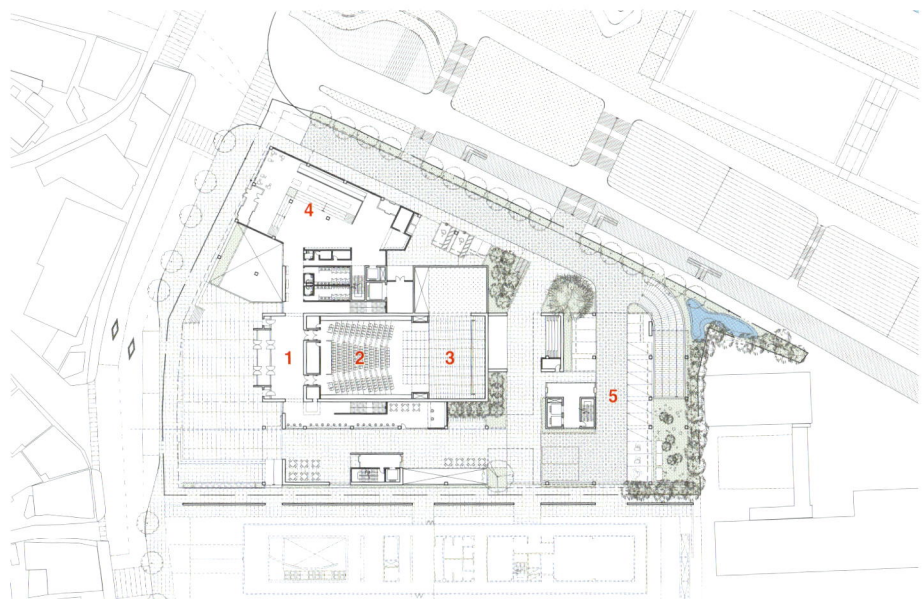

1st floor plan

1. Lobby
2. Auditorium
3. Stage
4. Arirang hall
5. Parking
6. Startup incubating center
7. Office sharing
8. Start-up office
9. Small library
10. Cafeteria
11. Staff welfare space
12. Meeting room
13. Office space
14. Total document library
15. Head office
16. Exhibition area
17. Exhibition office

Namdo Righteous Army History Museum
남도의병역사박물관

Hyunje Joo Baukunst / Hyunje Joo + I_architects / Nohwook Park
주현제 바우쿤스트 / 주현제 + 건축학동 건축사사무소 / 박노욱

Place of metaphor_Righteous army history museum that embraces the nature of the place

The reason why we built the museum is to make future generations remember the spirit and soul of the medics who ordinary people like my father / brother / brother did without hesitation when the country was in crisis, and to say that It is to instill a sense of unity in the heart. If so, what should be the architectural space that can instill that bold courage and unity in the heart? What should a space be like to contain the souls of their noble sacrifices?

Unexposed museum

The museum located in the historical park is not seen by topography and forests until visitors get off the parking lot and walk to the museum. Since it is not revealed, it remains unchanged after 100 years. The spirit of the righteous army is not in the exterior of the museum, but in the trees, in the fields, in the wind, and in the sky. The museum that we suggest is step back from the things that have a soul around them.

Facade as a carrier of messages

The only exposed part of the museum is the kinetic facade as a message carrier seen from the other side of the Yeongsan River. Normally, it acts as an awning to reduce the intensity of light, but it shakes when the wind blows, and makes sounds which could the museum's presence is showed from afar. It is just like ordinary people became righteous army oppose to national crisis.

Architectural space that embraces elements of nature

In order to become a place that contains the souls of righteous army and their beliefs, it requires length of flow as much as their weight, and spatial sequence. When visitors get out of the car and go to the museum, they experience the elements of nature infused with the spirit of righteous army, and a dramatic architectural space that embraces them.

은유의 장소_장소가 가진 자연을 포용한 의병 역사박물관

우리가 박물관을 건립하는 이유는, 우리 아버지 / 형 / 동생같은 평범한 사람들이, 나라가 위기에 처했을 때 주저 없이 나섰던 의병의 정신과 영혼을 미래세대가 기억하게 하고, '나와 의병이 다르지 않다'라는 동질감을 가슴 속에 심어주는데 있다. 그렇다면 그 담대한 용기와 동질감을 가슴에 심어줄 수 있는 건축공간은 어떠해야 할까? 숭고하게 희생한 그들의 영혼을 담을 수 있는 공간은 어떠해야 할까?

드러내지 않는 박물관

역사공원 내에 위치한 박물관은 관람객이 주차장에서 내려서 박물관으로 걸어오기까지, 지형과 숲에 의해 그 모습을 드러내지 않는다. 드러내지 않으니 100년이 지나도 변함이 없다. 의병의 정신은 박물관의 외관에 있지 않고, 나무와 들판, 그리고 바람, 하늘에 있다. 박물관은 주위의 영혼을 지닌 것들에서 한 발 물러나 있다.

메시지 전달체로서의 파사드

유일하게 박물관이 노출된 부분은, 영산강 건너편에서 보이는 메시지 전달체로서의 키네틱 파사드이다. 평상시에는 빛의 세기를 줄여주는 차양 역할을 하지만, 바람이 불면 흔들리며, 소리를 내어 멀리서도 박물관의 존재감을 발휘한다. 평범한 국민이 국난에 맞서 의병이 된 것처럼 말이다.

자연의 요소를 포용하는 건축공간

의병의 영혼과 그들의 영혼과 신념을 담아가는 장소가 되기 위해서는, 그 무게만큼의 동선 길이와 공간 시퀀스가 필요하다. 차에서 내린 관람객들이 박물관으로 들어와서 관람하고 나가기까지, 의병의 영혼이 스며있는 자연의 요소와 이를 포용하는 극적인 건축공간을 경험한다.

Location : 23-3, Singok-ri, Gongsan-myeon, Naju-si, Jeollanam-do, Korea I **Function** : Cultural facility I **Site area** : 363,686m I **Bldg. area** : 7,014m I **Total floor area** : 9,694m I **Stories** : B1, 2FL
Structure : Steel frame, Reinforced concrete rahmen I **Finish** : Exposed concrete

위치 : 전라남도 나주시 공산면 신곡리 23-3 일원 I **용도** : 문화시설 I **규모** : 지하1층, 지상2층 I **구조** : 철골, 철근콘크리트라멘 I **마감** : 노출콘크리트 I **건축주** : 전라남도 I **설계팀** : 서채린, 김병철, 오혜지

Site plan

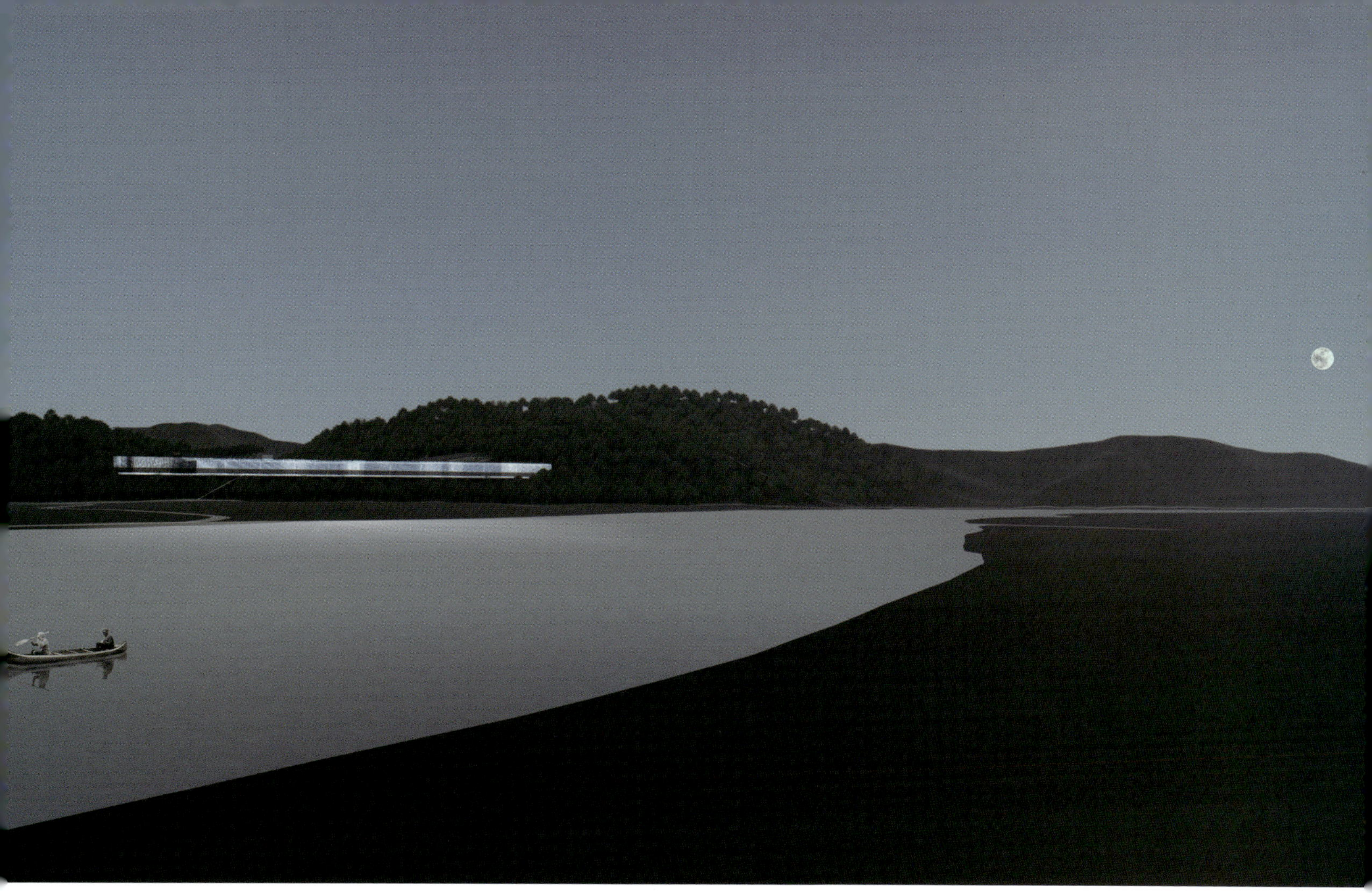

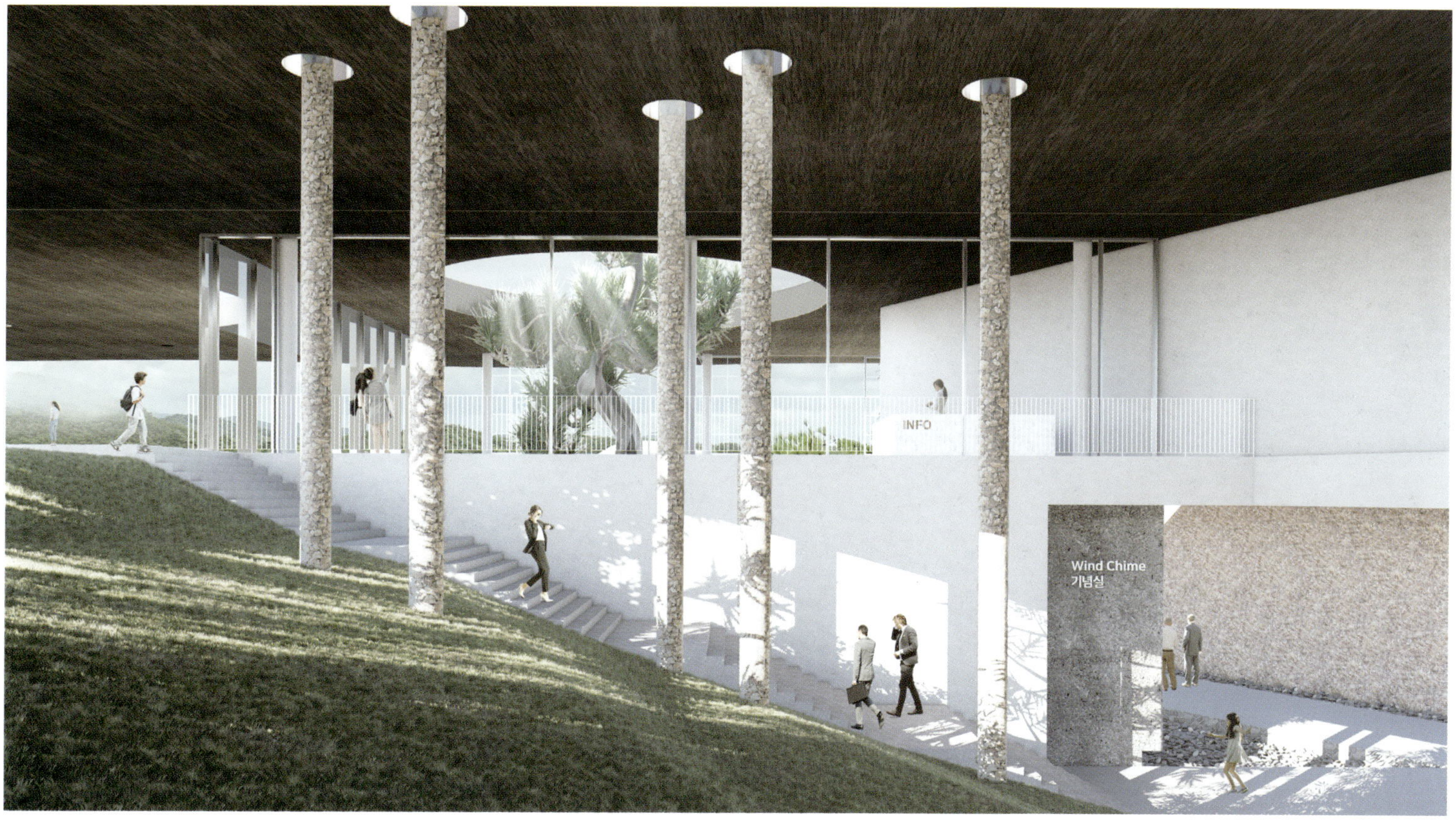
INFO
Wind Chime
기념실

Memorial Room Diagram

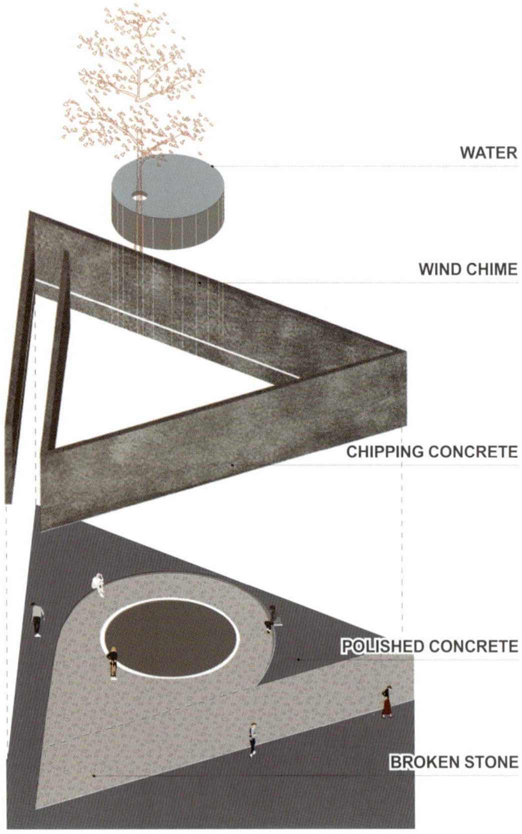

- WATER
- WIND CHIME
- CHIPPING CONCRETE
- POLISHED CONCRETE
- BROKEN STONE

Display Plan

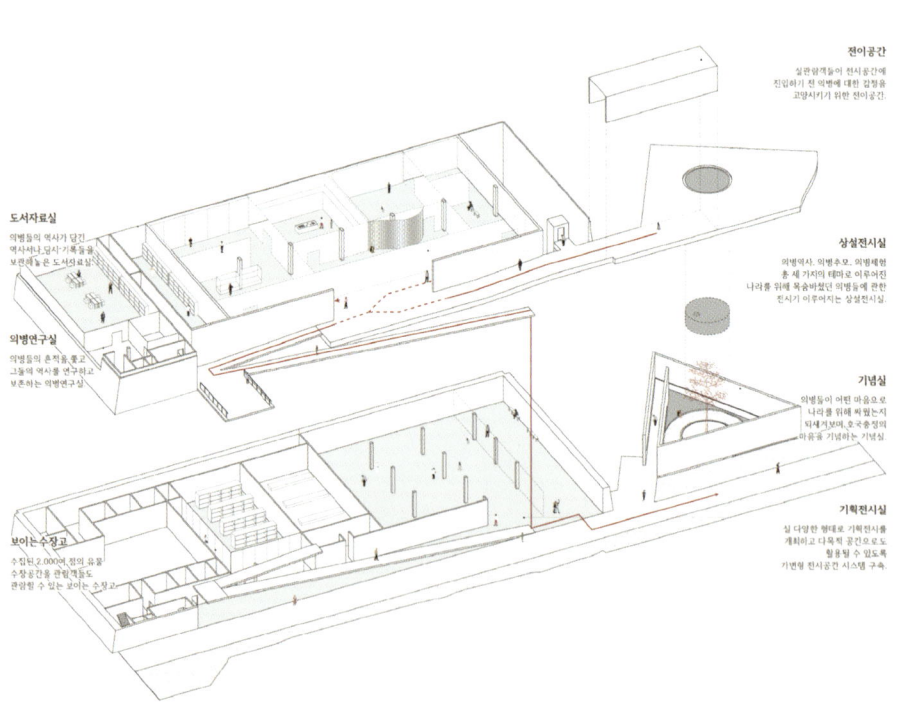

전이공간
전시관람객들이 전시공간에 진입하기 전 의병에 대한 감정을 고양시키기 위한 전이공간.

도서자료실
의병들의 역사가 담긴 역사서나 도서·기록물을 보관해놓은 도서자료실.

상설전시실
의병역사, 의병추모, 의병체험 총 세 가지의 테마로 이루어진 나라를 위해 목숨바쳤던 의병들에 관한 전시가 이루어지는 상설전시실.

의병연구실
의병들의 흔적들을 보고 그들의 역사를 연구하고 보존하는 의병연구실.

기념실
의병들이 어떤 마음으로 나라를 위해 싸웠는지 되새겨보며, 호국충정의 의병을 기념하는 기념실.

보이는수장고
수집된 2,000건의 유물 수장공간을 관람객들이 관람할 수 있는 보이는 수장고.

기획전시실
실 다양한 형태의 기획전시를 개최하고 다목적 공간으로도 활용할 수 있도록 가변형 전시공간 시스템 구축.

Namdo Righteous Army History Museum >>

■ Elevation Plan

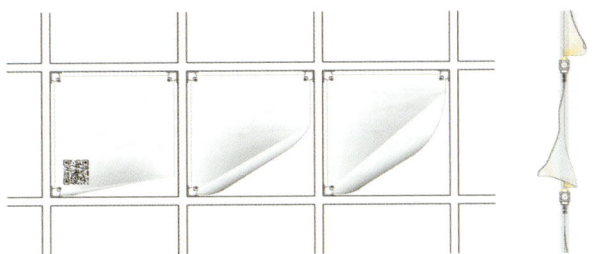

- 3-point fixed moving membrance structure

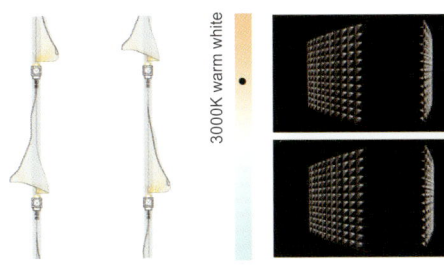

- Lighting for night seenery

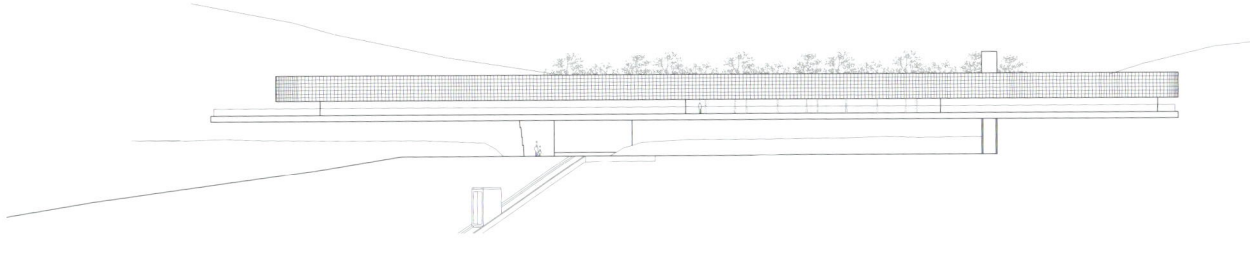

Elevation

Winner _당선작

190

Namdo Righteous Army History Museum >>

Winner _ 당선작

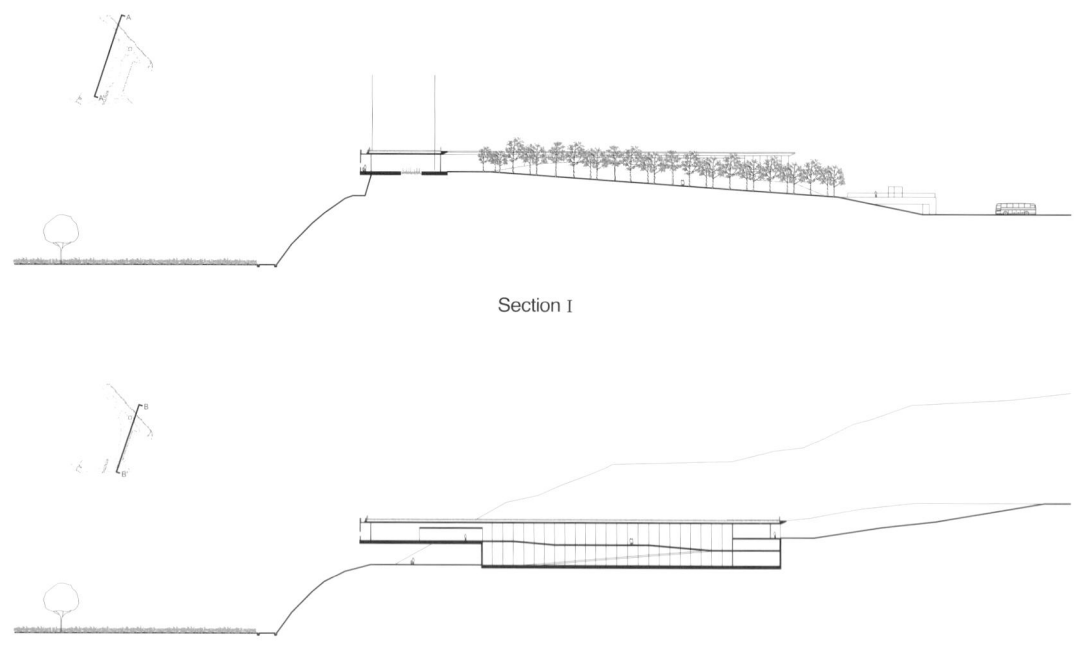

Section I

Section II

Namdo Righteous Army History Museum >>

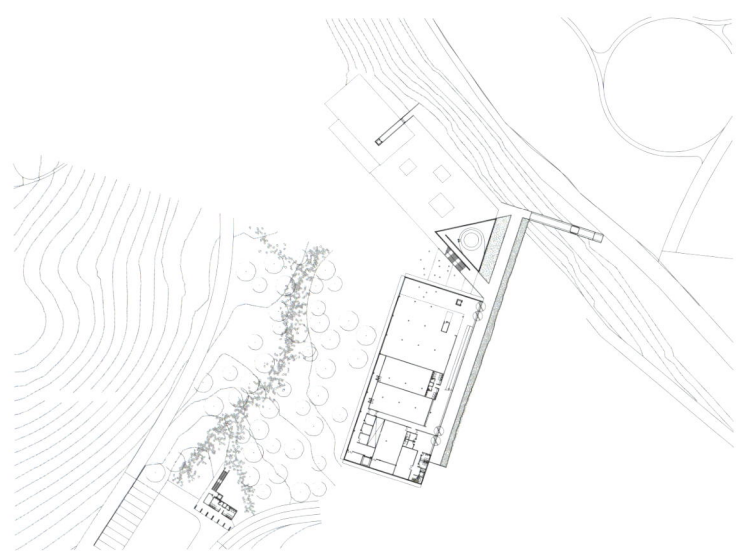

Basement floor plan

1st, 2nd floor plan

196	Boramae Hospital Safe Respirator Center 보라매병원 안심호흡기전문센터
202	Keflavik Airport 케플라비크 공항
206	Ferry Terminal Turku 투르쿠 페리 터미널
212	Chongqing Cuntan International Cruise Center 충칭 쿤탄 국제 크루즈 센터
218	Gyeongsangbuk-do Training Center 경상북도 수련원
222	Sinan-dong Sports Complex 신안동 복합스포츠타운
228	MZ Sports Plaza MZ 스포츠플라자
234	Sports Hall in Zatec 자테츠 스포츠홀
240	Barclay Tower 바클레이 타워
248	Ørestad Church 외레스타드 교회
252	Fragments of Nostalgia 노스탤지어의 조각
258	Magnifica Fabbrica 웅장한 공장
266	Yantai Seafront Garden 옌타이 해안 정원

Winner _ 당선작

Boramae Hospital Safe Respirator Center
보라매병원 안심호흡기전문센터

Baum Architects, Inc / Hyunchul Noh
㈜범건축종합건축사사무소 / 노현철

Natural walk
It is a three-dimensionally connected landscape with a newly transformed healing garden and a stacked resting deck, and it is a medium that connects the main hospital and the park. It becomes a healing space for medical staff, patients, and visitors by breathing with nature that crosses the boundaries of the inside and outside.

Arrangement plan
In harmony with the existing layout axis of the main hospital, the wards are placed in consideration of the south-facing and park-facing views. To preserve the characteristics of a hospital in Boramae Park, green areas of the second level of ecological nature are preserved, and a healing environment in nature is created with a healing garden connected to the park.

Floor and medical plan
Cross-infection is prevented by vertically separating the operating system that can be changed in stages during normal times and infection crisis and the access of infected patients. The pleasant lobby opened three floors is designed to organically connect to the external stairs and the topography while looking at the healing garden. The emergency medical center and the entrance of general patients are separated and a sufficient emergency parking area is provided. Also, a healing terrace is installed on each floor considering the user. An efficient ward that can operate a cohort isolation ward by stage is designed by clearly separating the medical staff and patient routes.

자연걸; 음[飮] (飮 : 마시다, 호흡하다)
새롭게 탈바꿈된 치유정원과 켜켜이 쌓인 휴게데크가 입체적으로 연계되는 랜드스케이프이며 본원과 공원을 이어주는 매개체이다. 내외부의 경계를 넘나드는 자연과 함께 호흡함으로써 의료진, 환자, 방문자 모두가 함께 치유되는 힐링공간이 된다.

배치계획
기존 본원의 배치축과 조화롭게, 병동부는 남향과 공원을 향한 조망을 고려하여 배치하였다. 보라매공원 내 위치하는 병원이라는 특성을 고려하여 생태자연도 2등급 녹지를 보존하고, 공원과 연계된 치유정원을 통해 자연 속 치유환경을 조성하였다.

평면 및 의료계획
평시 및 감염 위기 시 단계별 전환이 가능한 운영체계와 감염 환자의 출입을 수직으로 분리하여 교차감염을 방지하였다. 3개 층이 오픈된 쾌적한 로비는 치유정원을 바라보며 외부계단 및 지형에 유기적인 연결이 가능하도록 설계하였다. 응급의료센터와 일반환자의 출입구를 이격배치하고 충분한 응급주차영역을 확보하였다. 또한, 각층마다 사용자를 고려한 힐링 테라스를 설치하였다. 단계별 코호트 격리병동 운영이 가능한 효율적인 병동부는 의료진과 환자의 동선을 명확히 구분하여 설계하였다.

Location : 28, Boramae-ro 5-gil, Dongjak-gu, Seoul, Korea | **Function** : Hospital | **Site area** : 6,640m | **Bldg. area** : 2,214m | **Total floor area** : 9,986m | **Stories** : B3, 3FL | **Structure** : Reinforced concrete, Steel frame | **Finish** : Low-E pair glass, Ceramic sheet panel, Aluminium louver

위치 : 서울시 동작구 보라매로5길 28 (신대방동 722) | **용도** : 병원 | **규모** : 지하3층, 지상3층 | **구조** : 철근콘크리트, 철골 | **마감** : 로이복층유리, 박판세라믹패널, 알루미늄루버 | **설계팀** : 손기영, 김형석, 손재원, 정판기, 이인혁, 박효성, 이은채, 김소연, 김유라, 박다은, 전병재

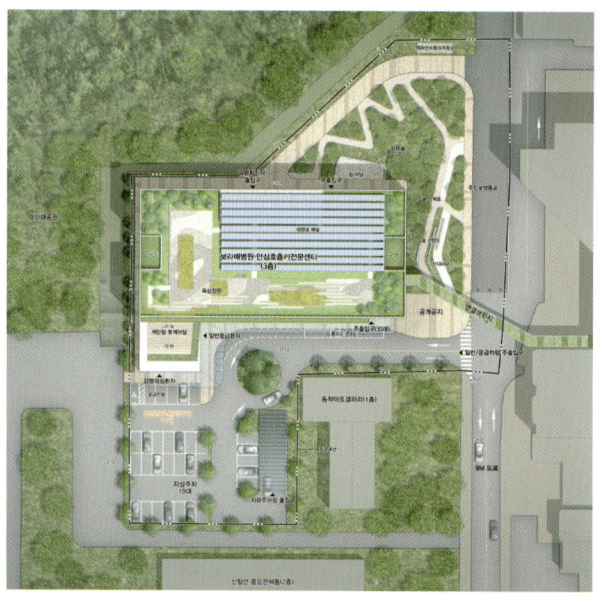

Site plan

Arrangement Plan

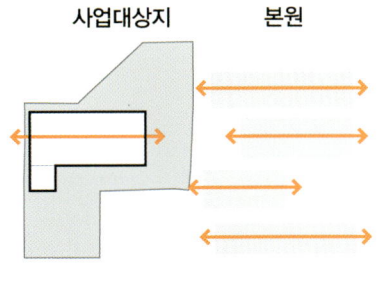

– 본원 배치축과의 조화

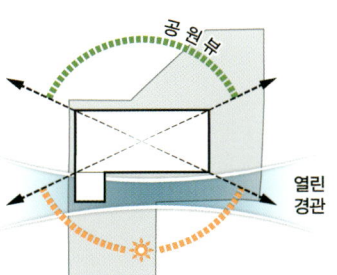

– 남향과 공원조망을 고려한 병동배치

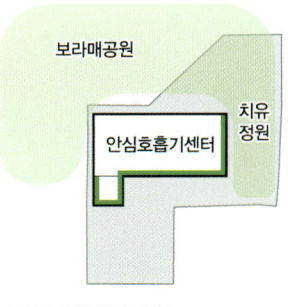

– 자연속의 치유환경 조성

Mass Process

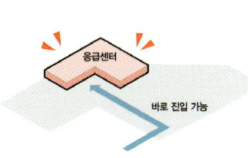

– 접근성과 인지성을 고려한 응급센터 배치

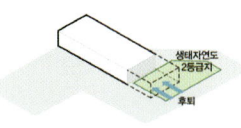

– 생태자연녹지 최대보존을 위한 매스 후퇴

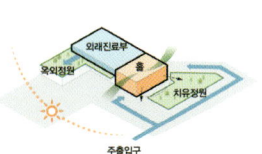

– 자연과 함께하는 외래진료부 및 홀 조성

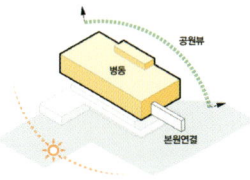

– 남향과 공원조망이 있는 병동배치

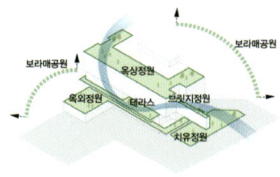

– 입체적인 옥외 휴게공간

Winner _ 당선작

Site Analysis

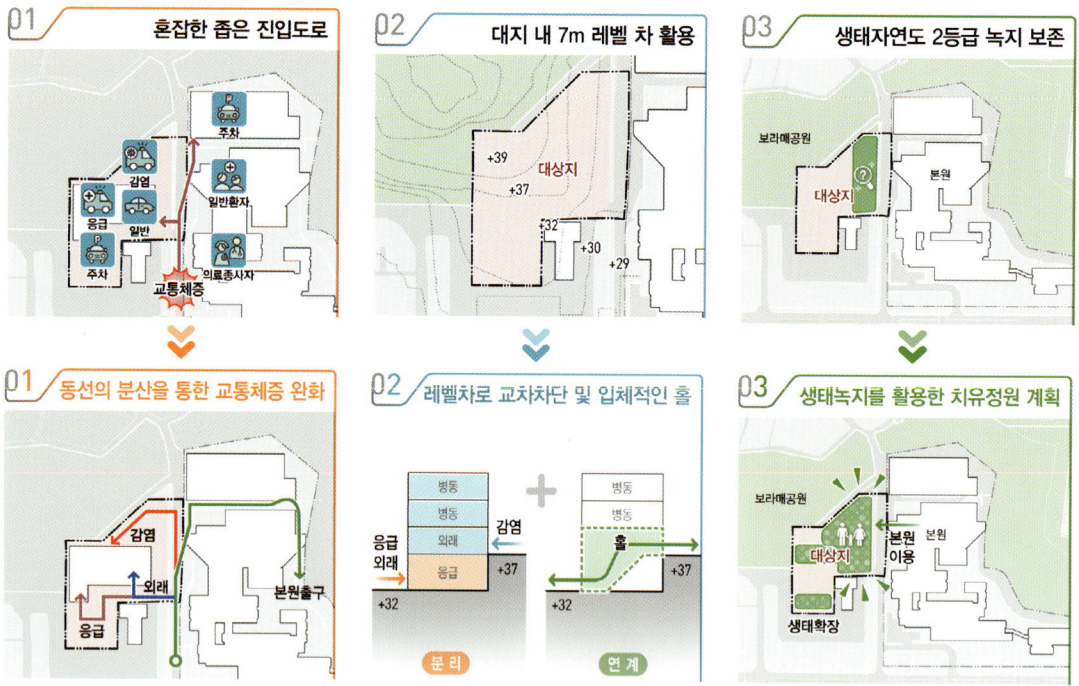

Program Zoning

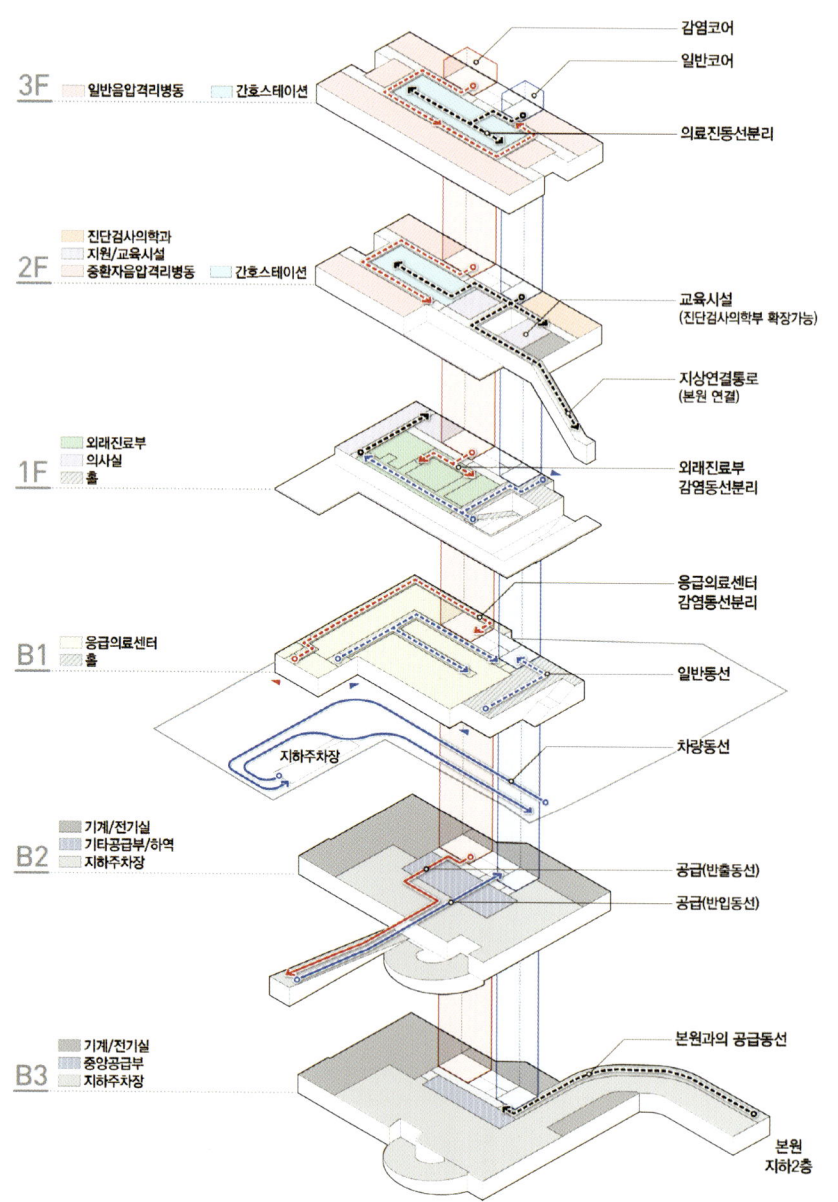

Boramae Hospital Safe Respirator Center >>

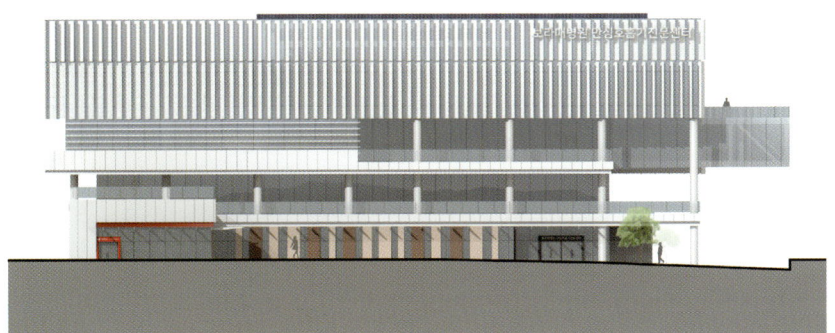

Elevation I

Elevation II

■ Section Plan

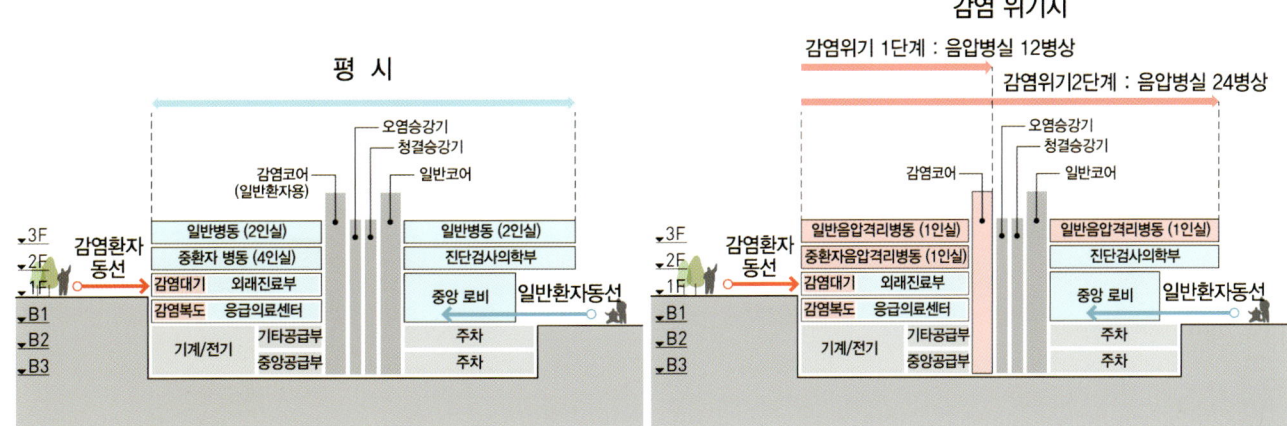

Winner _ 당선작

1. Single room
2. Nursing station
3. Nurse zone
4. Night duty room
5. Staff zone
6. Hall
7. Conference room
8. Air conditioning room
9. Doctor's residence
10. Bronchoscopy room
11. Lung function room
12. Lung center clinic
13. Infectious disease clinic
14. Infectious disease reception
15. Negative pressure isolation room
16. Medical department
17. ICU
18. Electrical room
19. Machine room
20. Laundry room
21. Waste storage
22. Linen storage
23. Central storage
24. Parking
25. Central supply center
26. Equipment storage
27. Wiring room
28. Day room
29. Lung center
30. Mild patient area
31. Wastewater tank

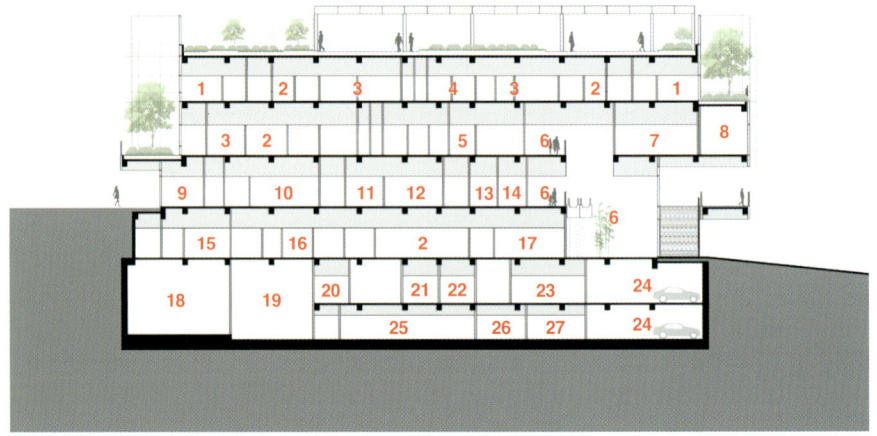

Section I

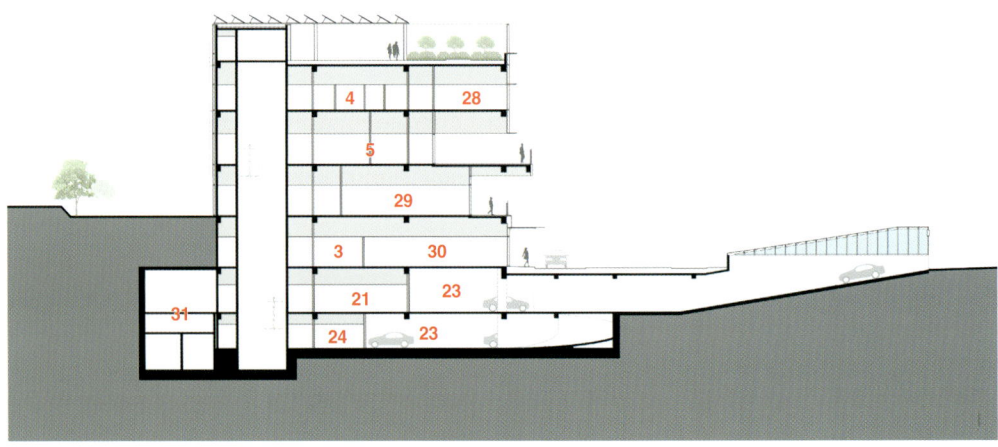

Section II

Boramae Hospital Safe Respirator Center >>

1. CT	14. Nursing station	27. Infectious disease clinic
2. X-ray	15. Pediatric beds	28. Infectious disease consultation office
3. Elevator hall	16. Mild patient beds	29. Infectious disease reception
4. Disaster prevention center	17. Critical beds	30. Blood collection room
5. Air conditioning room	18. CPR	31. Sterilization room
6. Hall	19. Doctor's residence	32. Outdoor garden
7. Negative pressure isolation room	20. Bronchoscopy room	33. Nurse zone
8. Operating room	21. Endoscopic recovery room	34. Staff zone
9. General containment room	22. Sound pressure room	35. Experiment zone
10. Pediatric clinic	23. Allergy testing room	36. Conference room
11. Clinic	24. Lung center clinic	37. Day room
12. Waiting room	25. Lung function room	
13. Medical department	26. Lung center counseling room	

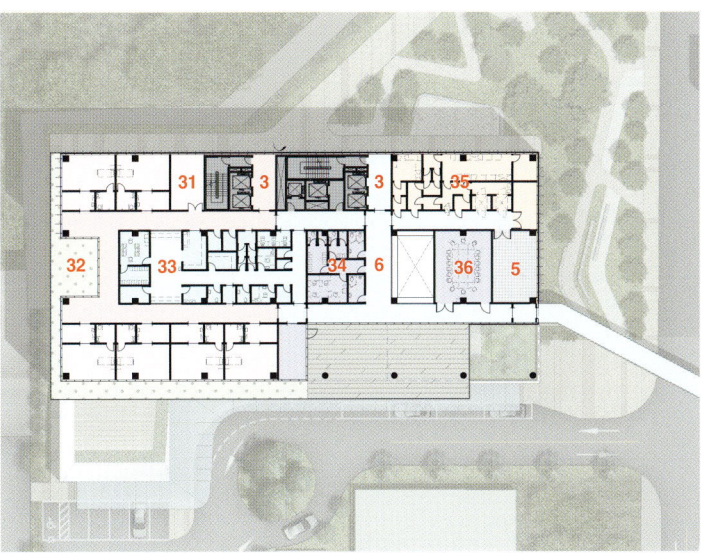

2nd floor plan

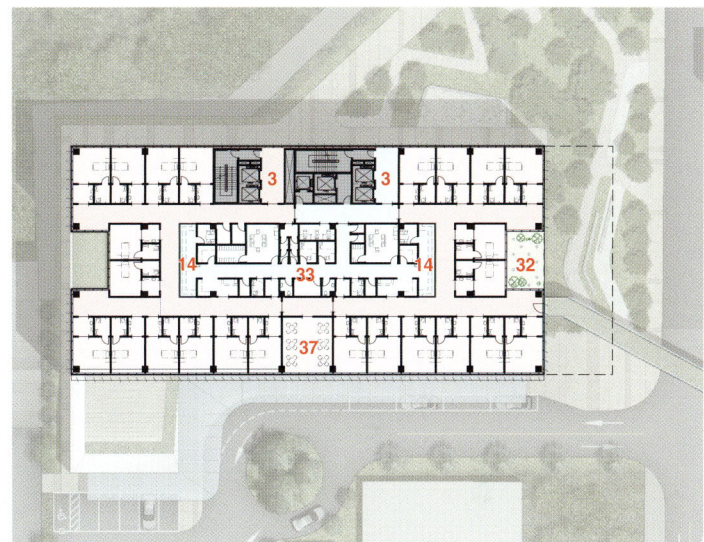

3rd floor plan

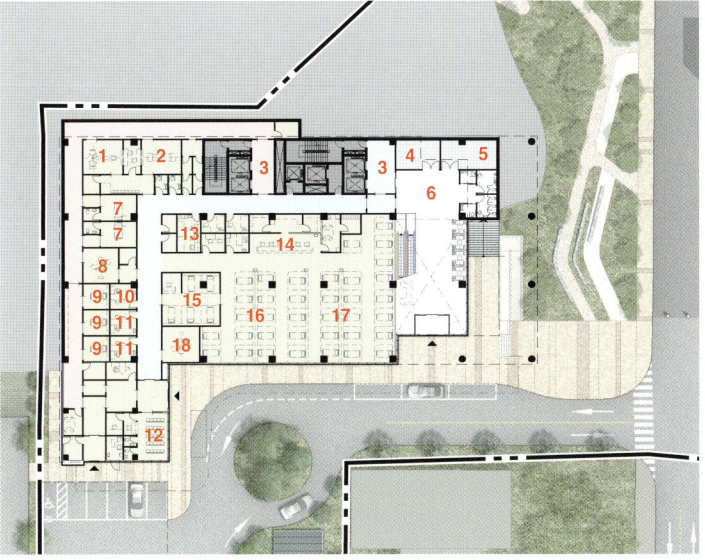

B1 floor plan

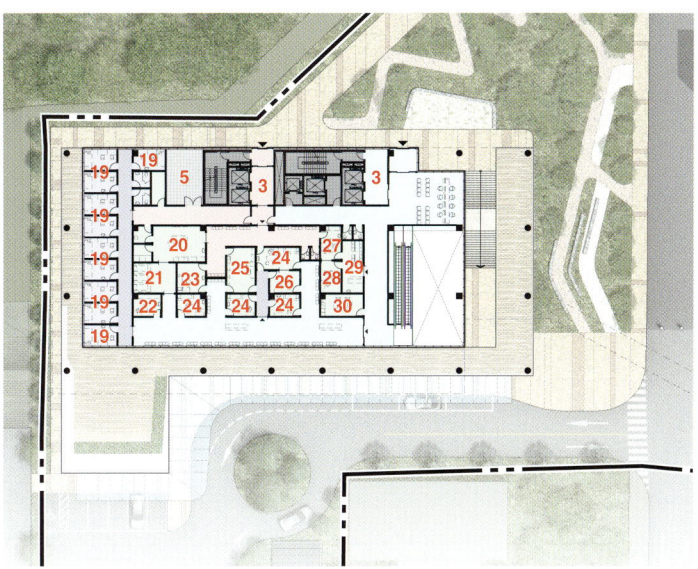

1st floor plan

Winner _ 당선작

Keflavik Airport
케플라비크 공항

KCAP
KCAP

The winning proposal aims to capitalise on Keflavik's strategic position between Europe and North America, its plentiful supply of renewable energy, and the captivating landscape of the Reykjanes UNESCO Geopark; it sets the path for long-term sustainable development and acts as a catalyst to Innovate the Icelandic economy. At the same time, the masterplan seeks to ensure that the airport and local communities of Reykjanesbær and Suðurnesjabær grow together in a mutually beneficial manner.

The team's strategy identifies catalyst sites that can achieve a high impact in terms of identity and value creation. These sites form a polycentric urban archipelago, connected by the landscape of the Airport Corridor. Landscape innovation will be deployed to integrate climate mitigation, and will play a fundamental role as the 'identity carrier' for the airport area. Tapping into the government's climate action plan that identifies reforestation as a top priority, the landscape concept connects the Airport Corridor into a national network of reforestation projects. A future oriented mobility vision comprising the Keflavík-Reykjavík Link (BRT), local public transport and active mobility linkages connects the airport area's urban nodes for travelers, workers and inhabitants alike.

이 제안의 목표는 유럽과 북미 간 케플라비크의 전략적 위치, 재생 에너지의 풍부한 공급, 레이캬네스 유네스코 지질공원의 웅장한 풍경을 활용하는 것이다. 장기적으로 지속 가능한 개발로 방향을 설정하고 아이슬란드 경제를 혁신하는 촉매제 역할을 한다. 동시에, 마스터플랜은 공항과 레이캬네스베르 및 수뒤르네스자베르의 지역사회가 서로 이익을 얻도록 함께 성장하고자 한다.

팀의 전략은 정체성과 가치 창출 측면에서 상당한 영향을 미치는 촉매 부지를 찾는다. 이 부지는 공항 통로의 풍경으로 연결된 다중심 도시 군도를 형성한다. 경관 혁신은 기후 완화를 위해 적용될 것이며, 공항 지역의 '정체성 전달자'로서 근본적인 역할을 할 것이다. 재식림을 최우선 과제로 하는 정부의 기후 행동 계획을 활용하여 경관 개념은 공항 통로를 재식림 프로젝트의 국가 네트워크로 연결한다. 케플라비크-레이캬비크 링크(BRT), 지역 대중 교통, 능동적인 이동성 연결이라는 미래 지향적인 이동성 비전은 여행자, 근로자, 주민 모두를 위해 공항 지역의 도시 결절점과 연계된다.

Location : Keflavik, Iceland I **Function** : Transportation facility I **Client** : Kadeco (Keflavík Airport Development Company)

위치 : 아이슬란드 케플라비크 I **용도** : 교통시설

Keflavik Airport >>

Master Plan

Integrated Development

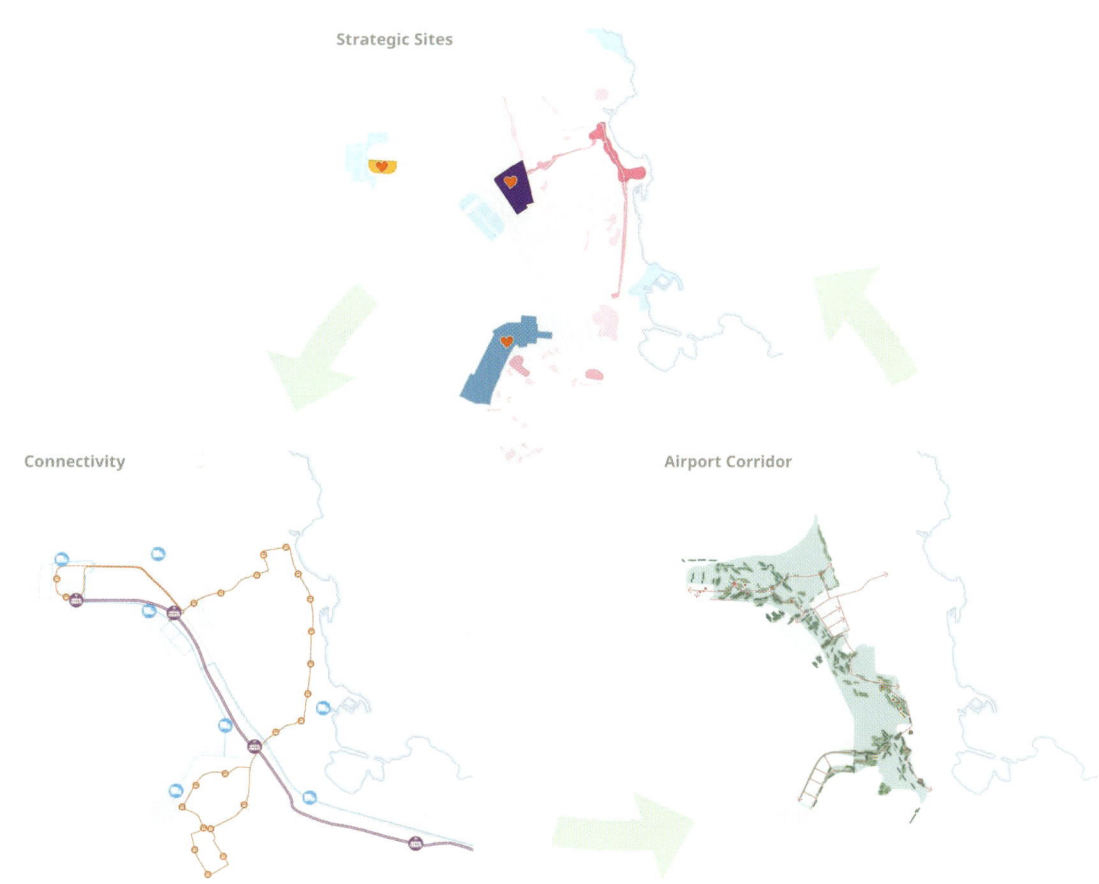

Strategic Sites

Connectivity

Airport Corridor

Winner _ 당선작

■ Design Concept

■ Circulation Plan

Keflavik Airport >>

■ Landscape Plan

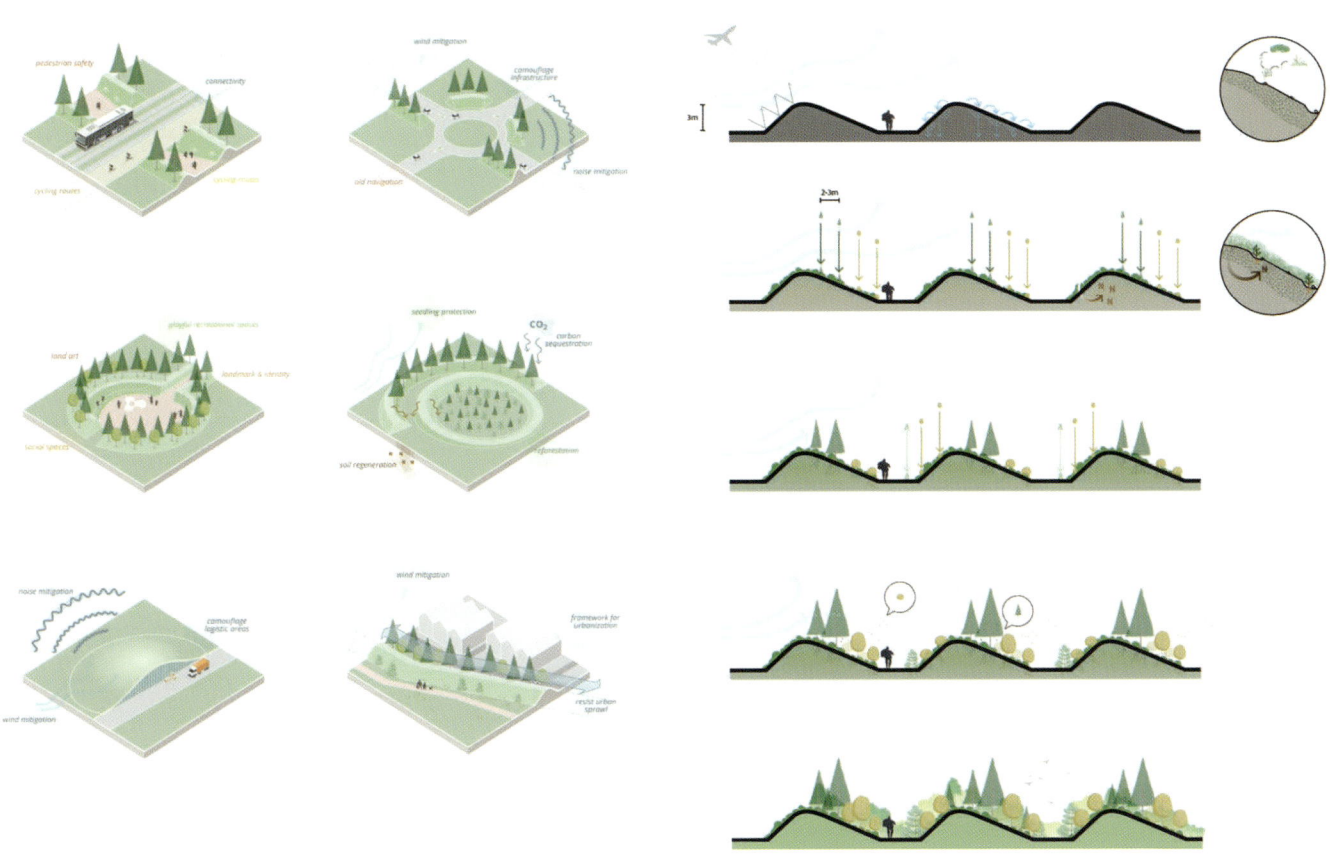

Winner _ 당선작

Ferry Terminal Turku
투르쿠 페리 터미널

PES-Architects / Tuomas Silvennoinen
PES-아키텍츠 / 투오마스 실벤노이넨

The Port of Turku in Finland's oldest city on the southwest coast is a busy cargo and passenger port with several daily sailings to Stockholm. The Ferry Terminal Turku project will upgrade the passenger harbour into a maritime hub that fulfils the requirements of modern, fast-paced and safe vessel traffic. The project also involves building a new joint passenger terminal to serve the two ferry lines operating out of the harbour.

In the international architecture competition for the terminal design, PES-Architects' proposal 'Origami' was selected as best fulfilling the goals set for functionality and fit in the urban context of the evolving Linnanniemi area around the harbour and the historic Turku Castle.

The architecture and functionality of the terminal are seamlessly integrated. A terminal is essentially a machine with a clear function, but alongside efficiency, the design aims to elevate and enhance the travel experience, whether for business or leisure. Ferry cruises often provide a break from the everyday, and the striking architecture of the terminal sets passengers in the right mood for a sea voyage.

The architecture is based on oblique triangular surfaces and a play with steel, glass and aluminium, delicately folded like a giant work of origami. The building's sculptural lines and shimmering surfaces evoke maritime images of a silvery salmon or steel ship.

The origami concept also refers to the multiple ways of folding a single piece of paper, just as the terminal can be modified to serve the different needs of its users. The flexible lobby space, for example, can easily be adapted to changing security control processes.

The terminal is situated as close as possible to the ships to minimise embarkation and disembarkation times. The layout is organised for optimally direct routes from the entrance to the departure hall on the second level and from the ships down to the exit. Incoming and outgoing passenger flows are separated but visually connected through glass walls. Spatial clarity, transparency and clear, straightforward routes make wayfinding easy and intuitive throughout the terminal.

Natural light floods the passenger areas, with warm wood complementing the steel, aluminium and glass structures. Scenic views of the city, the Turku Castle and sea can be enjoyed by both arriving and departing travellers through tall glass walls and terraces.

The terminal is easily accessible by various forms of transport, with public transport stops and a taxi stand immediately outside the terminal and a car park nearby.

Location : Port of Turku, Finland I **Function** : Transportation facility I **Site area** : 16,250m I **Bldg. area** : 5,575m I **Total floor area** : 13,625m I **Stories** : 3FL I **Structure** : Reinforced concrete, Steel frame I **Finish** : Sea aluminium, Glass I **Client** : Port of Turku I **Design team** : Eleanna Breza, Simon Richardus, Oskar Suomalainen, Martin Genet, Margarita Vodneva, Juho Jääskeläinen, Emanuel Lopes

위치 : 핀란드 투르쿠 항구 I **용도** : 교통시설 I **규모** : 지상3층 I **구조** : 철근콘크리트, 철골 I **마감** : 알루미늄, 유리

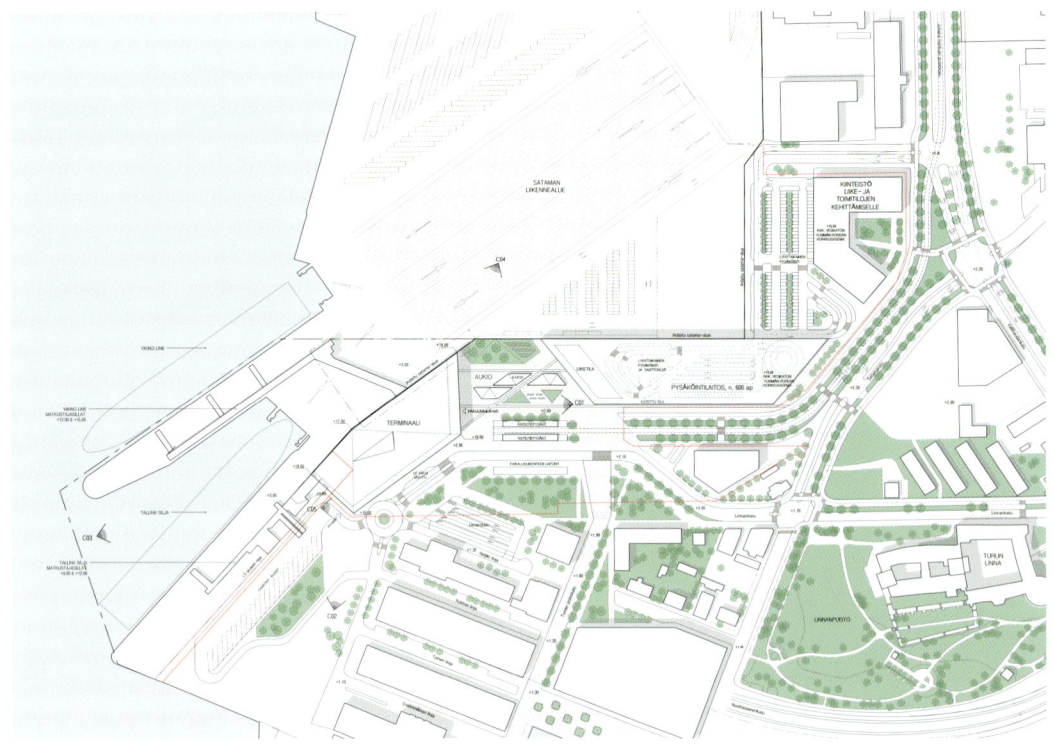

Site plan

핀란드의 남서 해안에 있는 가장 오래된 도시의 투르쿠 항구는 스톡홀름으로 매일 여러 번 항해하는 분주한 화물/여객 항구이다. 페리 터미널 투르쿠 프로젝트는 여객 항구를 현대적이고 빠르고 안전한 선박 교통의 요구 사항을 충족하는 해양 허브로 업그레이드할 것이다. 이 프로젝트에는 항구에서 운영되는 2개의 페리 라인을 위한 새로운 공동 여객 터미널 건설도 포함된다.

터미널 디자인을 위한 국제 현상 공모에서 PES-아키텍츠의 '종이접기' 제안이 기능적 목표를 달성하고 항구와 역사적인 투르쿠성 주변의 진화하는 린난니에미 지역의 도시 맥락에 부합하기 때문에 선정되었다.

터미널의 건축과 기능은 완벽하게 조화를 이룬다. 터미널은 본질적으로 명확한 기능을 갖춘 시스템이지만, 효율성과 함께 비즈니스 또는 레저 목적으로 여행 경험을 향상하고 개선하는 것을 목표로 한다. 페리 크루즈는 일반적으로 일상에서 휴식을 제공하며, 눈에 잘 띄는 터미널 건물은 승객에게 바다 여행 분위기를 선사한다.

이 건물은 삼각형 표면이 비스듬하게 있고 강철, 유리, 알루미늄이 사용되어 마치 거대한 종이접기 작품처럼 섬세하게 접혀 있다. 건물의 조각 같은 선과 반짝이는 표면은 은빛 연어나 강철 선박이라는 해안 이미지를 연상시킨다.

종이접기 개념은 한 장의 종이를 다양하게 접을 수 있는 것처럼, 사용자의 다양한 요구에 맞게 터미널을 변경하는 것이다. 예를 들어, 유연한 로비 공간은 변화하는 보안 제어 프로세스에 쉽게 적응할 수 있다.

터미널은 승하선 시간을 최소화하기 위해 선박과 최대한 가깝게 위치한다. 배치는 입구에서 2층 출발 홀까지 그리고 선박에서 출구까지 최적의 직선로로 구성되어 있다. 들어오고 나가는 승객 경로는 분리되지만 유리벽을 통해 시각적으로 연결된다. 공간적 명확성, 투명성, 분명한 직선로를 통해 터미널 전체에서 쉽고 직관적으로 길을 찾을 수 있다.

자연광이 여객 공간에 가득 채워지며, 따뜻한 목재가 강철, 알루미늄, 유리 구조를 보완한다. 높은 유리벽과 테라스를 통해 도착하고 출발하는 모든 여행자는 도시, 투르쿠성, 바다의 아름다운 풍경을 조망할 수 있다.

터미널을 나가자마자 대중 교통 정류장과 택시 승강장이 있고 인근에 주차장이 있어 다양한 형태의 교통 수단으로 터미널에 쉽게 접근할 수 있다.

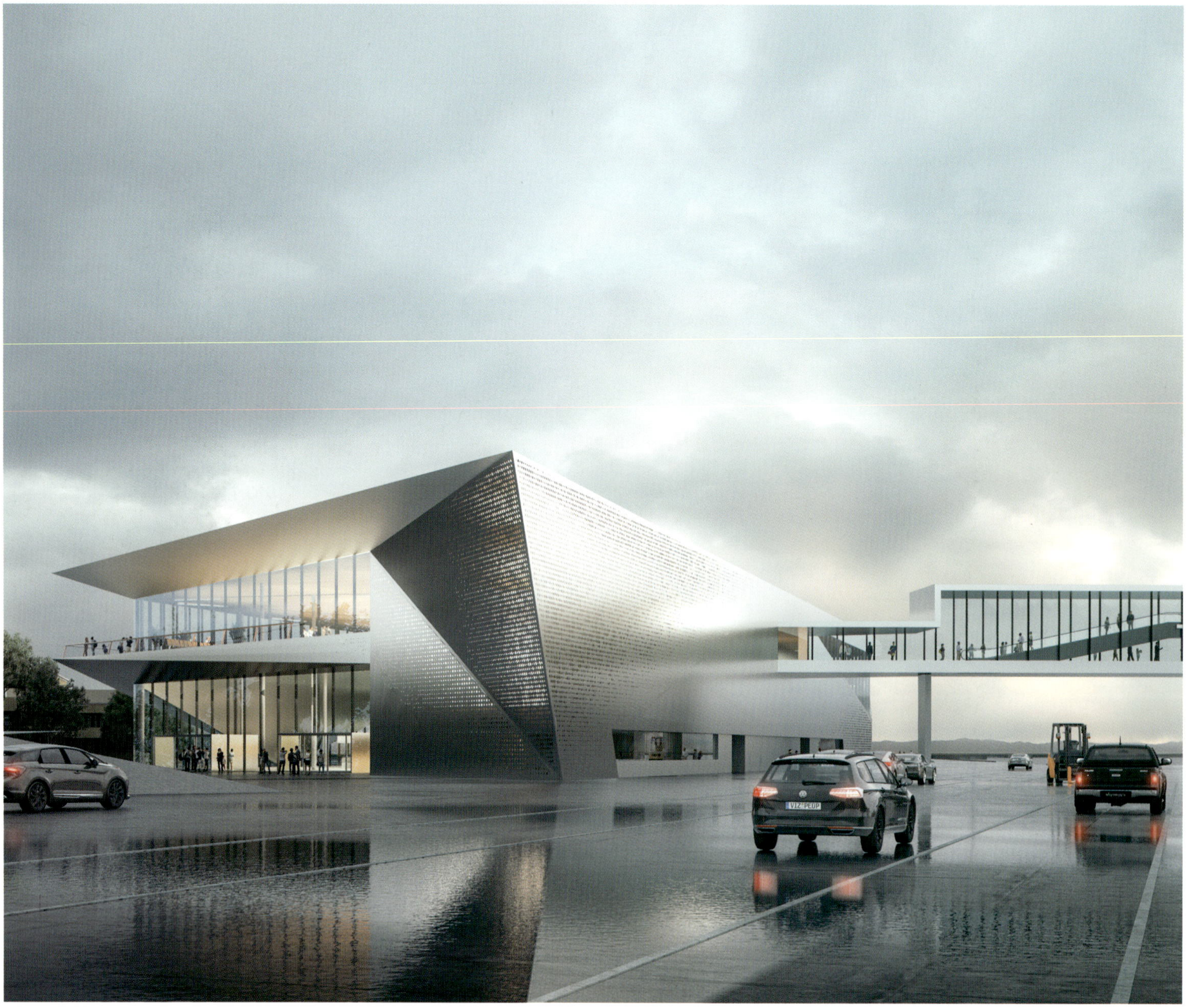

East elevation

West elevation

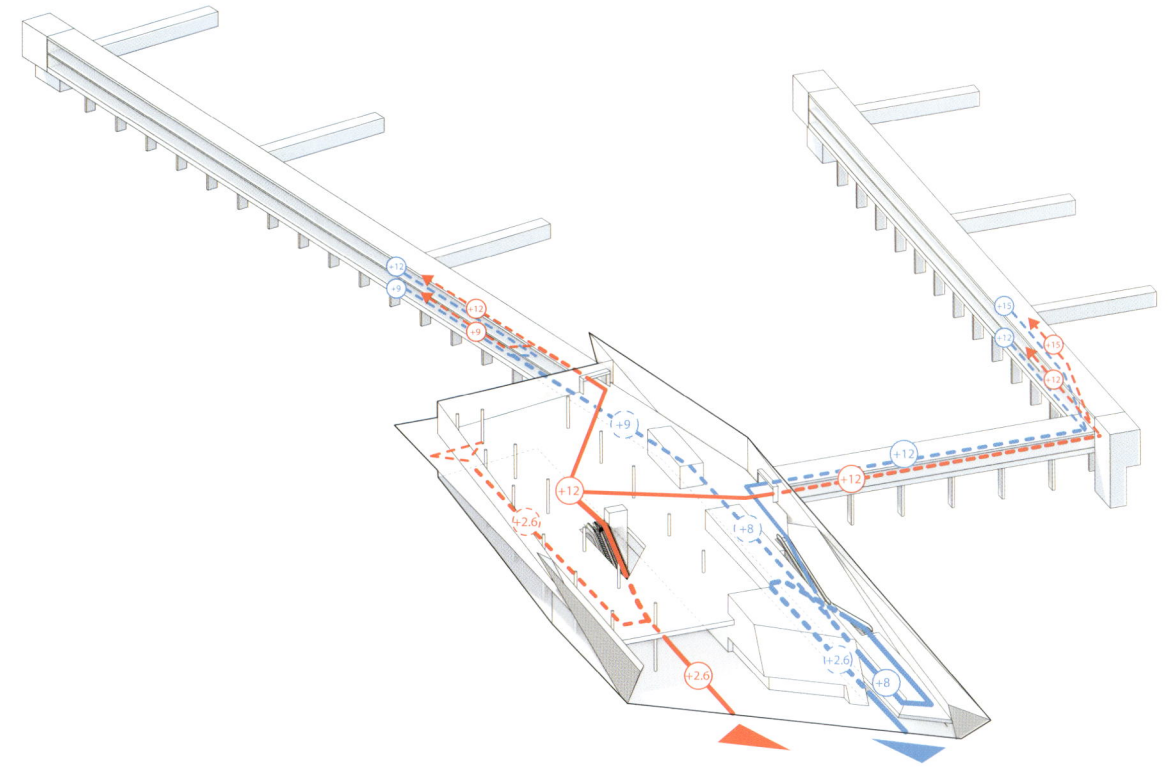

■ Circulation Plan

1. Waiting lounge / Departure hall
2. Terrace
3. Cleaning
4. Security control
5. Exit
6. Tour groups
7. Entrance lobby
8. Fresco
9. Roofed area
10. Plaza
11. Machine room
12. Customer service
13. Offices
14. Port area

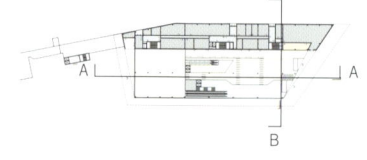

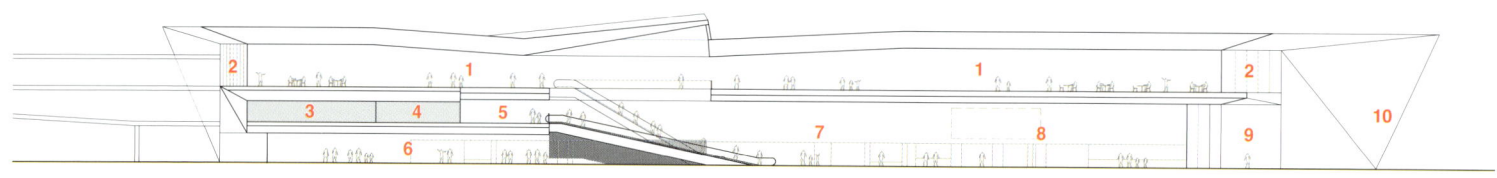

Section A

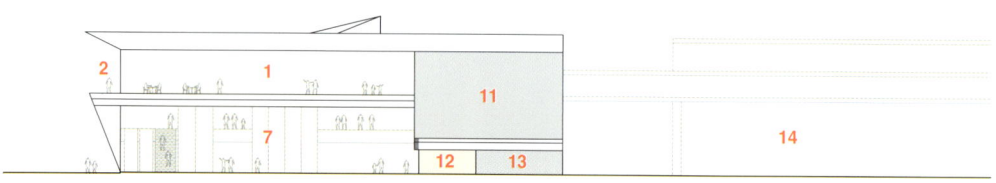

Section B

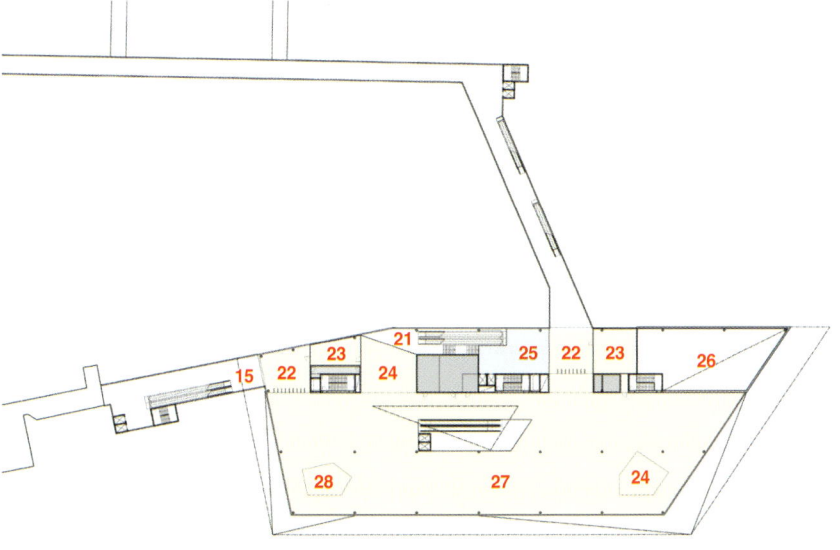

2nd floor plan

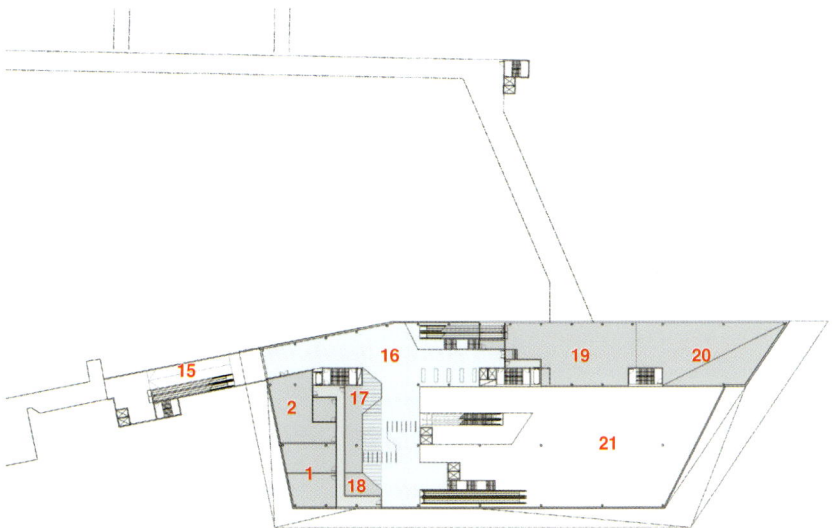

Intermediate floor plan

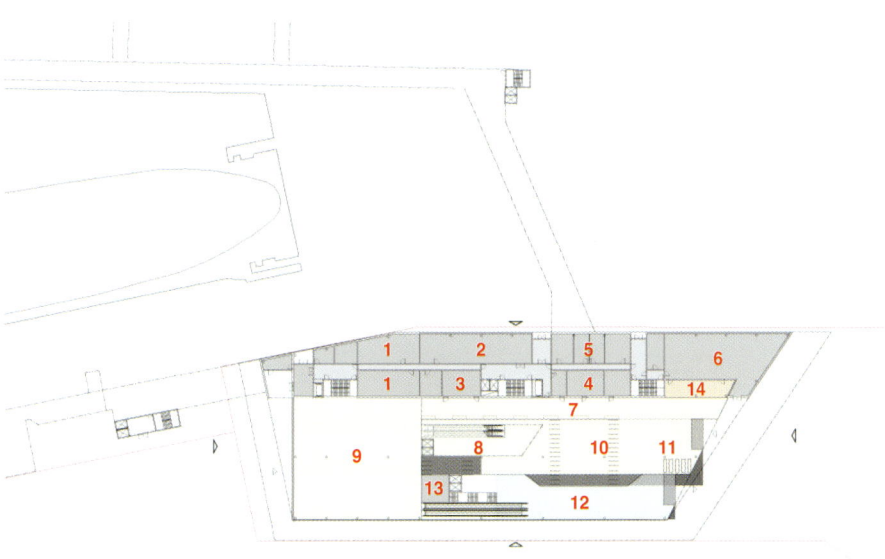

1st floor plan

1. Dressing room
2. Break room / Task assignment
3. Toilet
4. Luggage room
5. General cabling
6. Work premises / Shared
7. Transfer area from sales to security control
8. Transfer area to 2nd floor
9. Tour groups
10. Security control and check-in
11. Entrance lobby
12. Exit lobby
13. Auxiliary space
14. Customer service desks
15. Passenger bridges
16. Exit control area
17. Security control
18. Customs
19. Machine room area
20. High space
21. Opening
22. Embarking
23. VIP lounge
24. Restaurant
25. Exit corridor
26. Upper part of machine room
27. Waiting lounge / Departure hall
28. Bar

Chongqing Cuntan International Cruise Center
충칭 쿤탄 국제 크루즈 센터

MAD Architects / Ma Yansong
매드 아키텍츠 / 마 얀송

The project site, currently a cargo terminal covering 66,000m², is located in Chongqing's Liangjiang New Area. The site and associated cruise terminal sit within the Cuntan Port area, allowing access to the Yangtze River. Under MAD's plans, the site will become a 65,000m² international cruise terminal and city complex, hosting a 15,000m² cruise port and 50,000m² for commercial space. "Chongqing has mountains and waters," said Ma Yansong, reflecting on the vision behind the winning scheme. "However, the Yangtze River is more than just a natural landscape in Chongqing. Because of human activities such as shipping traffic and industrial transport, this mountain city is also full of energy and movement. We want to transform this energy in Chongqing from traces of industry into an energy that stimulates the imagination. People can feel the kinetic energy of the city here, but also imagine the public spaces of the future."

'Gantry Crane' – Science fiction, and a walking city

While visiting the site, MAD team was inspired by the large orange gantry cranes that dominate the freight terminal. "These gantry cranes became living alien creatures that gave a sense of surrealism," Ma noted. "The new scheme is therefore not only about reflecting the industrial colours of the past, but also about respecting this original surrealism. We have designed the elevated buildings as if they were a futuristic, free-walking city, seemingly arriving here from elsewhere, and perhaps travelling elsewhere once again someday."

MAD's scheme, named the Yangtze River Skywalk, is a 430m long complex comprised of six separate and interconnected elevated buildings inspired by the gantry cranes. From a distance, the buildings rise and fall, recreating the rhythm of the industrial freight terminal cranes in both form and colour, while their elevated position above the ground allows for open, unobstructed views of the river. The building's aluminium curtain wall offers the site a surreal feel, bringing a contemporary freshness to the building's mixed-use tenants, which include a parlour, shops, and restaurants.

Urban Public Space – A layering of urbanism, and a landscaped green axis

Underneath the 'floating' building complex is the new Cruise Ship Landscape Park and Cruise Ship Hall, designed by MAD to extend and enhance the Cuntan Central Golden Axis Pier. The urban interventions form a new link between the adjacent Pier Park and the Century Cuntan Park, creating a 100,000m² urban green space that blends naturally with the Central Golden Axis. The new scheme allows the public to fully experience the diversity of uses across the site, enjoying the scenery of the city and the Yangtze from a maritime perspective, a condition otherwise reserved for river vessels.

Located underneath the Cruise Landscape Park is the Cruise Centre Hub, providing access to the 'floating' complex and the ground level landscape park. The design of the Cruise Centre includes skylights to enhance natural light within the interior space, while an upper cantilevered building avoids overbearing direct sunlight.

When completed, the Cuntan International Cruise Centre will transform industrial memories into a fully realised piece of urban imagery, creating a unique urban landmark in Chongqing.

Location : Chongqing, China I **Function** : Transportation facility I **Site area** : 66,000m I **Total floor area** : 65,000m I **Client** : Chongqing Cuntan International Cruise Home Port Development Co. I **Consortium** : China Academy of Building Research Ltd. I **Principal partners in charge** : Ma Yansong, Dang Qun, Yosuke Hayano I **Associate in charge** : Liu Huiying I **Design team** : Yang Xuebing, Lei Kaiyun, Wang Ruipeng, Chen Wei, Ning Tong, Wang Yiding

위치 : 중국 충칭 I 용도 : 교통시설

Site plan

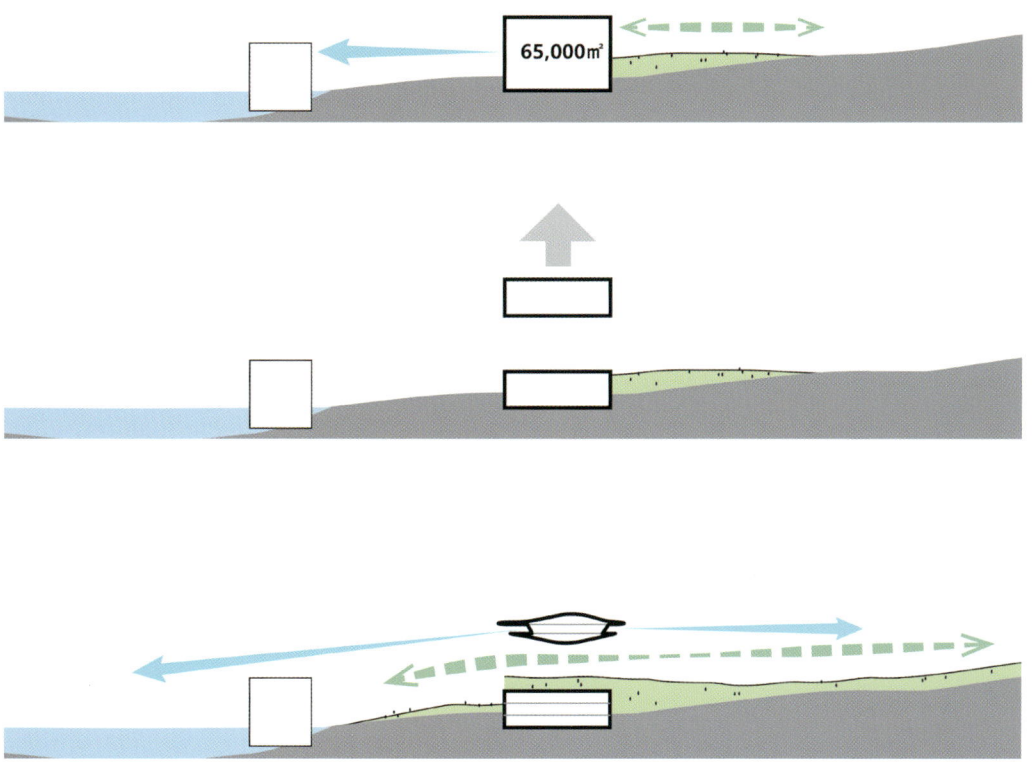

■ Design Process

Winner _ 당선작

현재 66,000㎡ 면적의 화물터미널인 사업 부지는 충칭의 량장 신규 구역에 있다. 대지와 관련 크루즈 터미널은 쿤탄항 지역 내에 위치하여 양쯔강에 접근할 수 있다. MAD의 계획에 따라 이 부지는 65,000㎡ 규모의 국제 크루즈 터미널과 도심 복합단지가 되며 15,000㎡의 크루즈 항구와 50,000㎡의 상업 구역이 들어설 예정이다.

마 얀송은 현상설계에서 선정된 계획의 비전을 회상하며 '충칭에는 산과 강이 있다. 양쯔강은 충칭의 단순한 자연 경관 그 이상의 의미를 갖는다. 해운과 산업 운송과 같은 인간 활동 때문에 이 산악 도시는 에너지와 생동감으로 가득 차 있다. 우리는 충칭의 이 에너지를 산업의 흔적에서 상상력을 자극하는 에너지로 바꾸고자 한다. 사람들은 여기에서 도시의 역동적인 에너지를 느낄 수 있을 뿐만 아니라 미래의 공공 공간도 상상할 수 있다'라고 말한다.

'갠트리 크레인' - 공상과학 소설과 걷는 도시

부지를 방문하는 동안 MAD 팀은 화물터미널의 큰 특징인 대형 주황색 갠트리 크레인에서 영감을 받았다. 마는 '이 갠트리 크레인은 초현실주의 느낌의 살아있는 외계 생명체가 되었다. 따라서 새로운 계획은 과거의 산업적 색상을 반영하는 것뿐만 아니라 이 독창적인 초현실주의를 존중하는 것이기도 하다. 우리는 이 고층 건물을 다른 곳에서 이곳에 도착하고 언젠가는 다시 다른 곳으로 여행할 수 있는 미래지향적이고 자유롭게 걷는 도시인 것처럼 설계했다'라고 말한다.

양쯔강 스카이워크라고 명명된 MAD의 계획은 갠트리 크레인에서 영감을 받은 6개의 분리되면서도 상호 연결된 고층 건물로 구성된 430m 길이의 복합 단지이다. 멀리서 보면 건물들이 오르락내리락하면서 형태와 색상 면에서 산업용 화물터미널 크레인의 리듬을 재현하고, 지상에서 높이 솟아 있어 양쯔강의 탁 트인 전망을 선사한다. 건물의 알루미늄 커튼월은 대지에 초현실적인 느낌을 부여하여 응접실, 상점, 레스토랑 등 건물의 복합용도 세입자에게 현대적인 신선함을 제공한다.

도시 공공 공간 - 겹쳐진 도시화와 조경된 녹지축

'수상' 복합단지 아래에는 MAD가 쿤탄 중앙 황금축 부두를 확장하고 개선하기 위해 설계한 새로운 크루즈선 조경 공원과 크루즈선 홀이 있다. 도시 개입은 인접한 부두 공원과 센추리 쿤탄 공원 사이에 새로운 연결을 형성하여 중앙 황금축과 자연스럽게 혼합되는 100,000㎡의 도시 녹지 공간을 만든다. 새로운 계획을 통해 사람들은 대지 전체의 다양한 용도를 완전히 경험하거나, 강 선박이 유보되어야 하는 해양 관점에서 도시와 양쯔강의 풍경을 즐길 수 있다.

크루즈 조경 공원 아래에 위치한 크루즈 센터 허브에서 '수상' 복합단지와 지상층 조경 공원에 접근할 수 있다. 크루즈 센터의 디자인에는 내부 공간에 자연 채광을 많이 끌어들이기 위한 채광창이 포함되어 있으며 상부 캔틸레버식 건물은 직사광선을 피한다.

완공 시 쿤탄 국제 크루즈 센터는 산업 흔적을 완전히 실현된 도시 이미지로 탈바꿈하여 충칭의 독특한 도시 랜드마크가 될 것이다.

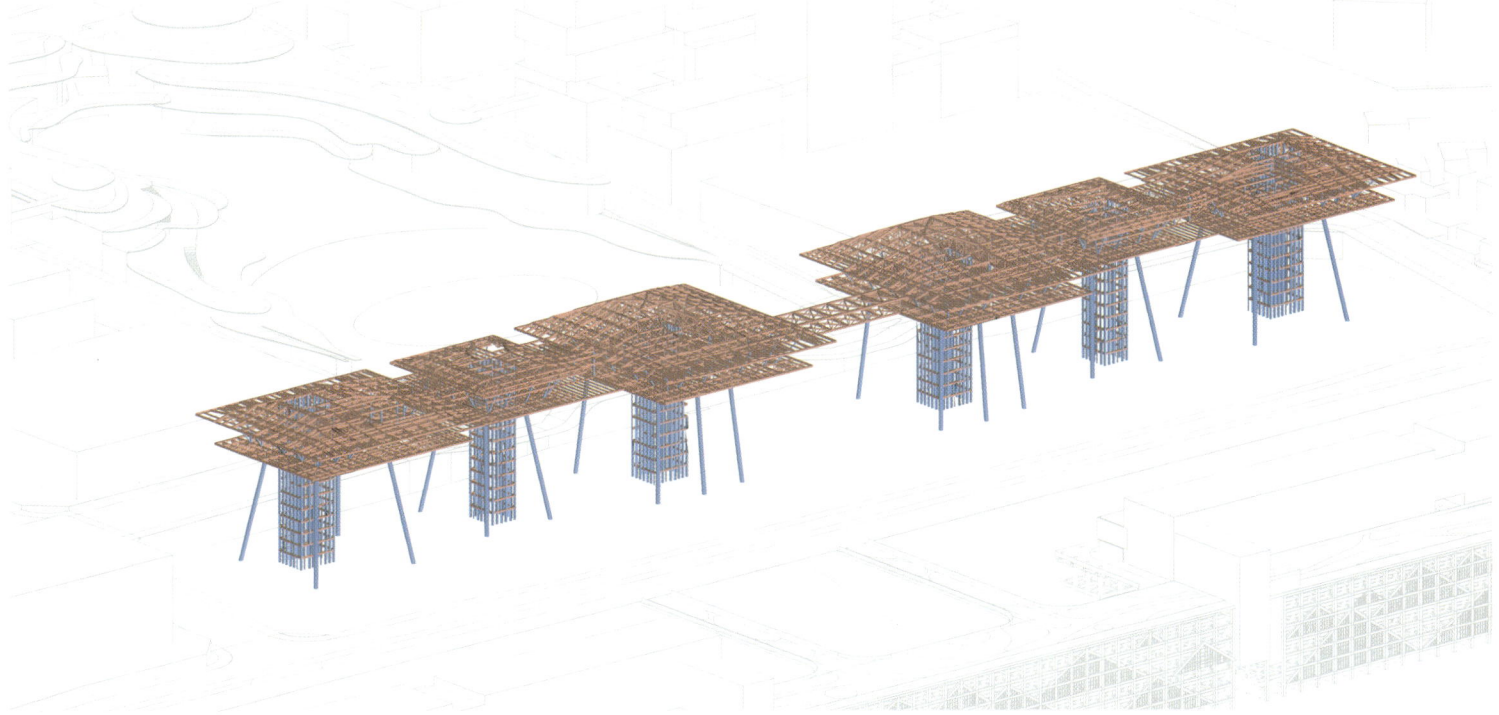

Structure

Winner _ 당선작

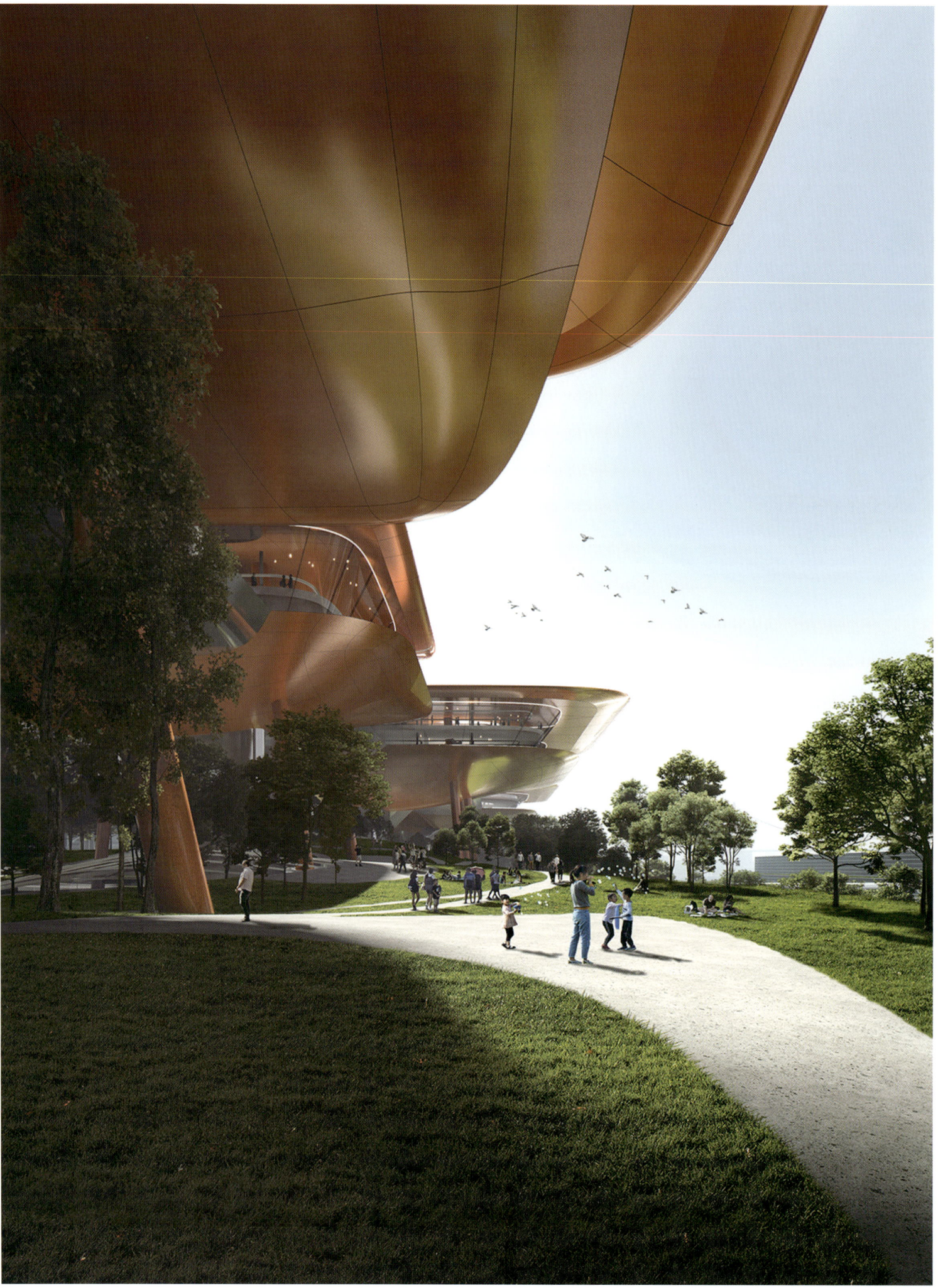

Chongqing Cuntan International Cruise Center >>

▍Zoning

Area map

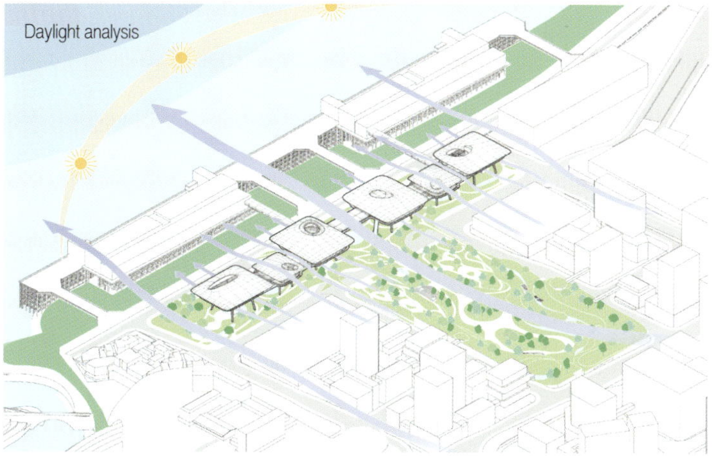

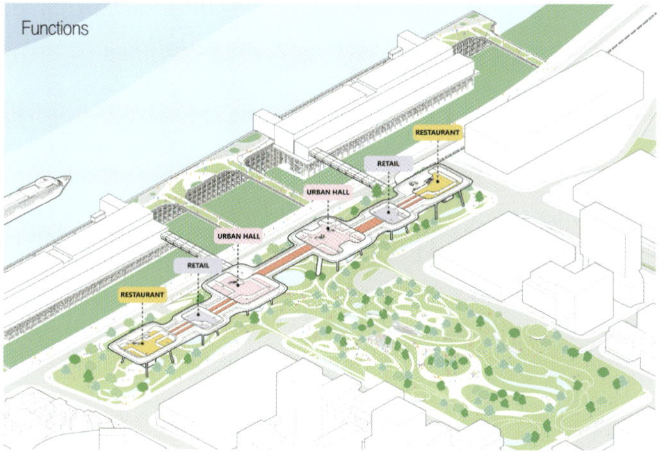

Daylight analysis

Functions

Public transportation

Winner _ 당선작

Gyeongsangbuk-do Training Center
경상북도 수련원

Shinhan Architects & Engineers / Inho Jeong, Sanghoon Kim + D&B PARTNERS architecture design group / Doyeun Cho, Kyunghwan Lee + HANBIT architecture / Sunghoon Han
㈜신한종합건축사사무소 / 정인호, 김상훈 + ㈜디엔비파트너스건축사사무소 / 조도연, 이경환 + ㈜한빛종합건축사사무소 / 한성훈

Although the business site within the Goraebul coastal tourist destination has beautiful sea views and excellent development conditions, the environment to enjoy nature was not provided due to the unstable ground and seawater along the coast. Through the plan, we aim to create a space where the natural ecosystem and humans coexist by restoring the value of nature to enjoy the great view of the beach, sandy beach, pine forest, and sea breeze.

Gyeongsangbuk-do Training Center is for relaxation, experience, education, and healing. The open view to the sea and the water space resembling the sea embrace both humans and nature. The long horizontal mass with gentle curves opens the field of view to capture the wide sea like a landscape painting, and harmonizes with the coastline of Goraebul. A long floating mass provides all rooms ocean views and comfortable privacy.

Various natural elements within the site, such as sand, pine forest, and wetland, were preserved and utilized. We created an external space where both people and nature can be healed through experiences that stimulate the five senses by looking at nature, hearing sounds, and smelling scents. In particular, various themed spaces such as indoor swimming pool, indoor hot spring, outdoor water play area, seawater hot spring, and family pool provide pleasure and healing with water.

고래불 해안 관광지에 속하는 사업 부지는 뛰어난 바다 전망과 우수한 개발 여건을 갖고 있음에도 불구하고, 해안가의 불안정한 지반 및 해수로 인해 자연을 즐길 수 있는 환경이 갖춰지지 않았다. 본 계획을 통해 바닷가의 멋진 전경과 모래사장, 송림 그리고 바닷바람을 만끽할 수 있도록 자연의 가치를 회복하여, 자연생태계와 인간이 상생하는 장소를 조성하고자 한다.

경상북도 수련원은 휴식, 체험, 교육, 힐링을 위한 곳으로 바다로 열린 조망과 바다를 닮은 수공간은 인간과 자연을 동시에 품고 있다. 완만한 곡선의 긴 수평 매스는 넓게 펼쳐진 바다를 풍경화처럼 담을 수 있도록 시야를 열어주며, 고래불의 해안선과 조화를 이룬다. 떠 있는 긴 매스를 통해 모든 객실에서의 바다 조망과 편안한 프라이버시를 제공한다.

모래, 송림, 습지 등 대지 내에 존재하는 다양한 자연 요소를 살리고 활용하였다. 자연을 바라보고, 소리를 듣고, 향기를 맡으며 오감을 자극하는 체험을 통해 사람과 자연이 같이 치유되는 외부공간으로 조성하였다. 특히 실내 수영장, 실내 온천, 야외 물놀이장, 해수 온천, 가족풀 등 다양한 테마 공간으로 물을 통한 즐거움과 힐링을 누릴 수 있도록 하였다.

Location : 5-95, Geomuyeok-ri, Byeonggok-myeon, Yeongdeok-gun, Gyeongsangbuk-do, Korea | **Function** : Training facility | **Site area** : 39,104m² | **Bldg. area** : 14,431m² | **Total floor area** : 13,511m² | **Stories** : B1, 4FL | **Structure** : Steel frame, Reinforced concrets | **Finish** : Exposed concrete, Pine board pattern exposed concrete, Low-E pair glass, Terra cotta, White louver

위치 : 경상북도 영덕군 병곡면 거무역리 5-95 | **용도** : 수련시설 | **규모** : 지하1층, 지상4층 | **구조** : 철골, 철근콘크리트 | **마감** : 노출콘크리트, 송판무늬 노출콘크리트, 로이복층유리, 테라코타, 백색루버 | **설계팀** : 오인환, 이지웅, 임연수, 박정원

Winner _ 당선작

Front elevation

1. Situation room
2. Communication / Server room
3. Toilet
4. Separate collection room
5. Lounge
6. Linen room
7. Goraebul yard
8. Laundry room
9. Warehouse
10. Kitchen
11. Conference hall
12. Waiting room
13. Goraebul valley
14. Infinity fool
15. Room
16. Hall
17. Outdoor water park
18. Baby fool
19. Roof garden

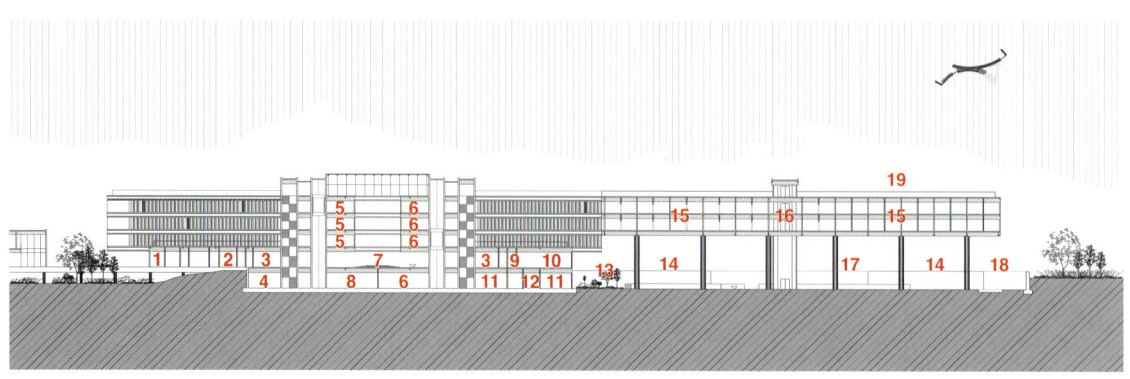

Cross section

Gyeongsangbuk-do Training Center >>

1. Two-person unshared house
2. Four-person unshared house
3. Six-person unshared house
4. Office
5. Lounge
6. Goraebul yard
7. Restaurant
8. Kitchen
9. Convenience store
10. Cafeteria
11. Waiting yard
12. Program parking
13. Botanic gardens cafe
14. Parking
15. Wetland experience center
16. Pool
17. Hall

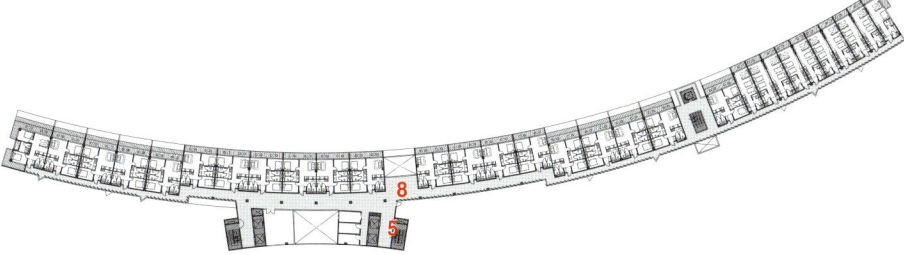

4th floor plan

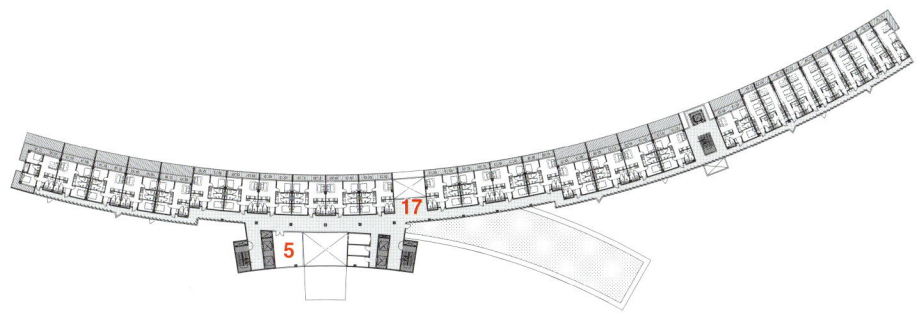

3rd floor plan

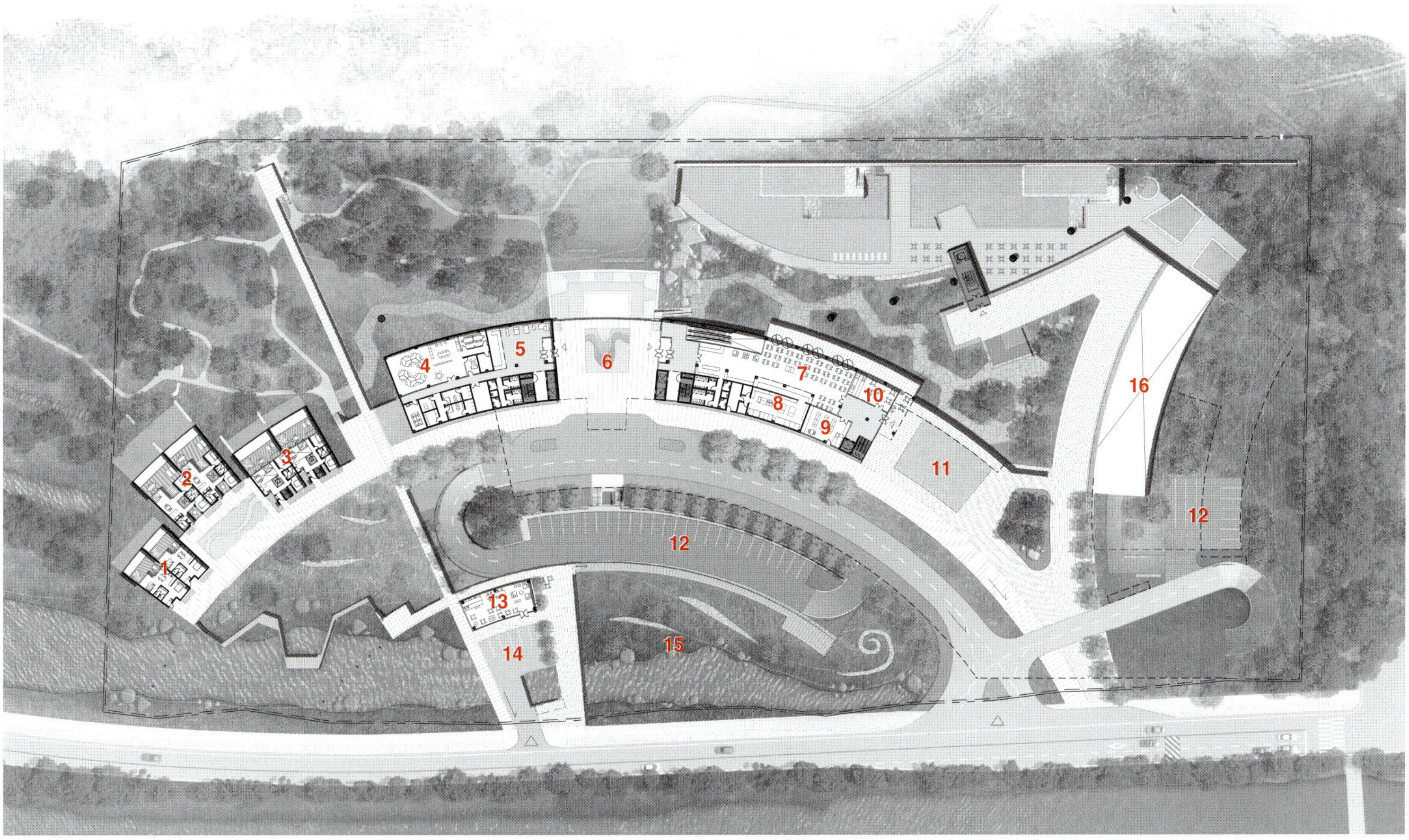

1st floor plan

Sinan-dong Sports Complex

신안동 복합스포츠타운

SPACE GROUP CO.LTD / Sangleem Lee
㈜공간종합건축사사무소 / 이상림

The target site is located near the original downtown of Jinju, which was developed based on Jinjuseong Fortress. In 1968, when the Jinju Public Stadium was built, there were only paddy fields around it, and as the city grew, the surrounding area gradually filled with low houses and stores, and the public stadium became the center of the city.

On the site of an attached stadium near the Jinju Public Stadium, which has been neglected for a long time since it was closed in 2016, a new complex sports town in Shinan-dong will be built as a facility for local residents. As a result, it is expected that the living environment of the residents will be greatly improved by enhancing the poor surrounding environment of the Jinju Public Stadium and providing the necessary sports infrastructure for the citizens of the northwest region. In addition, this place, which will experience rapid changes in the future, will have the historical significance of Jinju City, and be visited by many citizens due to its easy access. Therefore, it is planned to be an open space as a daily living sports space for local residents as well as a new place for communication. This will be connected to the surrounding environment beyond a simple sports facility and be a place for everyone as a base space without borders. Now, as part of the city, it is hoped that it becomes a new place for sports culture for the healthy and vibrant life of Jinju citizens as a complex sports town in Shinan-dong that can be used by all ages, and easily accessed by anyone through an expanded open space.

대상지는 진주성을 기반으로 발달한 진주 원도심 인근에 자리하고 있다. 진주 공설운동장이 지어질 1968년 주변은 논과 밭뿐이었고, 도시가 성장하면서 차츰 주변에 낮은 주택과 상가들로 채워지면서 공설운동장은 도시 중심부가 되었다.

2016년에 폐지된 이후로 오랜 기간 동안 방치되었던 진주공설운동장 인근 부속 운동장 부지에 신안동 복합스포츠타운이 지역주민들을 위한 시설로 새롭게 조성된다. 이로 인해 열악했던 진주공설운동장 주변 환경에 대한 개선과 동시에 서북권 시민들에게 그동안 부족했던 체육 인프라를 제공함으로써 주민 생활환경이 대폭 개선될 것으로 기대된다. 또한, 앞으로 급격한 변화를 겪게 될 이 장소는 진주시의 역사적 의미를 가지고 있을 뿐만 아니라, 접근성이 좋아 많은 시민들이 찾을 것으로 예상된다. 따라서 이곳을 지역주민들을 위한 일상의 생활체육공간 이자, 새로운 소통의 장으로서의 열린 장소로 계획하고자 하였다. 이는 단순한 체육시설을 넘어서 주변 환경과 연결되고, 경계가 없는 거점 공간으로서 모두를 위한 장소가 될 것이다. 이제 도시의 일부로서 모든 연령대가 이용할 수 있고, 확장된 오픈스페이스를 통해 누구나 쉽게 접근할 수 있는 신안동 복합스포츠타운으로 건강하고 활기찬 진주시민의 삶을 위한 스포츠 문화의 새로운 장소가 되길 희망한다.

Location : 1-85, Sinan-dong, Jinju-si, Gyeongsangnam-do, Korea I **Function** : Sports facility I **Site area** : 17,672m² I **Bldg. area** : 9,657m² I **Total floor area** : 8,948m² I **Stories** : B1, 2FL I **Structure** : Steel framed reinforced concrete, Steel frame I **Finish** : Metal panel, Bricks panel, Polycarbonate, Low-E pair glass

위치 : 경상남도 진주시 신안동 1-85번지 I **용도** : 운동시설 I **규모** : 지하1층, 지상2층 I **구조** : 철골철근콘크리트, 철골 I **마감** : 메탈패널, 브릭스패널, 폴리카보네이트, 로이복층유리 I **건축주** : 경상남도 진주시 I **설계팀** : 김승국, 백현욱, 백창현, 이중운, 박용순

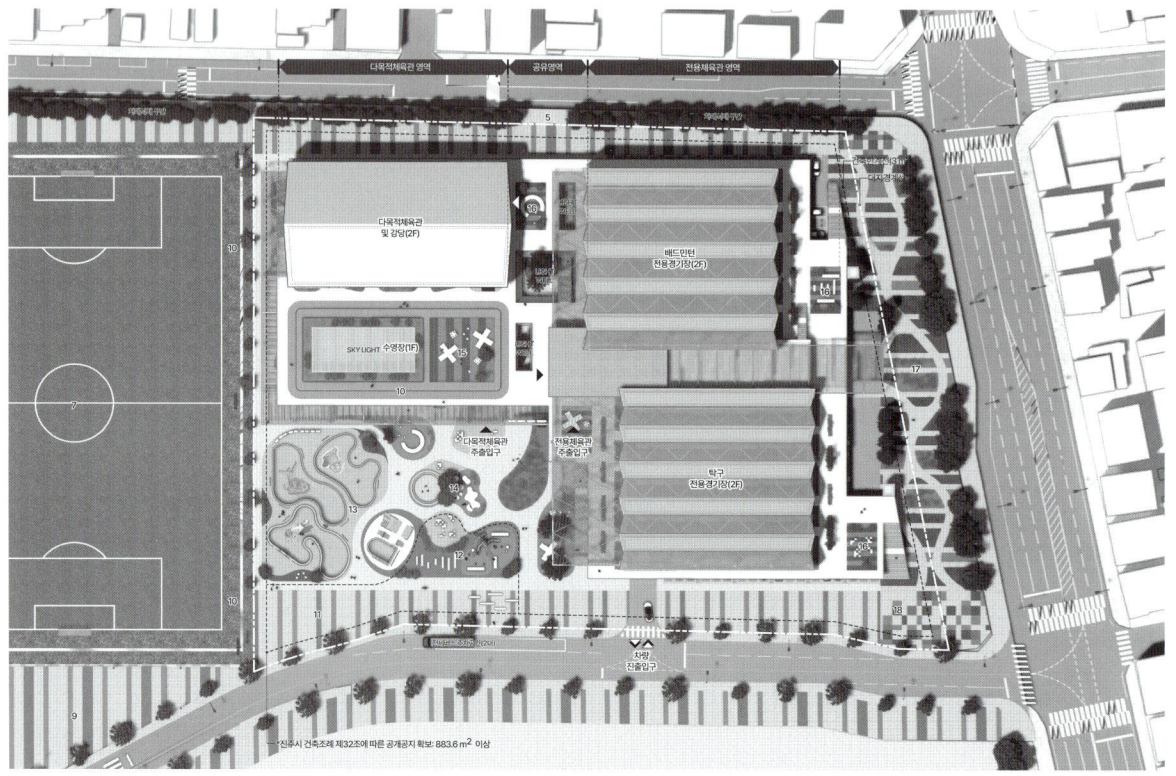

Site plan

Site Process

- Four programs

– 종목의 특성에 따라 각기 다른 면적과 천정고를 갖는 프로그램 매스

- Continuity of open space

– 외부체육시설 진입광장에서부터 확장된 오픈스페이스와 보행동선의 적극적 연결

- Pleasant piloti

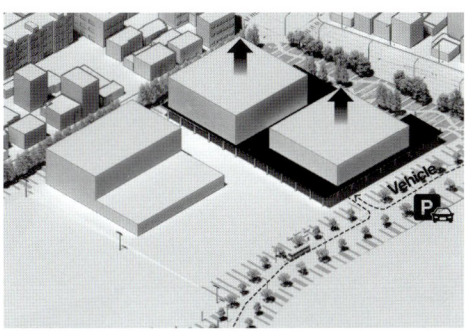

– 네면으로 접근이 용이하고 개방적인 필로티 공간을 활용한 주차장

- Plate expansion and light corridors

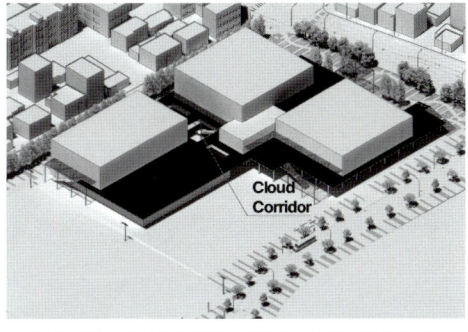

– 모든 공간은 입체적 플랫폼으로 연결되고 지상보행을 위한 가벼운 회랑

- Sports cloud

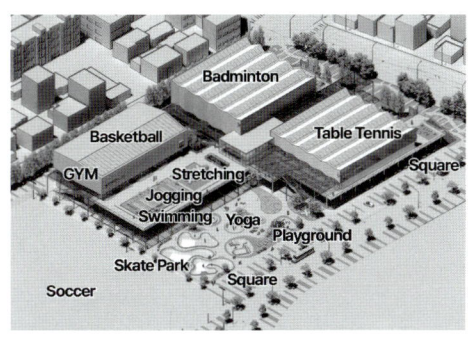

– 모든 연령대가 건강하게 교류하는 입체적 플랫폼의 SPORTS CLOUD

Winner _ 당선작

▌ Concept

- **Jinju scape : continuity of horizontal landscape**

- 진주시의 수평적 경관을 확장하는 토대로서 공공의 오픈스페이스 제공
- 증가하는 여가수요를 만족시킬 전문 운동시설을 수직적 변주요소로서 계획

- **Sports cloud : 3D platform**

- 지역민들이 운동을 매개로 자연스럽게 소통하고 교류할 수 있는 새로운 유형의 스포츠 복합시설

Sinan-dong Sports Complex >>

Open Space

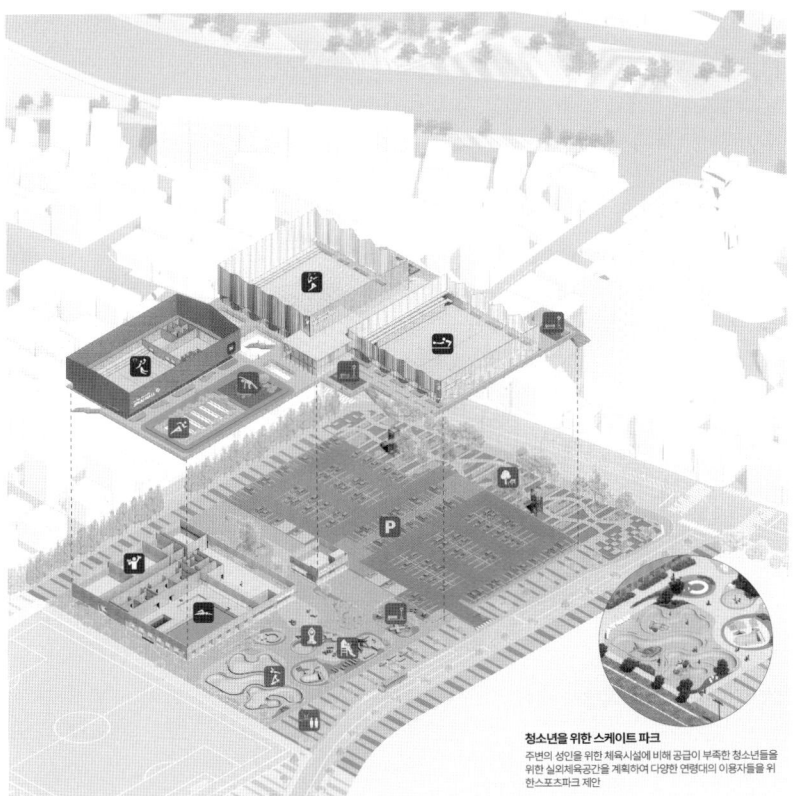

청소년을 위한 스케이트 파크
주변의 성인을 위한 체육시설에 비해 공급이 부족한 청소년들을 위한 실외체육공간을 계획하여 다양한 연령대의 이용자들을 위한 스포츠파크 제안

Indoor

Outdoor

Structure Plan

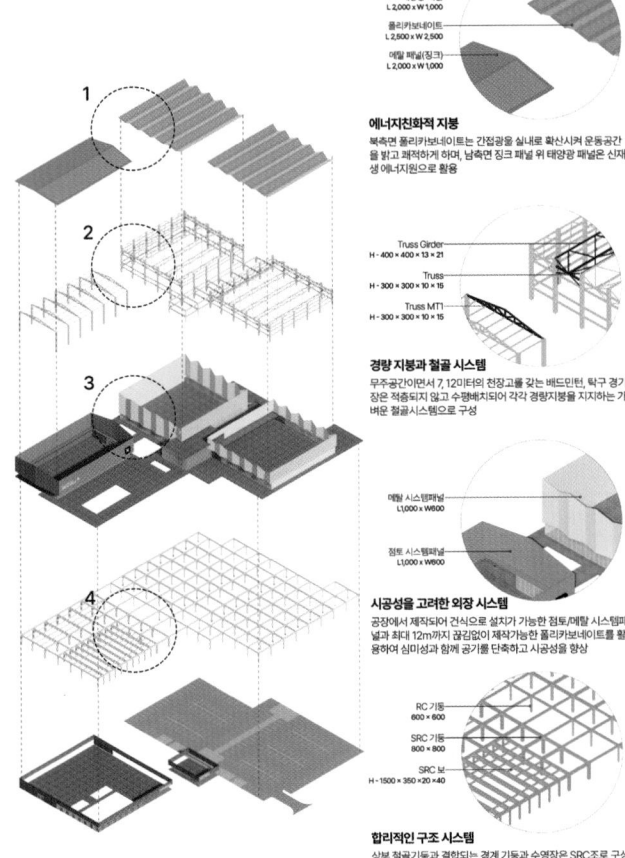

에너지친화적 지붕
북측면 폴리카보네이트는 간접광을 실내로 확산시켜 운동공간을 밝고 쾌적하게 하며, 남측면 징크 패널 위 태양광 패널은 신재생 에너지원으로 활용

경량 지붕과 철골 시스템
무주공간이면서 7, 12미터의 천장고를 갖는 배드민턴, 탁구 경기장은 적층되지 않고 수평배치되어 각각 경량지붕을 지지하는 가벼운 철골시스템으로 구성

시공성을 고려한 외장 시스템
공장에서 제작되어 건식으로 설치가 가능한 점토/메탈 시스템패널과 최대 12m까지 끊김없이 제작가능한 폴리카보네이트를 활용하여 심미성과 함께 공기를 단축하고 시공성을 향상

합리적인 구조 시스템
상부 철골기둥과 결합되는 경계 기둥과 수영장은 SRC조로 구성하고 내부 슬라브를 지지하는 경량 구간은 RC조로 구성

South elevation

West elevation

Winner _ 당선작

1. Gym parking
2. Multipurpose gym parking
3. Badminton gym
4. Pingpong gym
5. Hall
6. Powder room
7. Water tank
8. Electrical room
9. Dynamo room
10. Central monitor room
11. Fitting room
12. Shower room
13. Toilet
14. Multipurpose gym & Auditorium
15. Health counseling room & Operation office
16. Cloud corridor

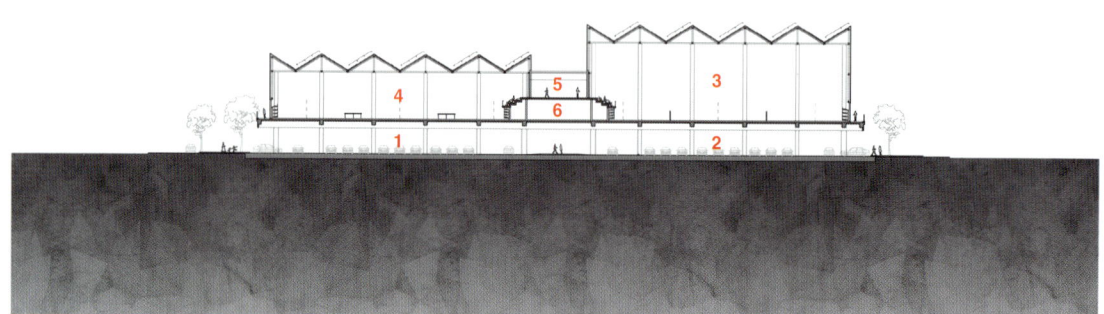

Longitudinal section

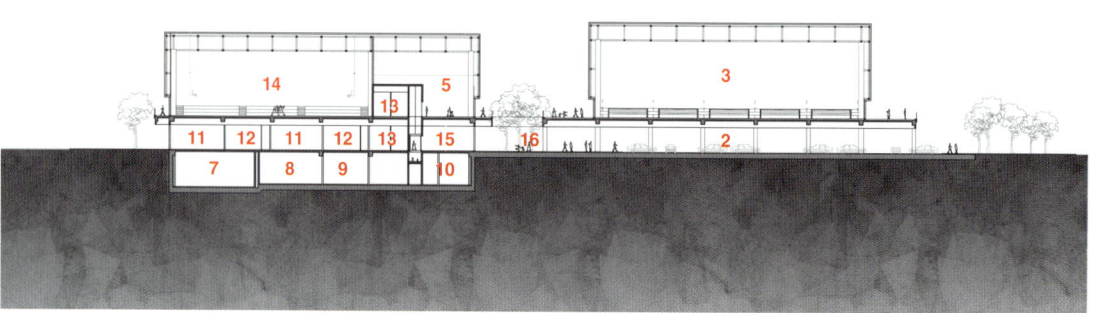

Cross section

Sinan-dong Sports Complex >>

1. Sport plaza
2. Park
3. Cloud lounge
4. Lobby
5. Viewing lounge
6. Fool
7. Faculty room
8. Medical service room
9. Fitting room
10. Shower room
11. GX room
12. Fitness center
13. Operation office
14. Health counseling room
15. Multipurpose class room
16. Cloud corridor
17. Jogging track
18. Hall
19. Auditorium
20. Waiting room
21. Badminton gym
22. Pingpong gym
23. Warehouse
24. Viewing garden

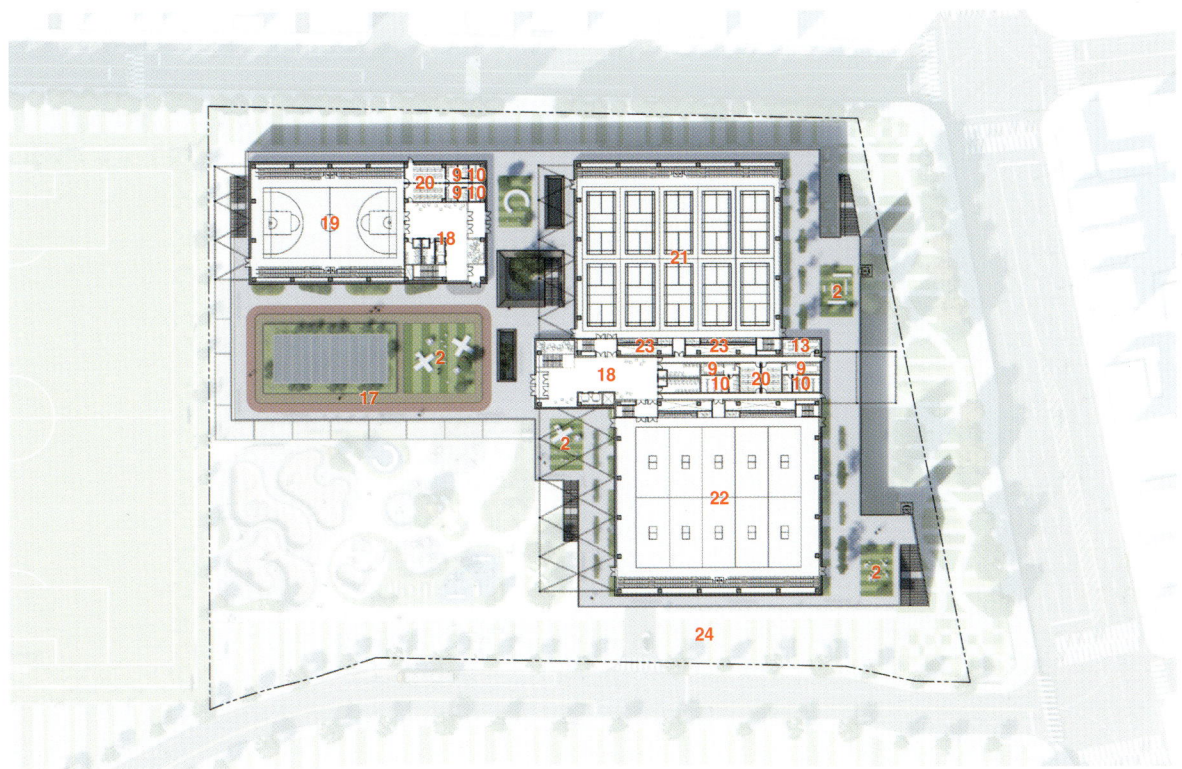

2nd floor plan

1st floor plan

227

Winner _ 당선작

MZ Sports Plaza
MZ 스포츠플라자

HEA / Jonghyun Baek + SLDESIGN / Sungjin Kim + MSGD / Juntaek Kim
㈜에이치이에이 / 백종현 + 에스엘디자인㈜ / 김성진 + 엠에스지디 / 김준택

'MZ Gather GO' is the starting point of experiencing the huge nature of Anyangcheon Stream, and it creates a sensuous cultural space for various generations, including the MZ generation.
First, the spaces under the overpass are three-dimensionally connected through community deck facilities such as net observatory shelter and deck shelter. Various experiences such as community activities and relaxation activities are provided depending on the height, and it realizes connection with surrounding cities and free flow in the interior space. Second, In addition to the multi-sport space, which is the main space, it seeks to revitalize the surrounding sub spaces. Sub-zones such as weight zones, artificial rock climbing grounds, and running tracks are made thick to increase accessibility and usability to the main exercise space. Lastly, using vivid colored outer fences and landscape lighting allow people to enjoy the feast of light and color not only during the daytime but also at night. As a space that contains the personality and characteristics of the MZ generation, it creates an artistic landscape through vivid light and color, and structures spaces to become a local landmark.

'MZ 모이자 GO'는 안양천이라는 거대한 자연을 접하는 시작점이자 MZ세대를 포함한 다계층의 사람들이 모이는 감각적인 문화 아지트를 만들고자 설계되었다.
첫째, 네트 전망쉼터, 데크 쉼터 등의 커뮤니티 데크 시설을 통해 고가 하부 공간을 입체적으로 연결했다. 높이에 따라 커뮤니티 활동, 휴식 활동 등 다양한 경험을 제공하며 주변 도시와의 연계와 내부 공간 속 자유로운 흐름을 계획했다. 둘째, 메인 공간인 멀티스포츠 공간뿐만 아니라 주변의 서브 공간에도 주목해 공간 활성화를 도모했다. 웨이트존, 인공암벽등반장, 러닝트랙 등의 서브존을 두텁게 조성하여 메인 운동공간으로의 접근성과 활용도를 높였다. 마지막으로 비비드한 색의 외곽 펜스와 경관조명을 활용하여 주간뿐만 아니라 야간에도 빛과 색의 향연을 즐길 수 있는 이색적인 공간을 계획하였다. MZ세대의 개성과 특징을 담은 공간으로서 선명한 빛과 색감을 통한 예술적인 경관을 연출하며, 지역의 랜드마크로 자리 잡도록 공간을 구조화하였다.

Location : 409-379, Mok-dong, Yangcheon-gu, Seoul, Korea I **Function** : Sports facility I **Total floor area** : 6,000㎡ I **Stories** : 2FL

위치 : 서울시 양천구 목동 409-379일대 I **용도** : 운동시설 I **규모** : 지상2층 I **설계팀** : ㈜에이치이에이 / 김영동, 이지선, 김라희, 전성연, 최진범, 박혜실, 노진우, 최차인, 서창모, 이성인 + 에스엘디자인㈜ / 이혜정, 최예린, 장하원 + 엠에스지디 / 이민형

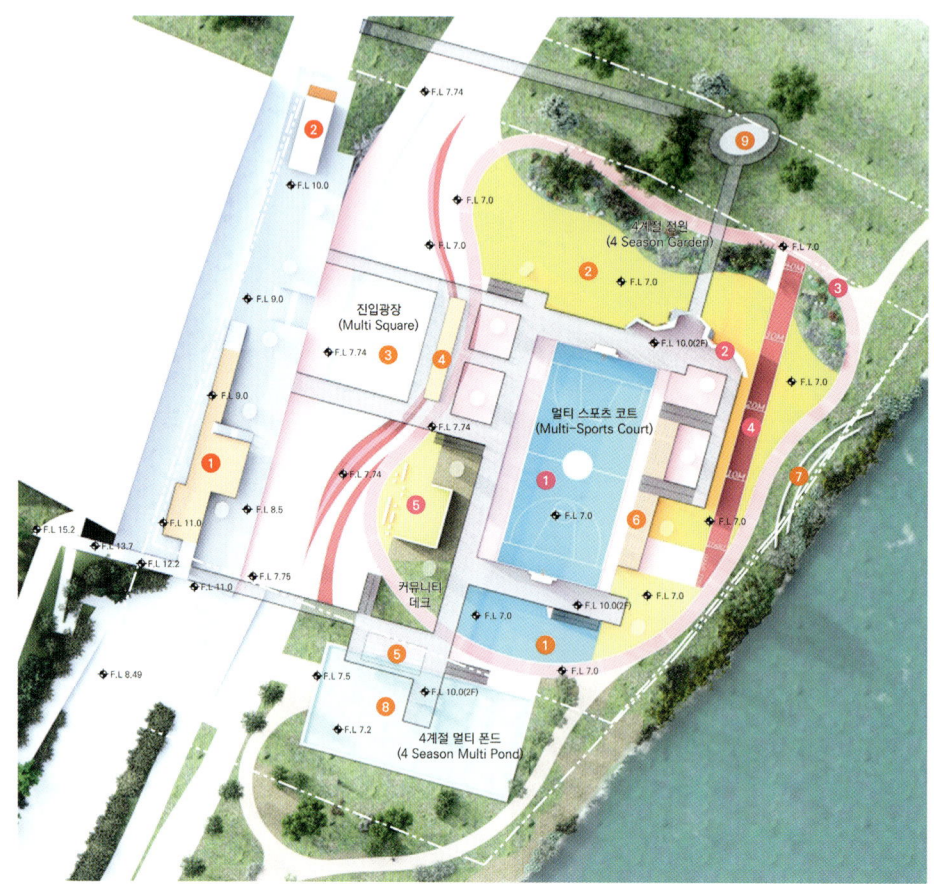

Site plan

▌Concept

- Subzone thickening

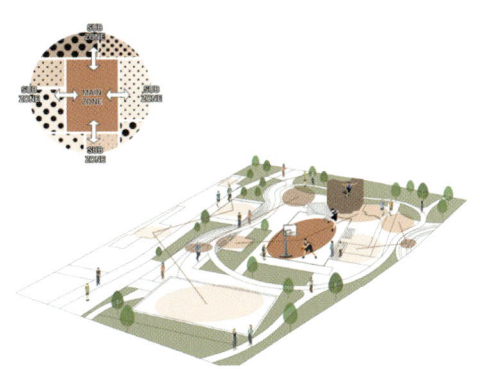

- 메인공간인 멀티스포츠 공간보다 주변의 서브공간에 주목하여 공간 활성화에 기여한다.
- 휴게기능, 대기존, 웨이팅존, 스트레칭 존, 정원공간 등에 더욱 집중함으로써 메인 운동공간의 접근성과 활용도를 높인다.

- Connect 3D

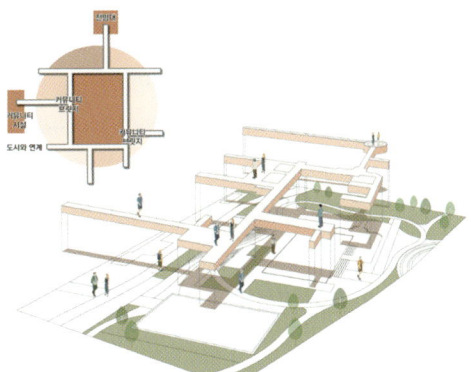

- 고가하부 공간을 입체적으로 연결하여 주변도시와의 연계와, 커뮤니티 시설, 메인공간과의 연결을 자유롭게 한다.
- 2개의 층위로 구성된 공간에서 일어나는 다양한 활동들을 기대한다.

- Light & Color

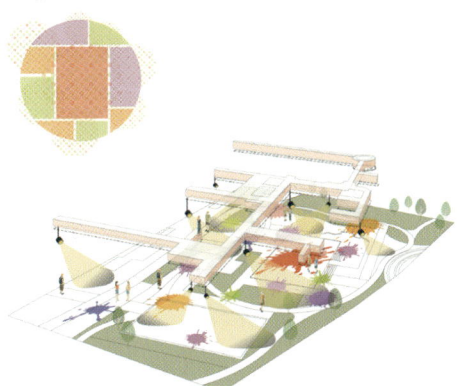

- 빛이 만들어내는 경관은 인상적인 경관을 연출하며, 공간을 구조화한다.
- 아름다운 빛깔의 향연을 즐기는 공간으로 야간경관 및 주간에도 빛과 색의 향연을 느낄 수 있도록 계획했다.
- MZ세대의 개성을 살린 이색적인 공간을 연출했다.

Winner _ 당선작

▌Concept

- 3D circulation plan

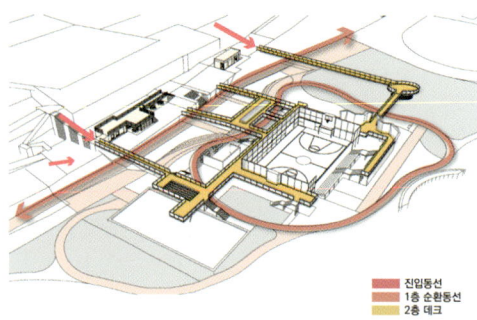

- 안양천 이용자를 위한 편리한 외부 유입 동선과 커뮤니티 데크를 통해 1층과 2층으로 분리한 동선계획

- Connection of space

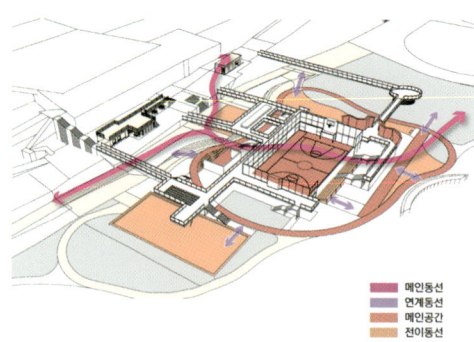

- 순환형의 메인 동선과 곳곳의 연계동선을 통해 다양한 프로그램이 발생 할 수 있는 공간계획

- Strengthening the connection between green areas and rivers

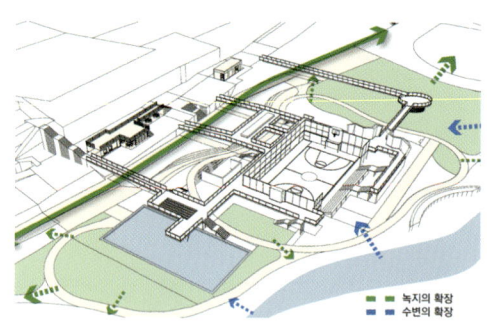

- 국회대로 공원화 계획을 따라 대상지를 가로지르는 녹지축과 안양천의 물길을 대상지내로 확장

▌Zoning

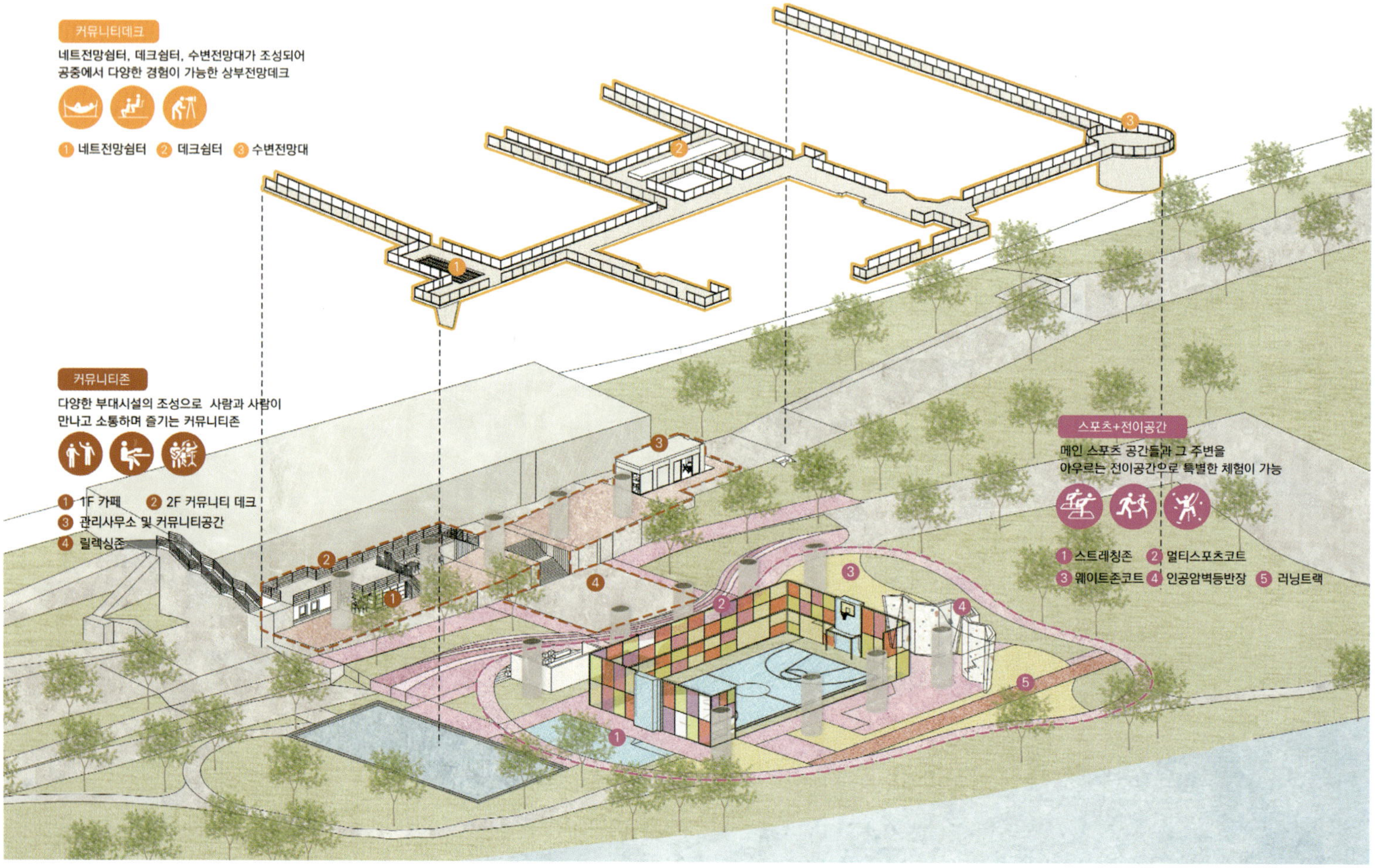

커뮤니티데크
네트전망쉼터, 데크쉼터, 수변전망대가 조성되어 공중에서 다양한 경험이 가능한 상부전망데크
1. 네트전망쉼터 2. 데크쉼터 3. 수변전망대

커뮤니티존
다양한 부대시설의 조성으로 사람과 사람이 만나고 소통하며 즐기는 커뮤니티존
1. 1F 카페 2. 2F 커뮤니티 데크 3. 관리사무소 및 커뮤니티공간 4. 릴렉싱존

스포츠+전이공간
메인 스포츠 공간들과 그 주변을 아우르는 전이공간으로 특별한 체험이 가능
1. 스트레칭존 2. 멀티스포츠코트 3. 웨이트존코트 4. 인공암벽등반장 5. 러닝트랙

MZ Sports Plaza >>

■ Spatial Detail

교각하부 메인공간

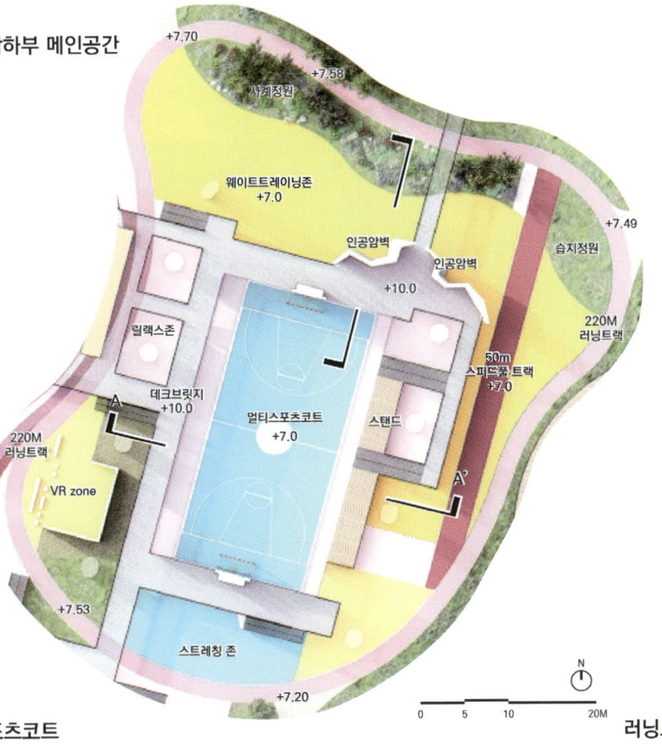

- 사계정원 +7.70 / +7.58
- 웨이트트레이닝존 +7.0
- 인공암벽
- 습지정원 +7.49
- 릴랙스존
- 인공암벽 +10.0
- 220M 러닝트랙
- 데크브릿지 +10.0
- 멀티스포츠코트 +7.0
- 50m 스피드풀트랙 +7.0
- 220M 러닝트랙
- VR zone
- 스탠드
- +7.53
- 스트레칭 존
- +7.20

멀티코트

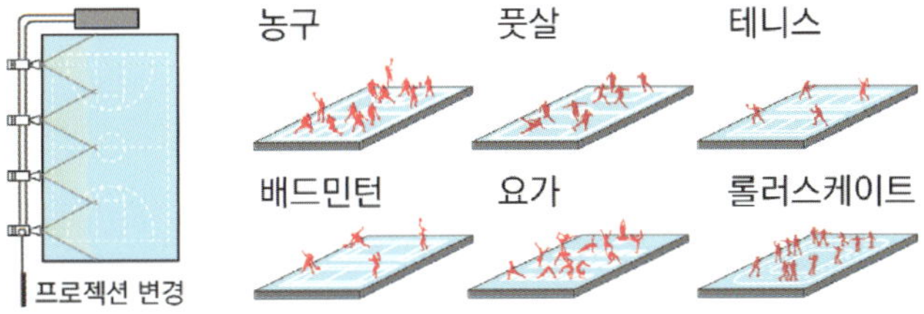

프로젝션 변경

농구 / 풋살 / 테니스 / 배드민턴 / 요가 / 롤러스케이트

zonning — 웨이트트레이닝존, 러닝트랙, 릴랙스존, 멀티 스포츠 코트, VR 존, 스트레칭존

bridge — 웨이트트레이닝존/릴랙스존 연결, VR존 연결, 멀티스포츠코트 연결, 커뮤니티시설 연결, 스트레칭 연결

track — 순환형 러닝트랙 (Running Track), 러닝트랙 (Fullspeed Track)

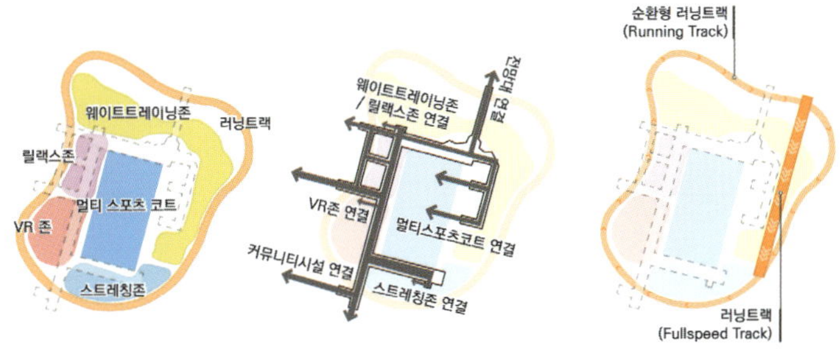

멀티스포츠코트 — 목통교, 데크 브릿지, 멀티 스포츠 코트, 스탠드, 러닝 트랙 (Fullspeed Track)

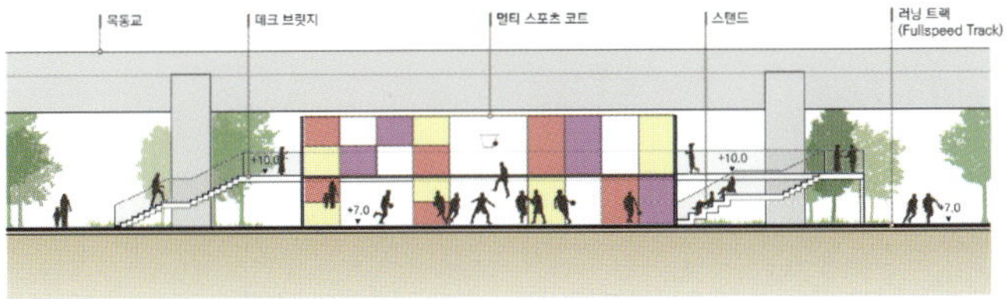

러닝트랙 — 데크 브릿지, 순환형 러닝트랙 (Running Track), 멀티 스포츠 코트, 스탠드, 러닝트랙 (Fullspeed Track), 목통교

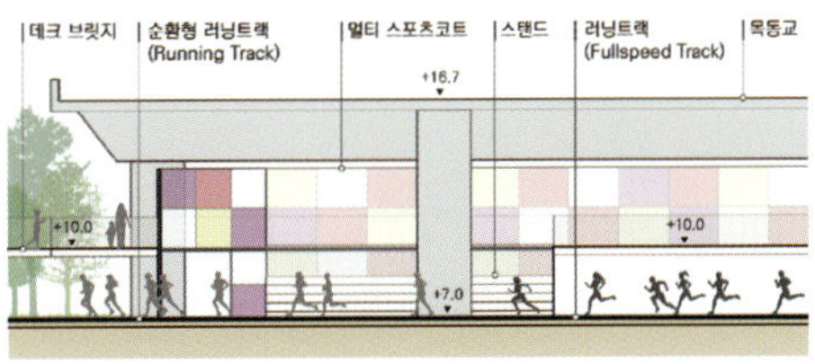

인공암벽등반 — 순환형 트랙 (Running Track), 사계기든, 웨이트존, 인공암벽, 데크 브릿지, 목통교

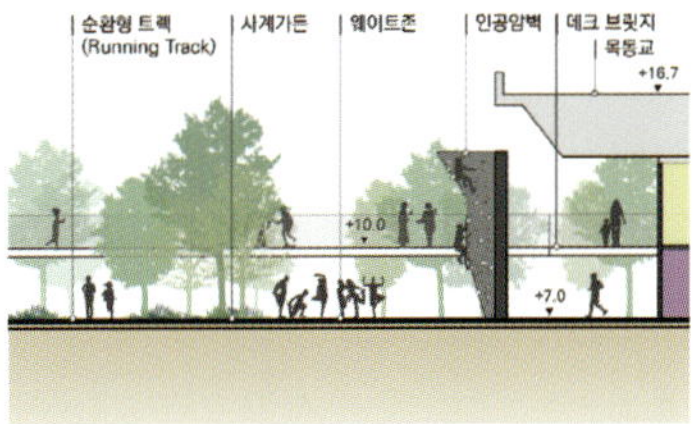

Winner _ 당선작

■ Additional Planning

- Gallery Pond

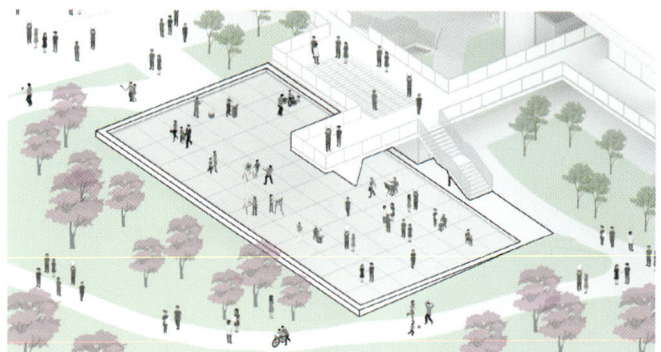

- 작품을 전시하고 이용자들에게 휴식과 예술을 쉽게 접할 수 있는 야외 전시공간 조성

- Water Pond

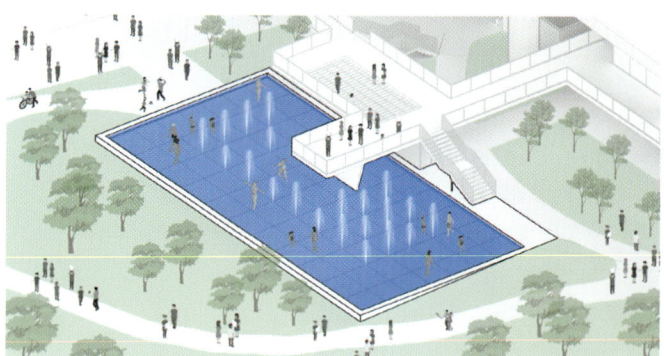

- 여름철 가족단위의 피서객들이 쉽게 이용할 수 있는 물놀이터 조성

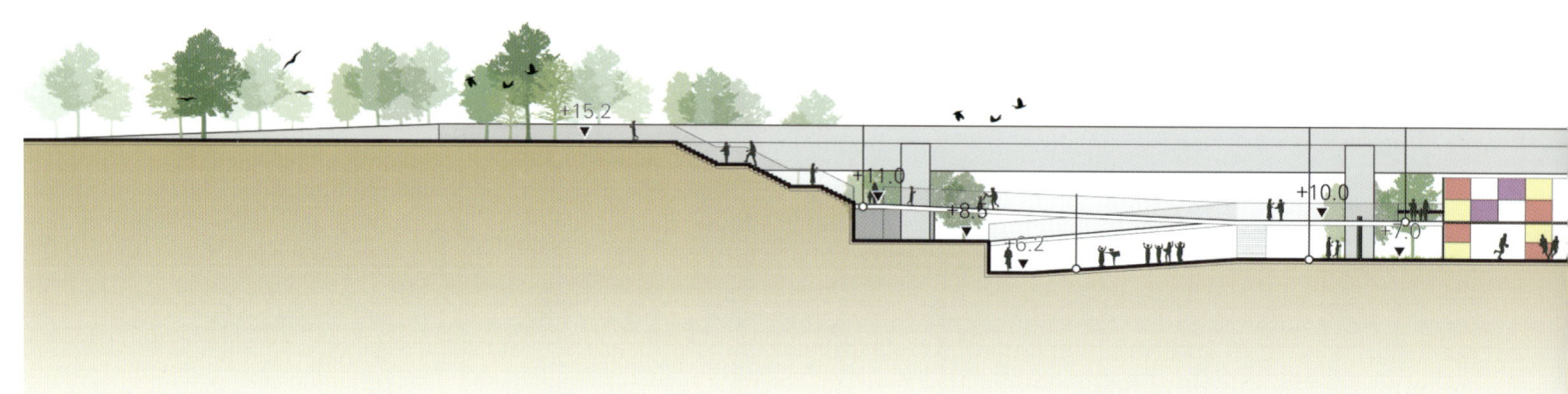

Section

■ Additional Planning

- Music Pond

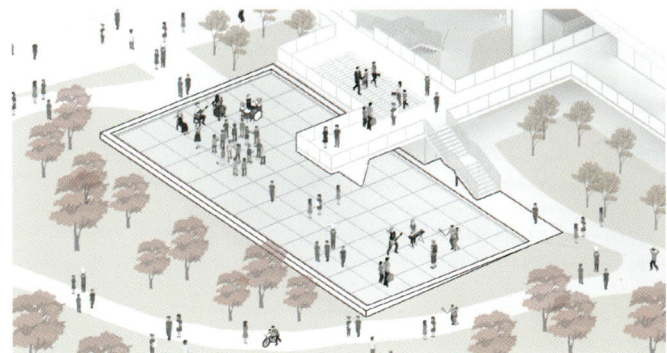

- 음악을 활용하여 이용객들의 흥미를 유발하는 야외 문화 공간 조성

- Snow Pond

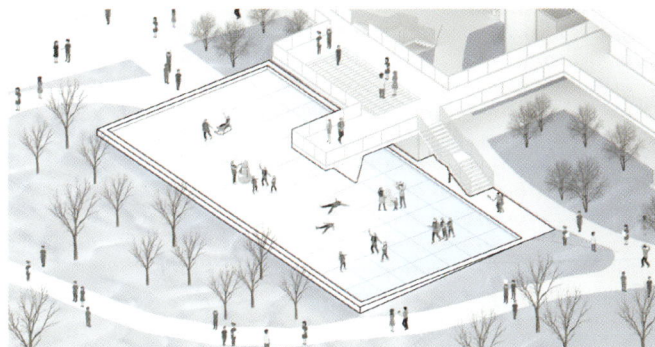

- 겨울에 얼어있는 빙판을 활용하여 다양한 활동을 할 수 있는 놀이공간 조성

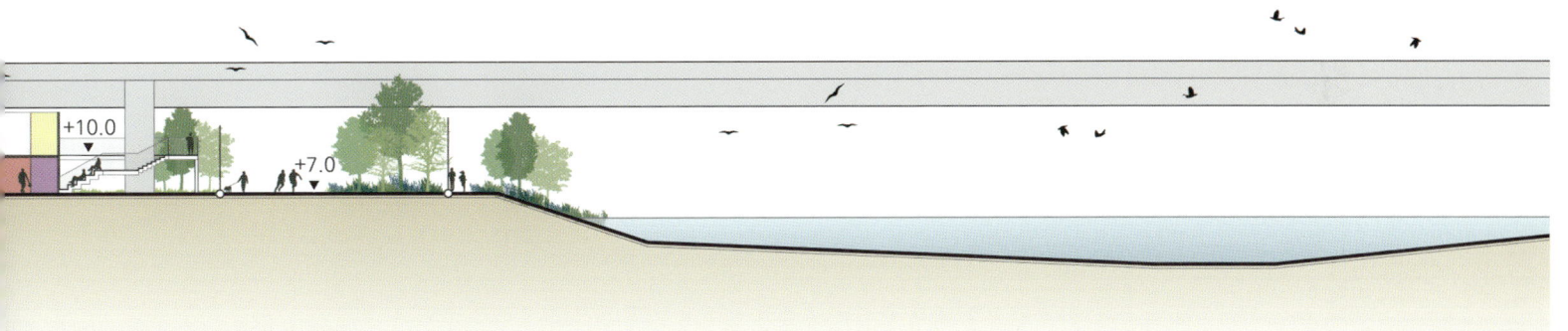

Winner _ 당선작

Sports Hall in Zatec
자테츠 스포츠홀

MACH + IDEA
MACH + IDEA

From the organiser
The town of Zatec announces an architectural competition for a new sports hall, which will be used mainly for ball games. The aim of the competition is to select the most suitable architectural solution designed with regard to the overall economy of the building. The new sports hall must be able to host competitive matches and training processes. Full use during school hours and in the evenings by both the public and local sports clubs is envisaged.

From author
We imagine the new Sports Hall of Zatec as a sustainable landmark settled inside a green park. The project is informed by old rural architecture by wrapping a complex sport program under one single pitched roof at this point of the city of Zatec. The building tries to bring domestic atmospheres in architecture to a large scale project by defining small and medium interstitial spaces between the outdoor park and the main training indoor area. By this means a unique and sensitive wooden roof underlines and confidently organizes the indoor spaces and programme.

주최자의 의견
자테츠 시는 주로 구기종목을 위한 새로운 스포츠홀의 설계공모를 발표했다. 설계공모의 목적은 건물의 전반적인 경제와 관련하여 설계된 가장 적합한 건축 솔루션을 선택하는 것이다. 새로운 스포츠홀은 경쟁 경기와 훈련 과정을 주최할 수 있어야 한다. 수업 시간과 저녁 시간에 공공 및 지역 스포츠 클럽 모두가 충분히 사용할 수 있다.

저자의 의견
새로운 자테츠 스포츠홀을 녹지 공원 안에 놓인 지속 가능한 랜드마크로 구상한다. 프로젝트는 자테츠 시의 이 지점에서 하나의 경사 지붕 아래에 복합 스포츠 프로그램을 감싸 옛날 시골 건물의 느낌을 자아낸다. 야외 공원과 주요 훈련 실내 영역 사이에 중소 규모의 사이 공간을 정의하여 건축의 가정적 분위기를 대규모 프로젝트에 담아내고자 한다. 따라서 독특하고 감성적인 목재 지붕은 실내 공간과 프로그램을 강조하고 대담하게 구성한다.

Location : Zatec, Czech Republic | **Function** : Sports facility | **Total floor area** : 6,700m² | **Structure** : Wood | **Client** : Zatec City Council | **Design team** : MACH / Laia Gelonch Llongarriu, Marc Subirana Ribera + IDEA / Carlos Maristany Ortiz, Luis Bellera Fernández de la Cruz, Eduardo Palao Valverde

위치 : 체코 자테츠 | **용도** : 운동시설 | **구조** : 목구조

Sports Hall in Zatec >>

Master Plan

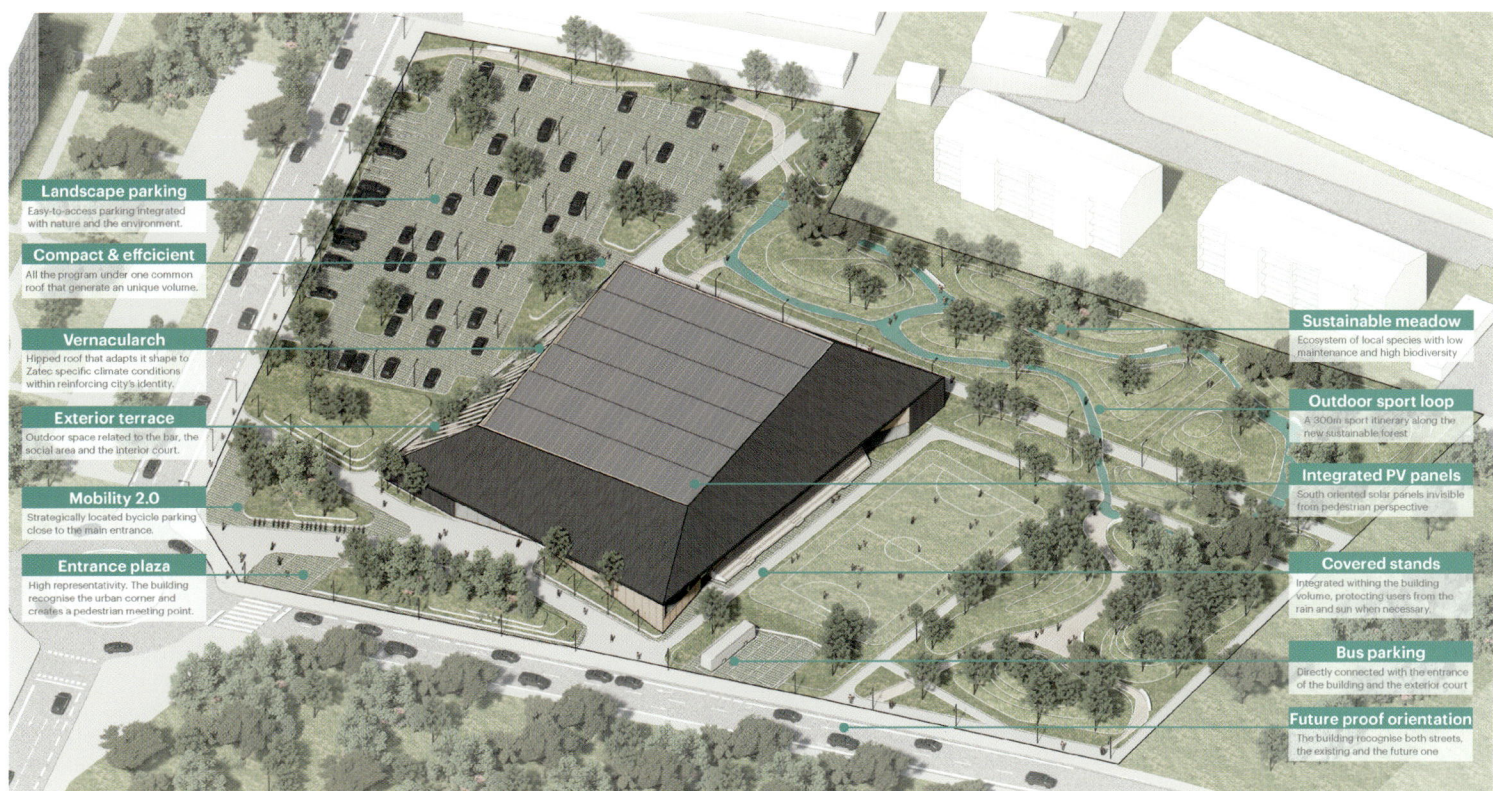

Design Concept

- Programme

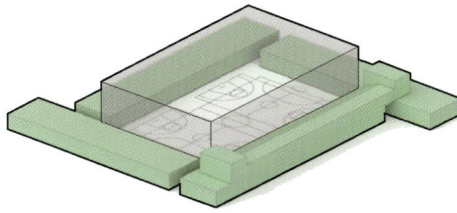

– We organize the main programme areas the central playing area, allowing each side of the hall to respond and resolve to the demands of the sports hall.

- Adaptive design

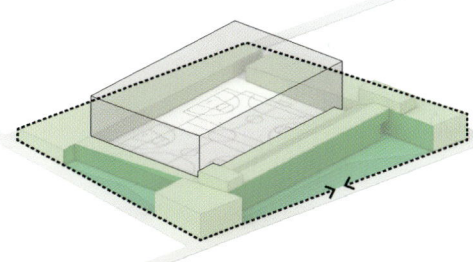

– We adapt the design in a clear and uniform geometry in order to pack all the programme in one single volume. This decision creates new opportunity spaces which function as a threshold.

- Under one roof

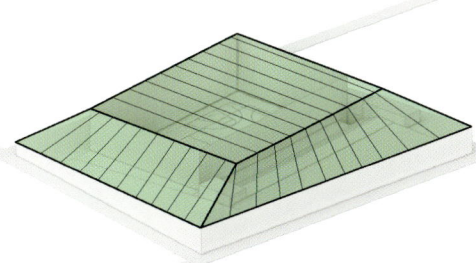

– An unique and sensitive wooden roof underlines and confidently wraps all the building in order to set the building in the context and to ease off the scale of the playing area box.

- New opportunities

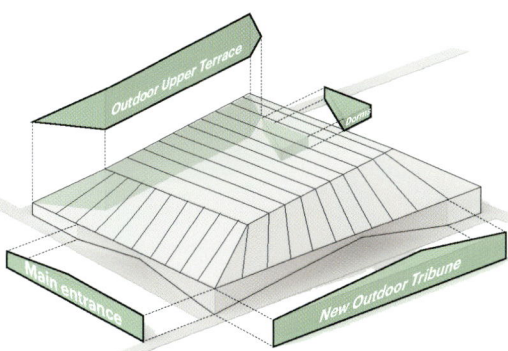

– The different entrances to the building are generated as cutouts in the overall geometry offering new opportunities such as shelter and welcoming spaces, including a new outdoor tribune on the east and an upper level exterior terrace on the west.

Winner _ 당선작

■ Axonometric

■ Detail

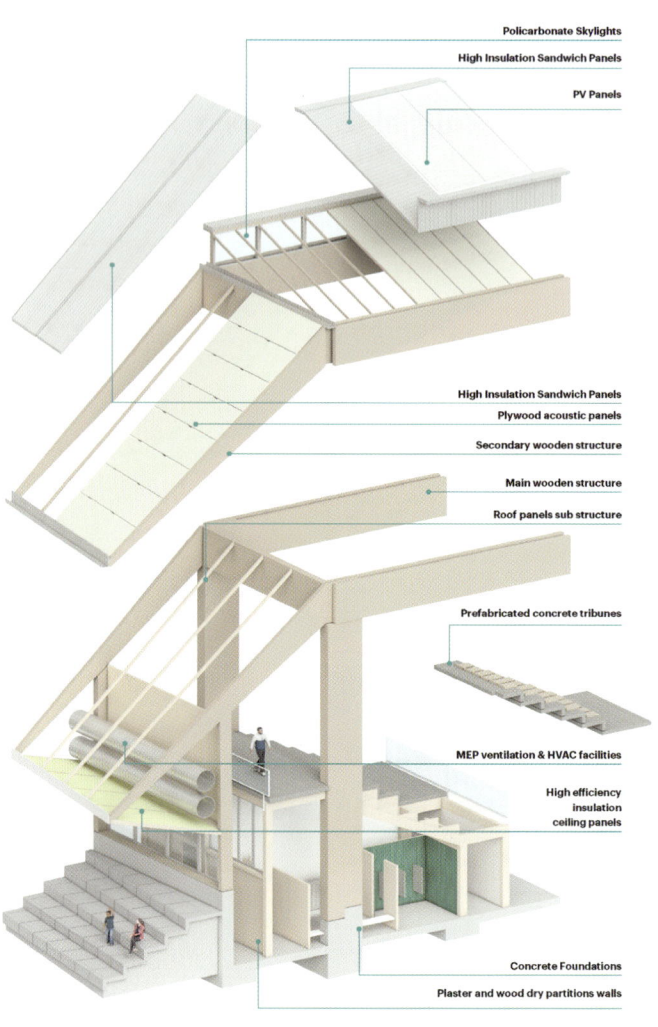

Sports Hall in Zatec >>

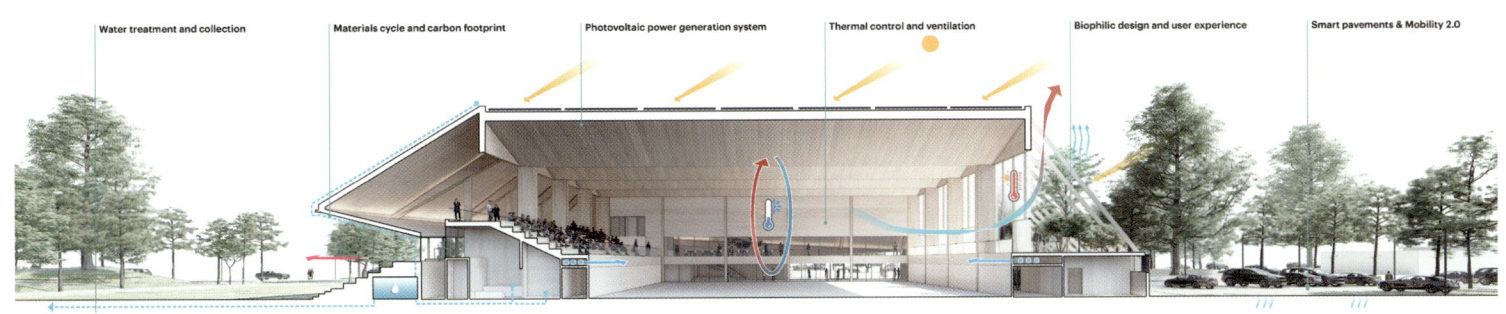

Section I

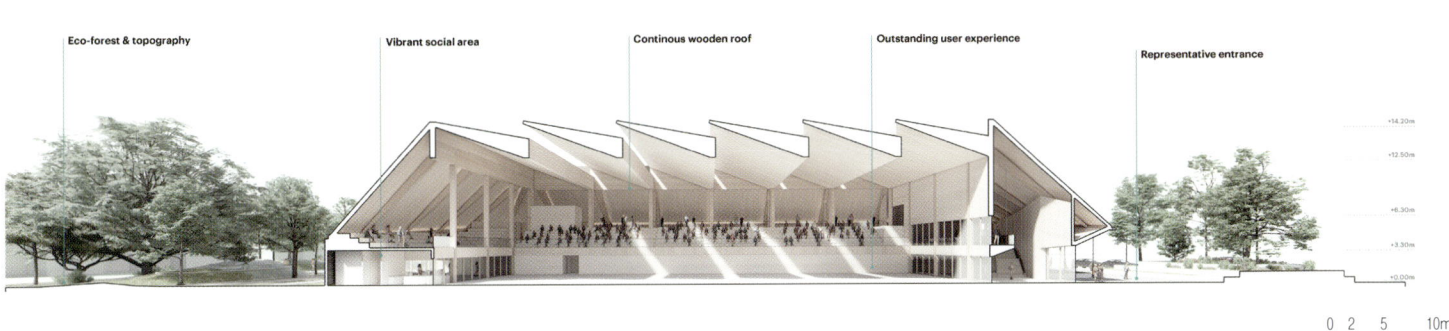

Section II

Winner _ 당선작

Sports Hall in Zatec >>

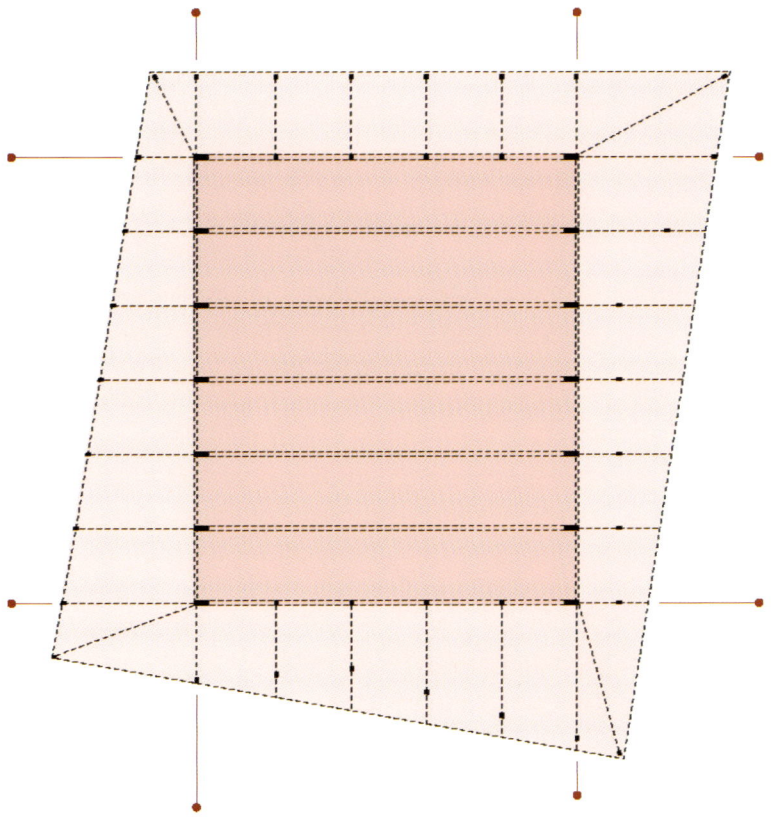

Structure plan

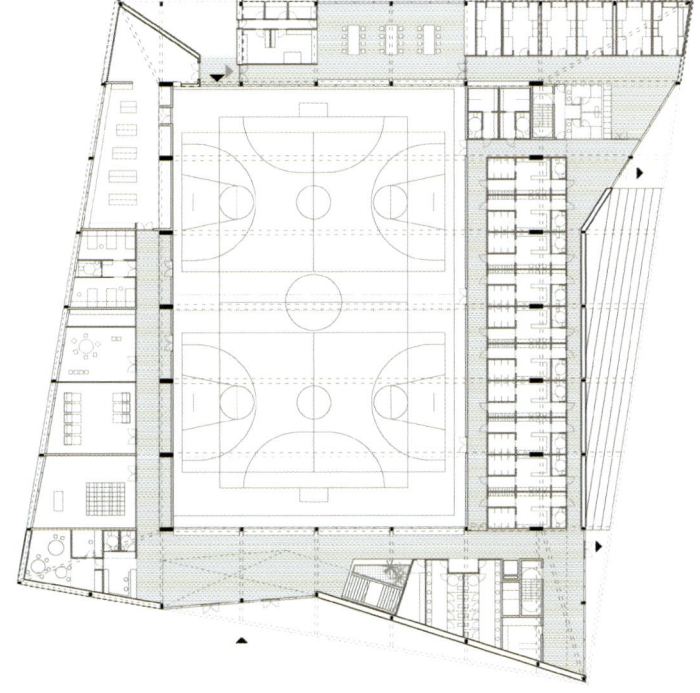

Ground floor plan

Winner _ 당선작

Barclay Tower
바클레이 타워

ACDF Architecture / Maxime-Alexis Frappier, Joan Renaud, Veronica Lalli, Laurent Bélisle, Pierre-Olivier Jacques
ACDF 아키텍처 / 막심 알렉시스 프레피에, 조안 르노, 베로니카 랄리, 로랑 벨리슬, 피에르 올리비에 자크

Located in the West End neighbourhood of Vancouver, the proposed Barclay development by ACDF Architecture is 48 storeys tall (464 feet) and is composed of 296 market housing and 81 social housing units. The tower is unique in its sophistication, with its soft curves and pale exterior that create a sense of peace and generosity in the city's dense urban fabric.

The design contrasts its surroundings while harmoniously complementing them, and is graceful in combining two different housing offerings without differentiating them or compromising quality for the overall design.

Drawing inspiration from the lines of nature, such as the towering trees of Vancouver's Stanley Park, Barclay provides a sense of familiarity with its organic yet highly mathematical form representative of the geometry in nature. Its innovative façade system has a structural concrete balcony system evocative of tulips and ginkgo leaves, as well as R-15 walls and triple glazing to meet the ambitious sustainability target of passive house. The tiered, scalloped balcony formation allows for optimal daylight penetration across the units, and also minimizes thermal bridging, a factor that is critical to treat in cold climates like British Columbia.

Barclay is modelled on a 'tower in a park' typology: a slender building surrounded by a landscaped base. A forest-like perimeter is composed of layered public and semi-private belts, planted with indigenous plants, while the corners of the site are open and connect to a larger network of public plazas. This integration of high-quality, landscaped public spaces and architectural promenades aims to promote positive livability.

At the centre of the site is the tower. Its southern facade is sheared upward, allowing the tower to follow the path of the sun, and an angled roof minimizes shadows on the neighbouring buildings and commercial strip. The east and west façades are detailed with its curvilinear balconies, while also cascade in a staggered pattern that follows the sun's angle.

The project seeks a timeless design, distinguishing itself in its elegance and contribution to a truly pleasant experience – for those passing by, visiting its landscaped and open plaza, or simply returning home.

Location : Vancouver, Canada I **Function** : Mixed-use facility I **Site area** : 1,608㎡ I **Bldg. area** : 972㎡ I **Total floor area** : 32,502㎡ I **Stories** : B10, 48FL I **Structure** : Reinforced concrete I **Finish** : Prefabricated structural concrete modules, Perforated ductile concrete shells, Triple glazed glass, Granite I **Client** : Grand Long Holdings – Pacific Norther Developments

위치 : 캐나다 밴쿠버 I 용도 : 복합시설 I 규모 : 지하10층, 지상48층 I 구조 : 철근콘크리트 I 마감 : 조립식 구조용 콘크리트 모듈, 천공 연성 콘크리트 쉘, 삼중 유리, 화강암

BARCLAY STREET TOWER

1063, 1069, AND 1075 BARCLAY STREET / WEST END, VANCOUVER, CANADA
48-STOREY RESIDENTIAL TOWER / 296 MARKET HOUSING UNITS + 81 SOCIAL HOUSING UNITS
SUSTAINABILITY TARGET OF PASSIVE HOUSE STANDARD

SENSE OF BELONGING

The objective of the project is to create a strong sense of belonging for the building occupants and the public alike that connects to the natural context of the city, while responding to current needs. The intriguing form of the tower creates a sense of belonging and identity within the skyline and streetscape of the city.

5 ARCHITECTURAL FACTORS CONTRIBUTING TO A SENSE OF BELONGING

1 LIVABILITY
TO CREATE A PROJECT THAT FOCUSES ON LIVABILITY IN ALL ASPECTS IN TERMS OF FUNCTIONALITY, COMFORT, QUALITY AND DURABILITY OF MATERIALS, THE GENEROSITY OF OUTDOOR SPACES DEDICATED TO THE SOCIAL HOUSING COMPONENT, THE MARKET HOUSING COMPONENT, AND THE PUBLIC ALIKE.

2 UNIQUE ARCHITECTURE
TO CREATE A UNIQUE ARCHITECTURE THAT IMMEDIATELY INSPIRES EMOTION, CAPTURES ONE'S ATTENTION AND ENGAGES IN A DIALOG WITH THE OBSERVER, FROM AFAR, AS AN ICON WITHIN THE SKYLINE AND FROM NEAR, AS A CONTRIBUTOR TO THE PEDESTRIAN EXPERIENCE.

3 QUALITY OF PUBLIC SPACES
TO PAY PARTICULAR ATTENTION TO THE QUALITY OF THE PUBLIC SPACES IN ORDER TO CREATE A RICH AND TIGHTLY BONDED COLLECTIVE SPIRIT WITHIN THE NEIGHBOURHOOD, BY ATTRIBUTING A HIGH QUALITY PROGRAMMED DESIGN TO THE GROUND PLANE AND THE INTERFACE WITH THE PUBLIC REALM.

4 ARCHITECTURAL PROMENADE
TO SCRIPT THE ARCHITECTURAL PROMENADE FROM THE PUBLIC STREETSCAPE TO THE PRIVATE LIVING SPACES BY THE USE OF A MULTITUDE OF ARCHITECTURAL AND LANDSCAPE LAYERS WHICH ENCOURAGE THE HUMAN CONNECTION TO NATURE AND PROMOTE A NETWORK OF GREEN SPACES AND PUBLIC PLAZAS.

Vancouver character

The site is located in the West End neighbourhood of Vancouver. The area, inhabited by many families, has a strong tradition of walkable streets along rich topographies, such as tree-lined Barclay Street. The project draws inspiration from the lines of nature, which innately provide a sense of belonging and familiarity for the public. This reinterpretation of the essence of nature refers to the beauty and serenity of the organic forms that so strongly characterize the city of Vancouver, and more specifically, the West End.

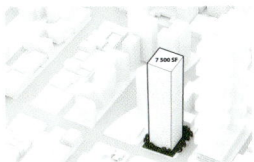
RECTANGULAR VOLUME
THE MAXIMUM BUILDABLE AREA, 43 STOREYS OF 7500 SF EACH, IS PLACED ON THE SITE, WITHIN THE REQUIRED SETBACKS, ALLOWING FOR A VEGETAL BELT TO WRAP THE TOWER.

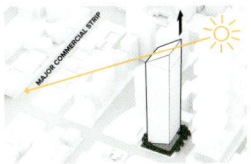
SHEARED VOLUME
THE SOUTHERN FACADE OF THE VOLUME IS SHEARED UPWARD SO THAT THE ROOF PLANE CORRESPONDS TO THE ANGLE OF THE SUN AT THE EQUINOX. THE GROUND LEVEL IS OPEN ALONG BARCLAY STREET, CREATING A GRANDIOSE LOBBY WITH THE ILLUSION OF AN OBJECT HOVERING ABOVE.

BALCONY SYSTEM
THE ORGANIC FORM OF THE PARK CLIMBS UP THE TOWERS' FACADES IN THE FORM OF CURVILINEAR BALCONIES. THIS SECONDARY SYSTEM OF BALCONIES ANCHORS ONTO THE EAST AND WEST FAÇADES, CASCADING IN A STAGGERED MANNER THAT FOLLOWS THE ANGLE OF THE SUN.

Tower form and tectonics

The balcony structural units are anchored to the tower and supported by the vertical balcony divider, minimizing thermal bridging (an important factor that contributes to meeting the environmental target of passive house in the extreme Canadian climate). The balconies take the shape as a floral tectonic – a delicate, but yet rational form that creates a three-dimensional system that grows form the forested site.

YELLOW TULIPS SYMBOLIZE CHEERFUL THOUGHTS AND SUNSHINE.

ROSES SYMBOLIZE BALANCE, PROMISE, HOPE, AND NEW BEGINNINGS.

PINK LOTUS SYMBOLIZE PURITY AND DEVOTION.

GINKGO SYMBOLIZES THE UNITY OF OPPOSITES, CHANGELESSNESS, HOPE AND LOVE.

TOWER TECTONICS
THE DELICATE, BUT YET RATIONAL, FORM OF THE GINKGO LEAF AND ITS SYMBOLISM ARE ABSTRACTED INTO A JOYOUS THREE-DIMENSIONAL SYSTEM THAT GROWS FROM THE FORESTED SITE.

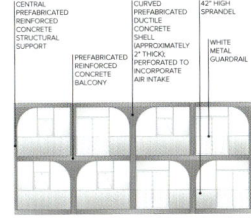
FACADE SYSTEM
THIS FACADE SCHEME ALLOWS FOR OPTIMAL DAYLIGHT PENETRATION AND RESPECTS THE MAXIMUM GLAZING REQUIREMENTS TO MEET THE SUSTAINABILITY TARGET OF PASSIVE HOUSE, INCLUDING R-15 WALLS, TRIPLE GLAZING AND MINIMAL THERMAL BRIDGING.

Fifth facade

Not only is the tower engaging at the scale of the city, but it also has a 'fifth façade' that engages pedestrians as they approach it, contributing to the experience at a human scale. The base of the tower, including the lobby and the landscaping, activates the public realm and, moreover, as one looks up at the tower from street level, the alternating balconies and curved soffits create an intriguing texture that is in dialogue with pedestrians, activating the pedestrian scale.

Public realm

Within the skyline, the tower will become an icon in the city with its delicate nature-inspired design, its angled roof that maximizes sunlight on neighbouring streets, and its rational design that follows the overall city grid and structure. The tower interacts with the public realm via games of light and shadow along the curvilinear balcony facades, in a uniquely graceful and elegant manner that is traditionally unseen in standard tower design. The balcony forms are poetically inspired by the Ginkgo leaf and speak to the unique quality of Vancouver as an urban city set in nature.

Layers

A layered conceptual treatment of the ground plane emphasizes the 'tower in the park' typology, while simultaneously activating the public realm. The semi-private vegetal belt wraps the limits of Thurlow Street and Barclay Street. It is composed of indigenous species, with access paths carved in. The corners of the site are opened and connect to a larger network of public plazas that line the site, extending toward the neighbouring public parks. Surrounded by greenery, these plazas evoke a quality similar to that of the edge of a forest, animating the streetscape and the public realm. They are designed to be inhabited by interactive objects, potentially a public art component that is utilized by and interacts with the public.

VIEW OF TOWER (THURLOW STREET)

VIEW OF MARKET HOUSING ENTRANCE

VIEW OF COMMON OUTDOOR SPACES

Winner _ 당선작

■ Sketch

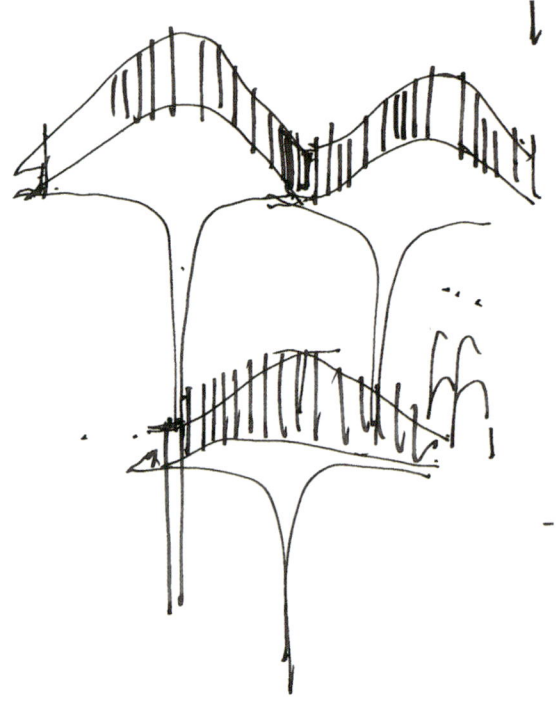

밴쿠버의 웨스트 엔드 지역에 위치한 ACDF 아키텍처가 제안한 바클레이 타워 개발은 48층(약 141미터) 높이이며 296개의 마켓 하우징과 81개의 소셜 하우징으로 구성된다. 타워는 도시의 빽빽한 도시 구조에서 평화와 관대함의 감각을 만들어내는 부드러운 곡선과 헬쑥한 외관의 정교함이 특징이다.

이 디자인은 주변 환경과 대조를 이루면서도 조화롭게 보완하고, 전체 디자인에 대한 차별화나 품질 저하 없이 두 개의 서로 다른 주택 제품을 우아하게 결합했다.

밴쿠버 스탠리 공원의 우뚝 솟은 나무와 같은 자연의 선에서 영감을 얻은 바클레이는 자연의 기하학을 대표하는 유기적이지만 고도로 수학적 형태로 친숙함을 제공한다. 혁신적인 파사드 시스템은 튤립과 은행나무 잎을 연상시키는 구조적 콘크리트 발코니 시스템뿐만 아니라 패시브 하우스의 야심찬 지속 가능성 목표를 충족시키기 위해 R-15 벽과 삼중 유리를 갖췄다. 계층화 된 가리비 모양의 발코니 형성은 유닛 전체에 최적의 일광을 허용하고 브리티시 컬럼비아와 같은 추운 기후에서 중요한 요소인 열교를 최소화한다.

바클레이는 '공원 속의 타워' 유형을 모델로 했다. 조경계획된 기지로 둘러싸인 가느다란 건물이다. 숲과 같은 주위는 토착 식물이 심어진 층으로 된 공공 및 반 사적 공간인 벨트로 구성되며 대지 모서리는 열려 있고 더 큰 공공 광장 네트워크에 연결된다. 높은 수준의 조경계획된 공공공간과 건축적 산책로의 이러한 통합은 긍정적인 거주 가능성을 촉진하는 것을 목표로 한다.

대지의 중심에는 타워가 있다. 남쪽 파사드는 위쪽으로 기울어져 타워가 태양의 경로를 따라갈 수 있도록 하고 각진 지붕은 이웃 건물과 상업 지구에 대한 그림자를 최소화한다. 동쪽과 서쪽의 파사드는 곡선형 발코니로 자세히 묘사되어 있으며 태양의 각도를 따라 엇갈린 패턴으로 계단식으로 배열되어 있다.

이 프로젝트는 세월이 흘러도 변치 않는 디자인을 추구하며, 우아함과 진정으로 즐거운 경험(지나간 사람들, 조경된 개방된 광장을 방문하거나 단순히 집으로 돌아오는 사람들)을 제공한다.

Winner _ 당선작

East elevation

West elevation

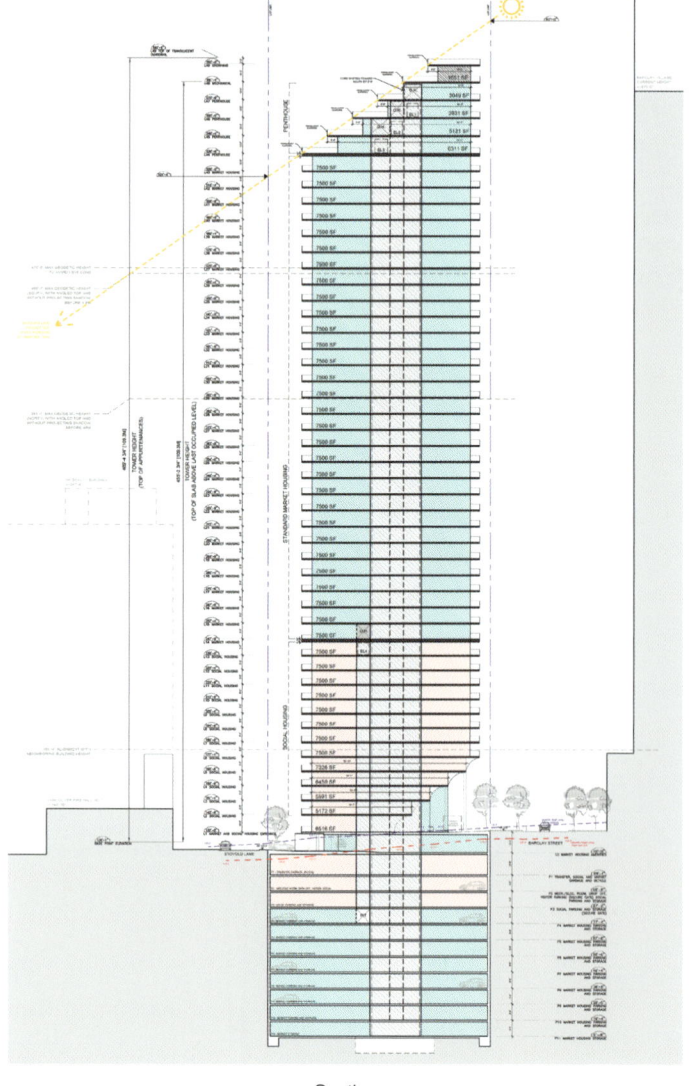

Section

Barclay Tower >>

Winner _ 당선작

Barclay Tower >>

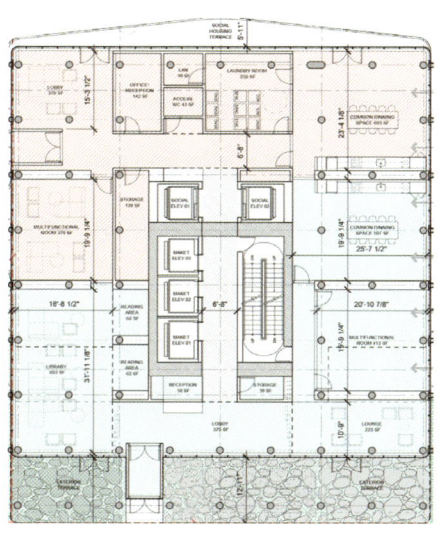

1st floor plan

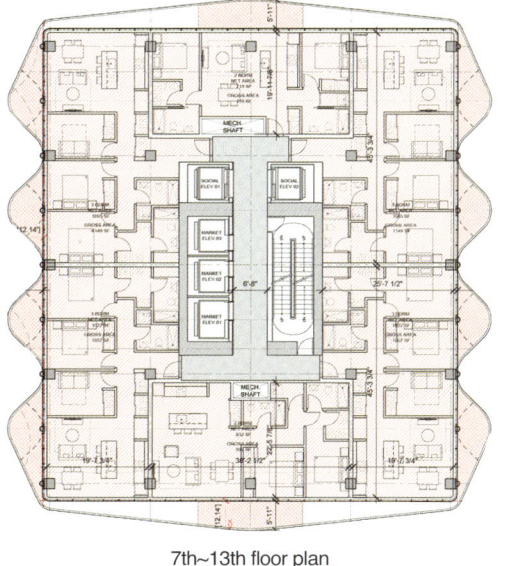

7th~13th floor plan

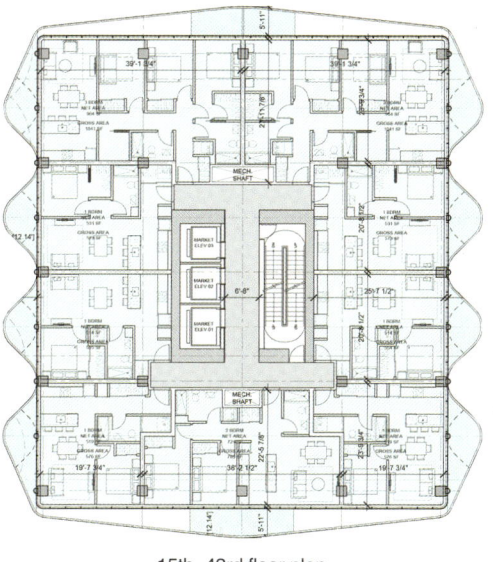

15th~43rd floor plan

247

Winner _ 당선작

Ørestad Church
외레스타드 교회

Henning Larsen Architects / Henning Larsen
헤닝 라센 아키텍츠 / 헤닝 라센

A modern monument for Denmark's capital, the Ørestad Church evokes a meeting place at a clearing in the trees. Built in wood and wood shingle, the church reflects the nature of Ørestad's open landscape, embracing the community and its surroundings – an inverted facade design creates protrusions within the deep church walls, an extroverted space for the community.

A striking sculptural roof becomes a new sustainable landmark and marks a natural meeting place for the local community.

Conjuring the sensation of standing under a canopy of trees in a forest, Henning Larsen's design for Ørestad Church features wooden roof domes through which light cascades.

The facade of the church is rough, like bark on a tree and changes character through the seasons and over time. The church is connected to its surroundings by a continuous brick floor of various tones and glazing, referencing fallen leaves, that rises to become benches, sitting niches and podiums – the path from the city and the common lead directly into the church.

덴마크 수도의 현대적인 기념물인 외레스타드 교회는 나무가 우거진 공터에서 만남의 장소를 연상시킨다. 나무와 너와판으로 지어진 교회는 외레스타드의 탁 트인 풍경을 반영하여 지역사회와 그 주변 환경을 포용한다. 뒤집힌 파사드 디자인은 지역사회를 위한 외향적인 공간인 깊은 교회 벽 안에 돌출부를 만든다.

눈에 띄는 조각 지붕은 새로운 지속 가능한 랜드마크가 되고, 지역사회를 위한 자연스러운 만남의 장소가 된다.

숲 속의 나무 캐노피 아래에 서 있는 것 같은 헤닝 라센의 외레스타드 교회 디자인은 빛이 쏟아지는 나무 지붕 돔이 돋보인다.

교회의 파사드는 나무 껍질처럼 거칠고, 계절과 시간에 따라 특성이 바뀐다. 교회는 낙엽에서 영감을 얻어 다양한 색조와 유리로 된 연속적인 벽돌 바닥으로 주변 환경과 연결되어 있으며, 벤치, 좌석, 연단으로 사용되도록 융기되어 있다. 이는 도시와 공유지에서 교회로 직결되는 통로가 된다.

Location : Copenhagen, Denmark I **Function** : Religious facility I **Total floor area** : 2,100m² I **Client** : Islands Brygges Parish

위치 : 덴마크 코펜하겐 | 용도 : 종교시설

▌ Design Concept

 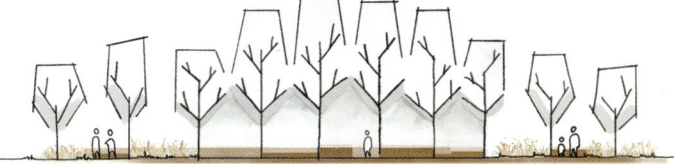

▌ Axonometric

Winner _ 당선작

South elevation

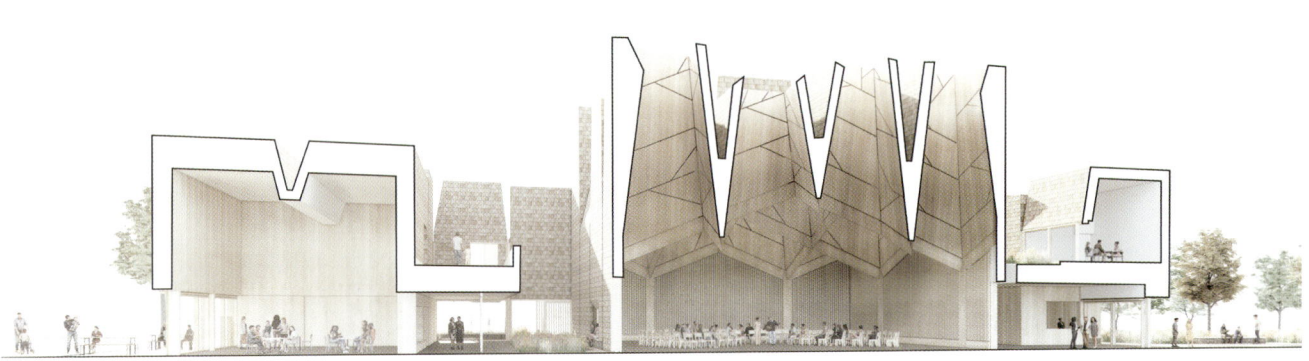
Section

Ørestad Church

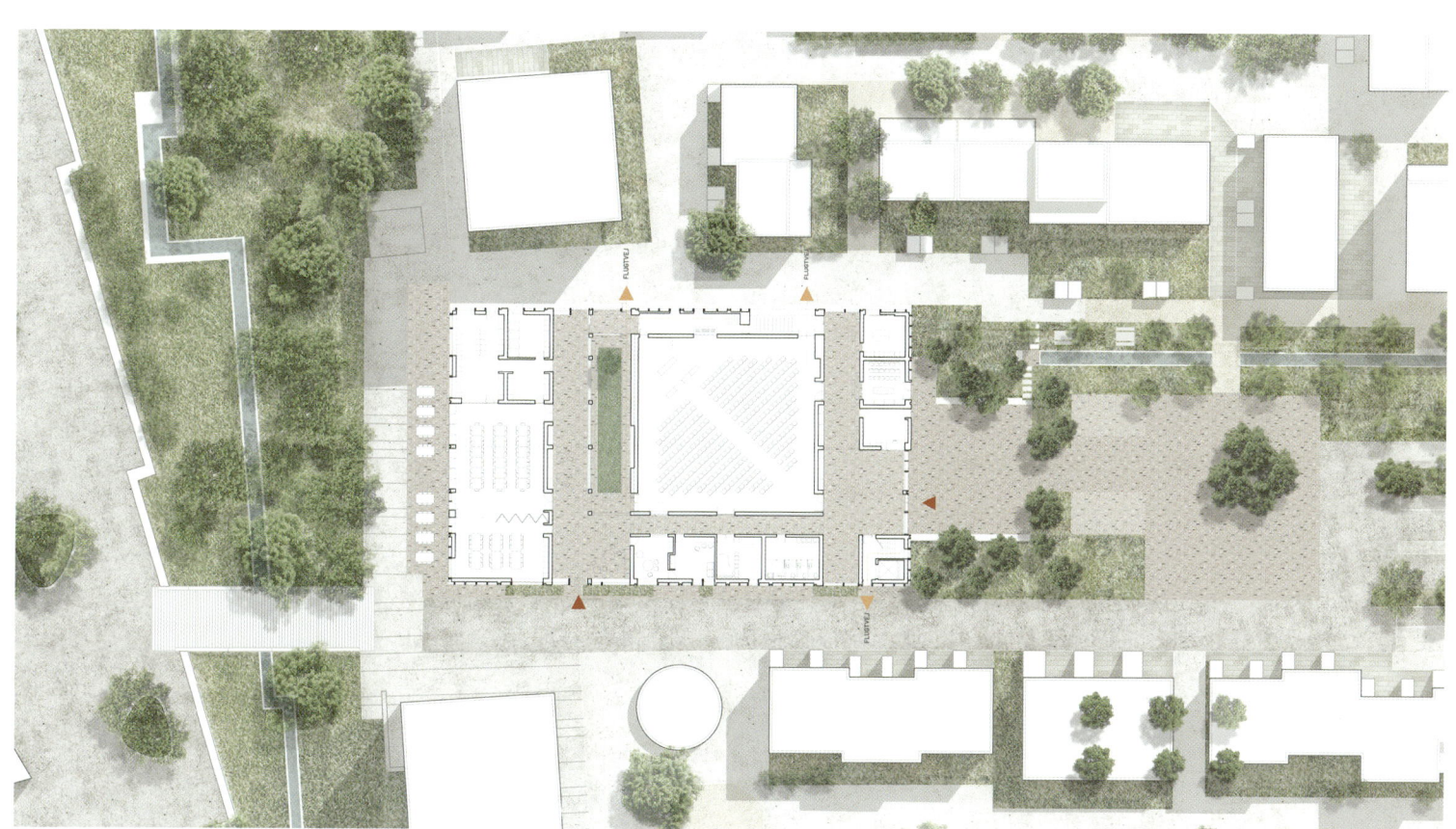

Ground floor plan

Winner _ 당선작

Fragments of Nostalgia
노스탤지어의 조각

alcolea+tárrago / Rubén Alcolea, Jorge Tárrago
알콜레아+타라고 / 루벤 알콜레아, 호르헤 타라고

Adaptive reuse of a sugar factory hall into conference and community center

The proposal 'Fragments of Nostalgia Mo2025' has been awarded the first prize in the national open and anonymous competition for the restoration and adaptive reuse of the factory 'La Azucarera' in Monzon, Spain. The project will preserve and transform the main sugar industrial hall into a conference center and will include additional co-working and shared spaces for local associations. The factory will have restored back its original morphology to accommodate a multifunctional open space. What once was the main driving force in the regional economy will be now a space for cultural activities and multigenerational interactions, reclaiming its presence in the city as a public center.

The project defines a mixed strategy, in which different interventions could be addressed separately, alternatively and over time but with a defined and clear master plan, to avoid interferences and optimize resources. The proposal will also transform the adjacent area into a public park with community gardens and a farmers' market.

The set of interventions will be partially funded as one of the strategic projects in the Urban Agenda program for the EU.

적응형 재활용을 통해 설탕 공장을 컨퍼런스 및 커뮤니티 센터로 탈바꿈

'노스탤지어 조각' 제안은 스페인 몬존에 있는 '라 아스카레로' 공장의 복원과 적응형 재활용을 위한 공개 및 익명 설계공모에서 1위를 수상했다. 이 프로젝트는 설탕산업 본당을 보존하고 컨퍼런스 센터로 탈바꿈하며, 지역협회를 위한 추가 공동 작업 및 공유 공간을 포함할 것이다. 공장은 개방형 다기능 공간을 수용하기 위해 원래의 형태를 복원할 것이다. 한때 지역 경제의 주요 원동력이었던 곳이 이제 문화 활동과 세대 간 교류의 장이 되어 공공 중심지로서 존재감을 다시 드러낼 것이다.

이 프로젝트는 간섭을 피하고 자원을 최적화하기 위해, 명확하게 정의된 기본 계획을 통해 다양한 개입을 개별적으로, 대안적으로 또는 시간이 지남에 따라 해결할 수 있는 통합 전략을 수립한다. 또한 인접 지역을 커뮤니티 정원과 농산물 시장이 있는 공공 공원으로 만드는 것을 제안한다.

일련의 개입은 EU를 위한 도시 의제 프로그램의 전략적 프로젝트 중 하나로 자금이 일부 지원될 것이다.

Location : Monzón, Spain I **Function** : Cultural, Mixed use facility I **Total floor area** : 5,000m I **Finish** : Concrete, Metal I **Client** : Ayuntamiento de Monzón

위치 : 스페인 몬존 I **용도** : 문화, 복합시설 I **마감** : 콘크리트, 금속

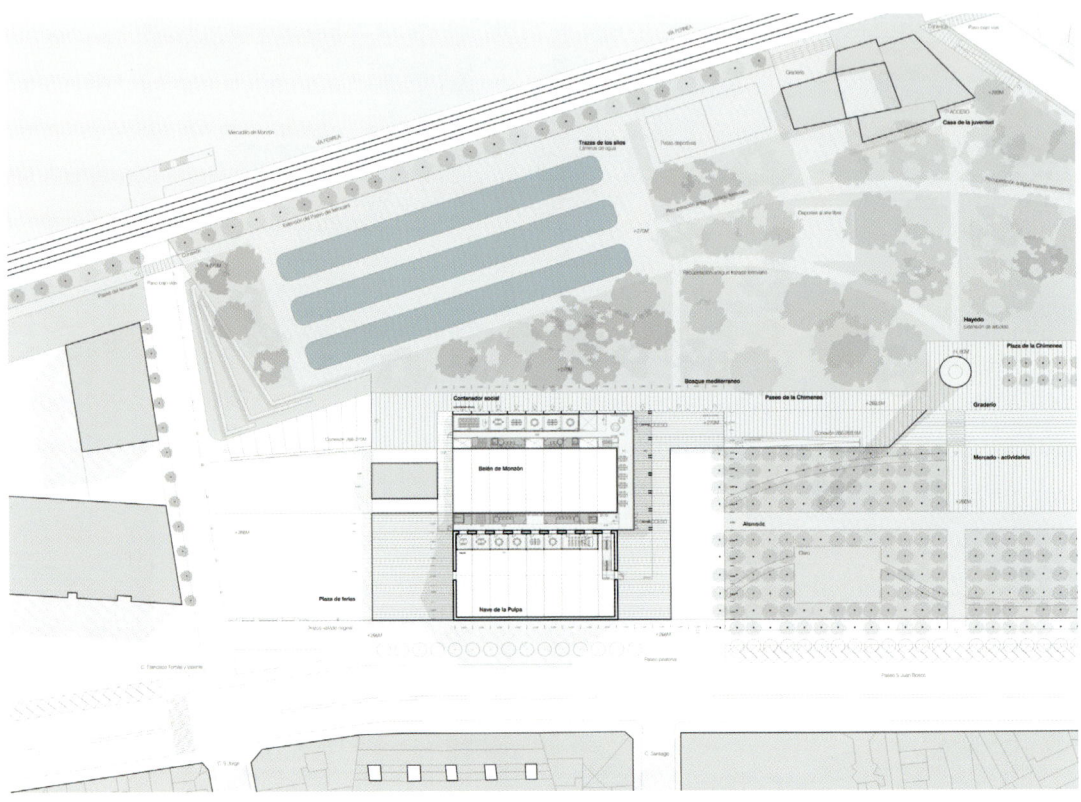

Master plan

Site Analysis

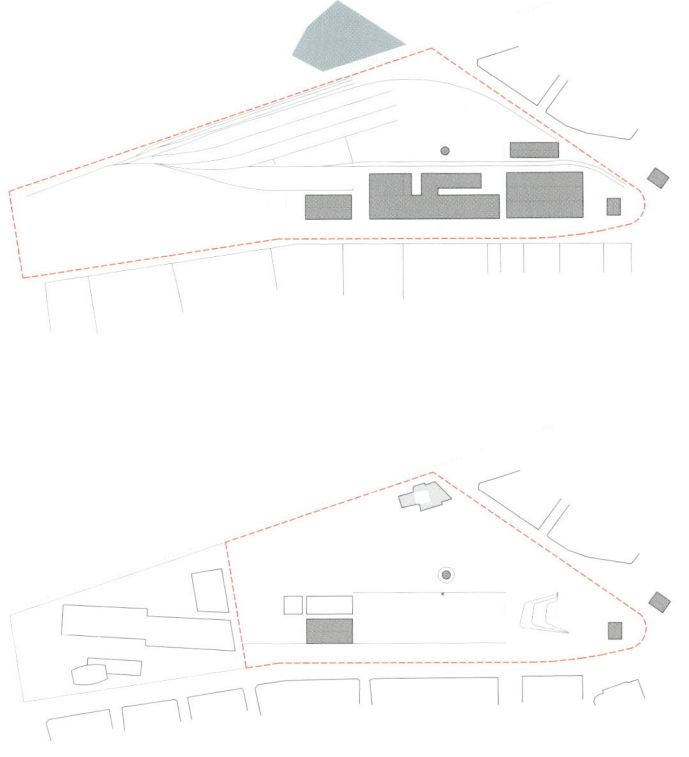
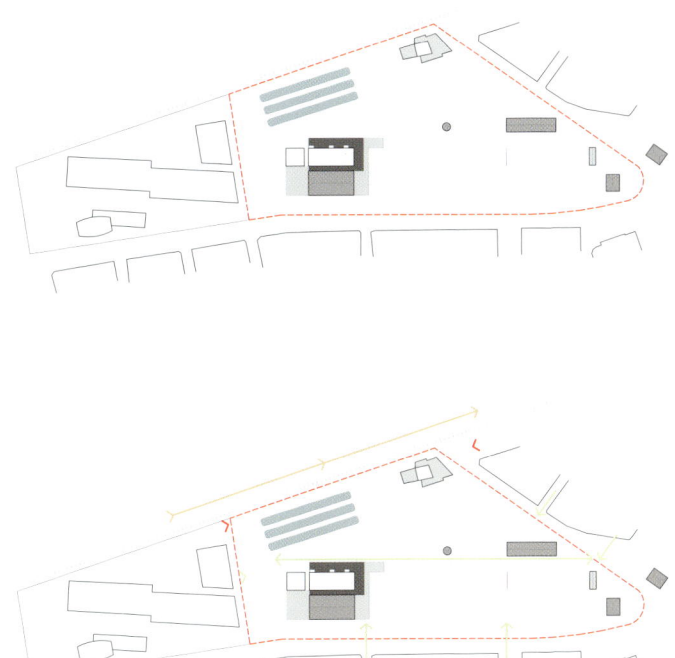

■ Model

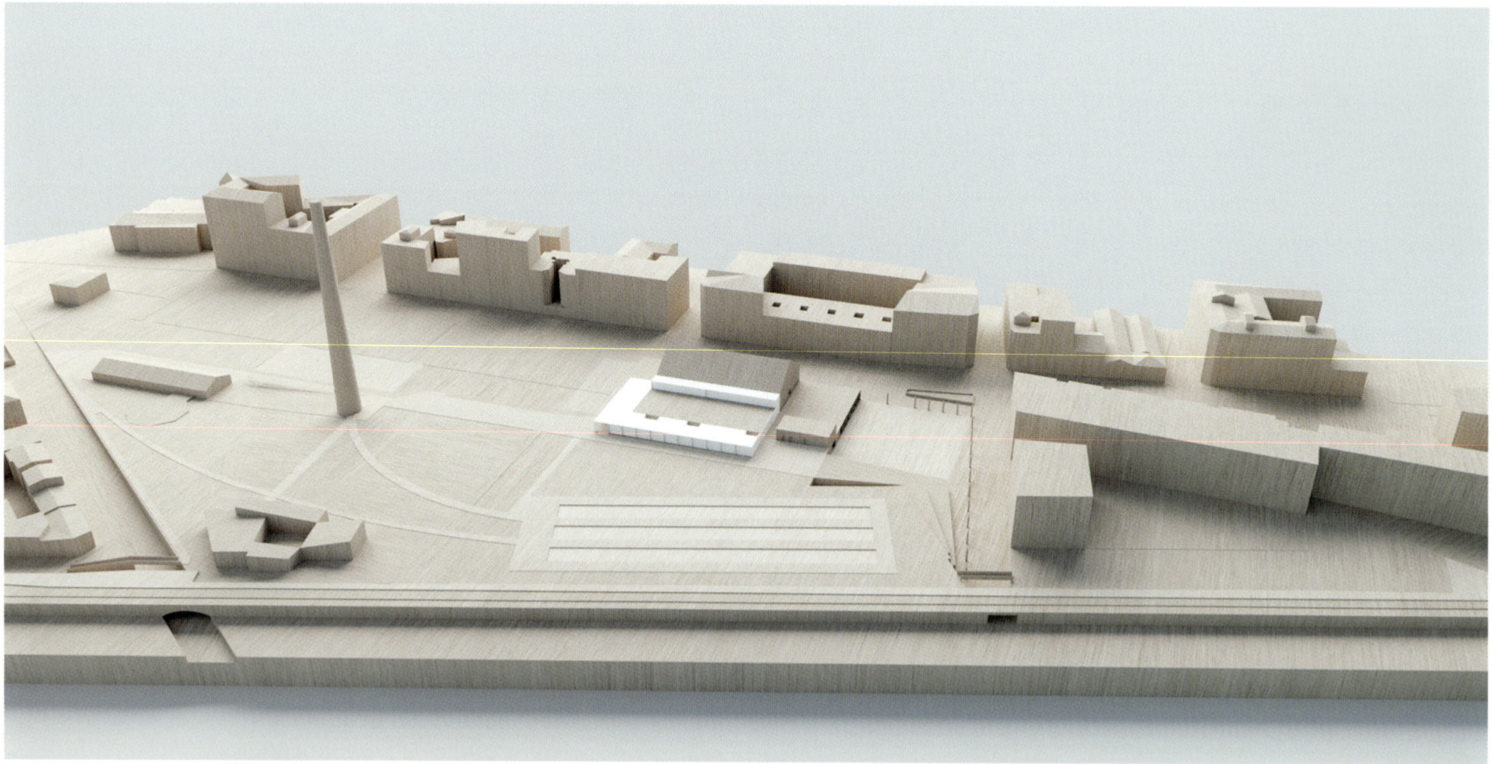

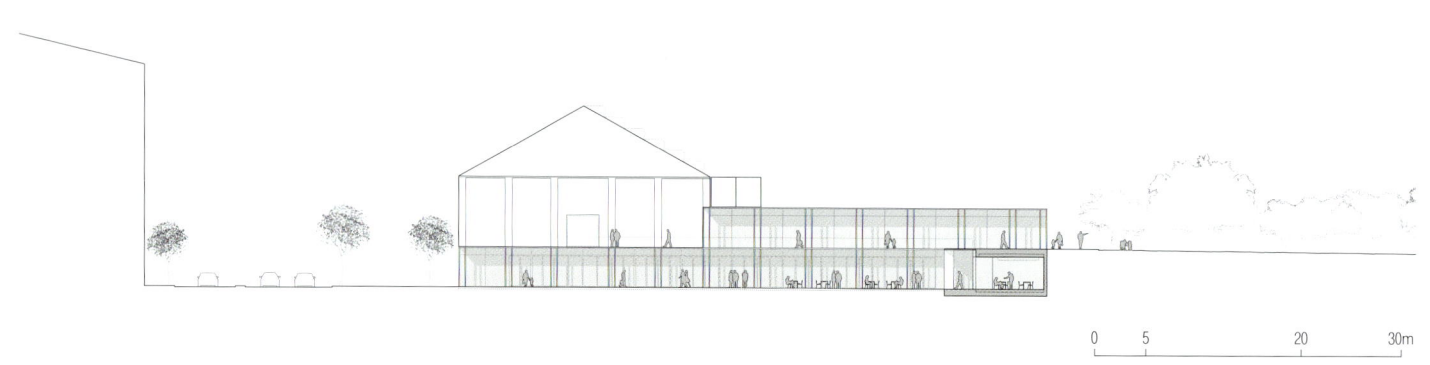

South elevation

Fragments of Nostalgia >>

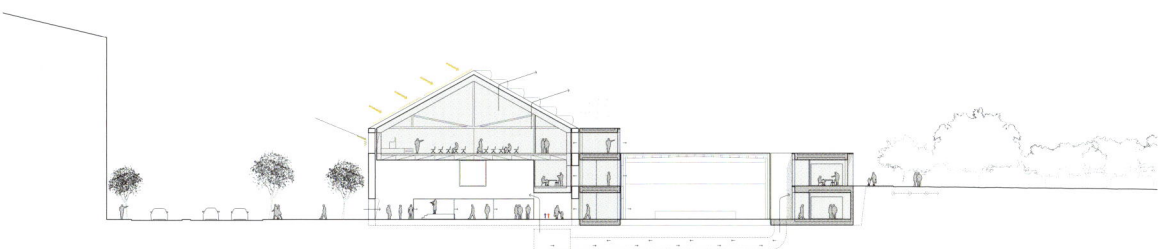

Cross section

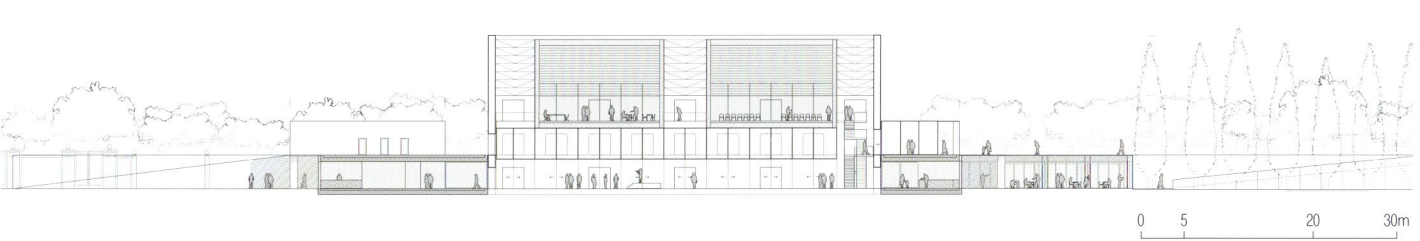

Longitudinal section

Program & Entries

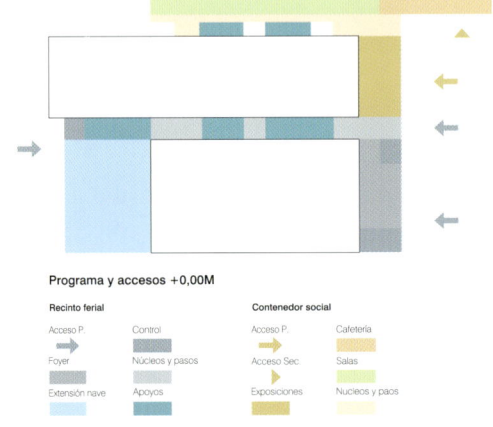

Programa y accesos +0,00M

Recinto ferial
- Acceso P.
- Foyer
- Extensión nave
- Control
- Núcleos y pasos
- Apoyos

Contenedor social
- Acceso P.
- Acceso Sec.
- Exposiciones
- Cafetería
- Salas
- Núcleos y paos

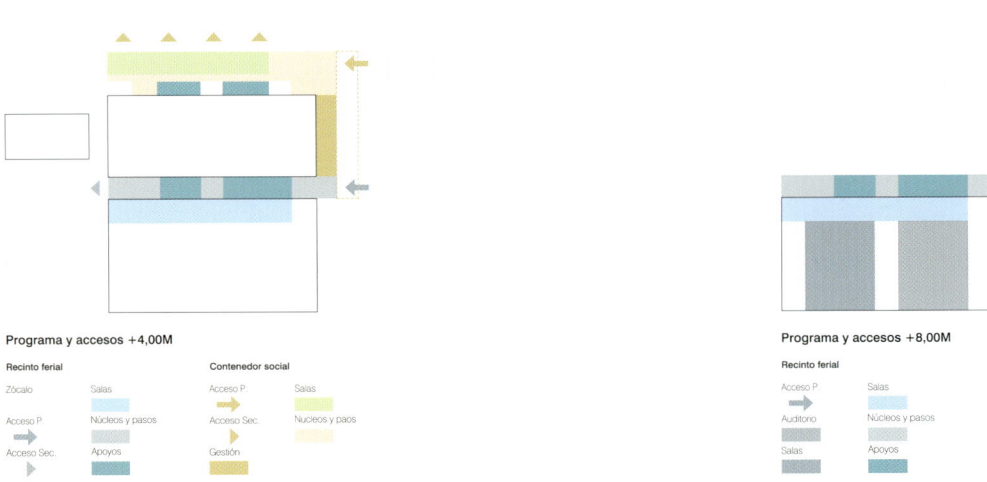

Programa y accesos +4,00M

Recinto ferial
- Zócalo
- Acceso P.
- Acceso Sec.
- Salas
- Núcleos y pasos
- Apoyos

Contenedor social
- Acceso P.
- Acceso Sec.
- Gestión
- Salas
- Núcleos y paos

Programa y accesos +8,00M

Recinto ferial
- Acceso P.
- Auditorio
- Salas
- Salas
- Núcleos y pasos
- Apoyos

Fragments of Nostalgia >>

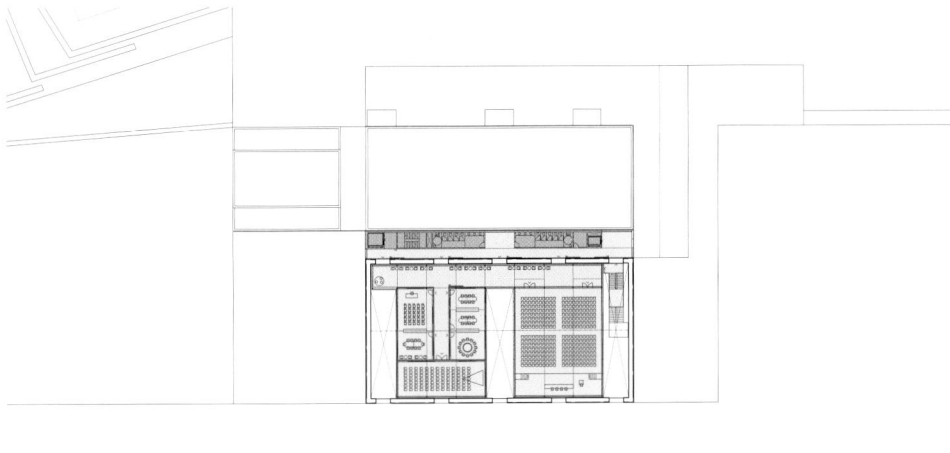

Upper floor plan

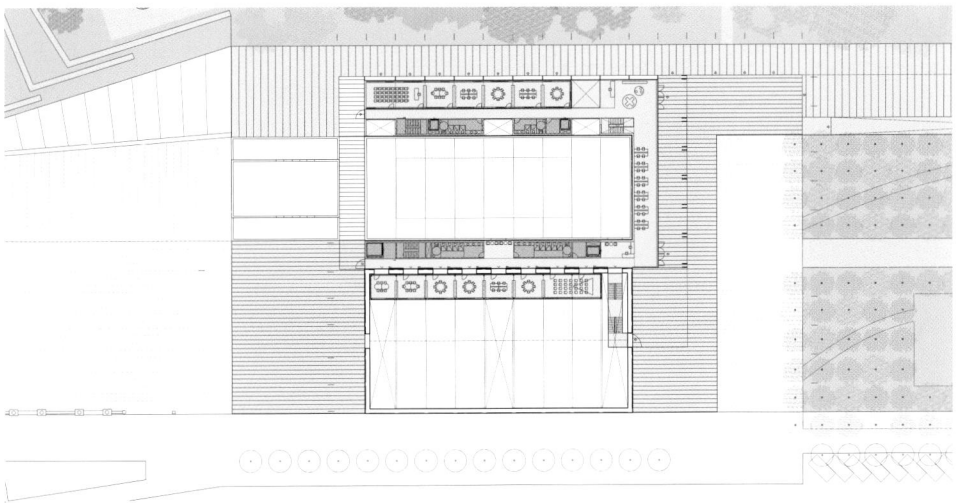

Park floor plan

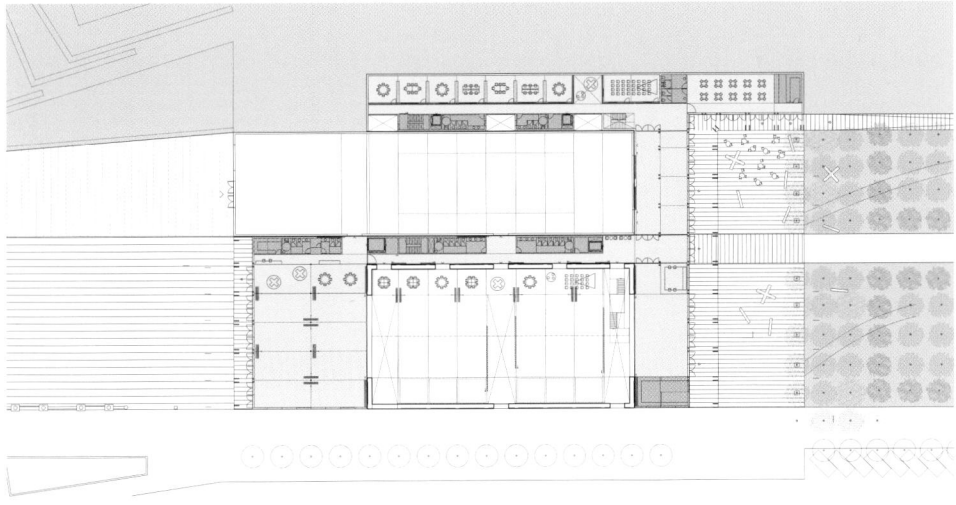

Floor plan

257

Winner _ 당선작

Magnifica Fabbrica
웅장한 공장

FRPO + WALK + SD Partners
FRPO + WALK + SD 파트너스

Milan vibrating city
Milan, a metropolis in constant movement, dynamic and productive, always reinventing itself, today opens its secret heart. A heart in which others beat, with different rhythms, moved by skillful hands that forge treasures invisible to most. Vibrating center of the city and meeting point between industriousness and genius, La Scala and the Magnifica Fabbrica, are called today to take a new step in the history of the city.

Technology, culture and landscape
Magnificent for its size and location, the new Fabbrica will open its treasures to the public view to reveal the work of artists who create true works of art for La Scala shows. The Magnifica Fabbrica will thus be able to reveal its light and illuminate the renewal of the city, rebuilding a new balance between technology, culture and landscape, caring for the environment as a home, and reconnecting the living fabric of the city.

Biodiversity Green Factory
The expansion of the Parco della Lambretta will accompany the new Magnifica Fabbrica as the driving force behind the transformation of the ex-Innocenti area, whose soul will be the recovery of the grandiose structure of the Palazzo di Cristallo, the former production site for millions of lambrettas that have filled the streets of Italy and the world, and that will become a Green Biodiversity Factory, a place for citizens to participate in a great collective work of art, to fill the future of the city with green.

Canals, meadows, and rows of trees
The Parco della Lambretta project exemplifies a balanced relationship between natural resources and human activities, conceiving the landscape as a resilient green infrastructure. It is based on a circular concept of the water cycle, an element that has characterized the development of the city, inspired by the traditional elements of Milan's agricultural heritage: canals, ditches, meadows, pedestrian paths and rows of trees, offering a natural environment and a usable public space for all. Water regains its leading role and becomes an indissoluble link between the Parco and the Fabbrica through the Water Gardens, a large green infrastructure for natural phytopurification, which will offer unique educational and sensory experiences.

Location : Milano, Italy I **Function** : Mixed-use facility I **Site area** : 168,310㎡ I **Bldg. area** : 66,450㎡ I **Client** : Comune di Milano e Teatro alla Scala I **Project leader** : FRPO / Pablo Oriol, Fernando Rodríguez + WALK / Juan Tur Mc Glone + SD Partners / Massimo Giuliani, Alessandro Viganò, Beatrice Meroni

위치 : 이탈리아 밀라노 I **용도** : 복합시설

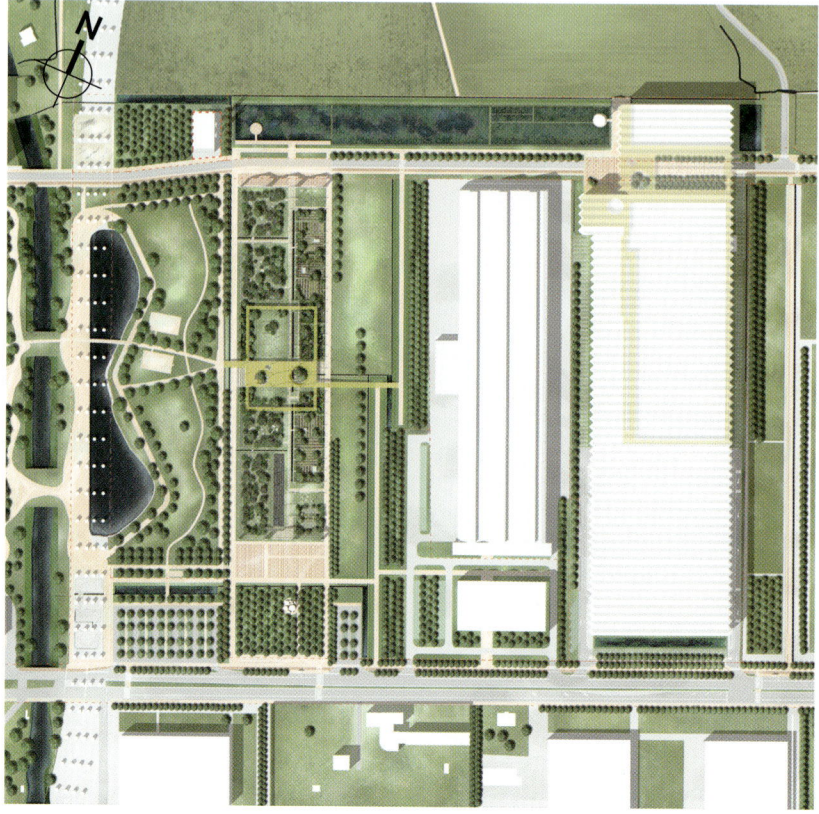

Master plan

밀라노, 생동감 넘치는 도시
끊임없이 변화하고 역동적이고 생산적이고 항상 스스로 재창조하는 대도시 밀라노가 드디어 감추어 두었던 심장을 드러낸다. 서로 다른 리듬으로 뛰는 심장은 눈에 보이지 않는 보물을 만드는 숙련된 손에 의해 움직인다. 활기찬 도시 중심지이자 근면함과 천재성의 만나는 곳인 라스칼라와 웅장한 공장은 도시의 역사에서 새로운 발걸음을 내딛는다.

기술, 문화, 경관
규모와 위치가 웅장한 새로운 공장은 라스칼라 쇼를 위한 예술 작품을 만드는 아티스트들에 의해 보물을 공개할 것이다. 웅장한 공장은 기술, 문화, 경관 사이의 새로운 균형을 다시 맞추고 집처럼 환경을 관리하고 도시 구조를 다시 연결하여 도시의 빛을 밝히고 활성화할 수 있다.

다양한 생물이 살아가는 녹지 공장
파르코 델라 람브레타의 확장은 새로운 웅장한 공장과 함께 이전 이노센티지역의 혁신을 이끄는 원동력이 될 것이며, 이탈리아와 전 세계의 거리를 채운 수백만 개의 스쿠터를 생산했던 곳인 팔라조 디 크리스탈로의 웅장한 구조를 복구하고자 한다. 시민이 함께 훌륭한 집단 예술 작품에 참여하여 도시의 미래를 푸르게 채우는 녹지 공장이 될 것이다.

운하, 초원, 가로수
파르코 델라 람브레타 프로젝트는 자연 자원과 인간 활동 사이의 균형 있는 관계를 보여주며 경관을 탄력적인 녹지 인프라로 구상한다. 운하, 도랑, 초원, 산책로, 가로수 등 밀라노 농업 유산의 전통적인 요소에서 영감을 받아 도시 발전에 중요한 요소인 물 순환의 개념을 기반으로 자연 환경과 누구나 이용 가능한 공공 공간을 제공한다. 물은 주도적인 역할을 되찾고, 자연 식물 정화를 위한 대규모 녹지 인프라인 수공원을 통해 파크코와 공장 사이의 불가분의 연결이 되어 독특한 교육 및 감각 경험을 제공한다.

Winner _ 당선작

Section - Park

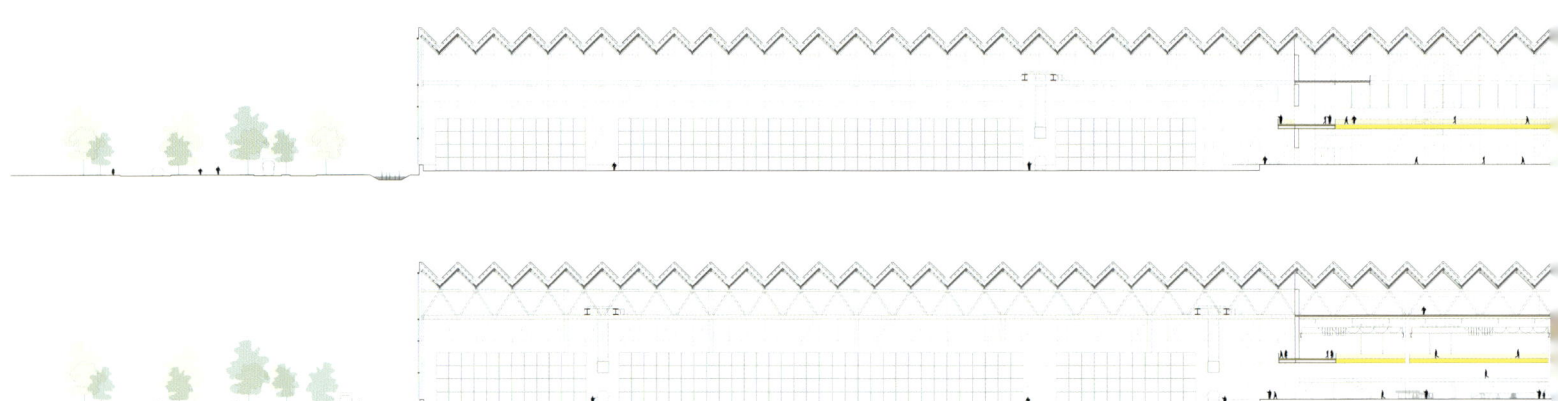

Longitudinal section

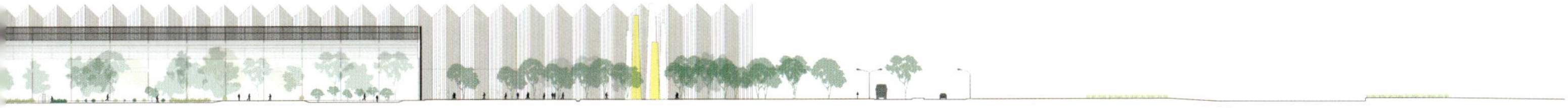

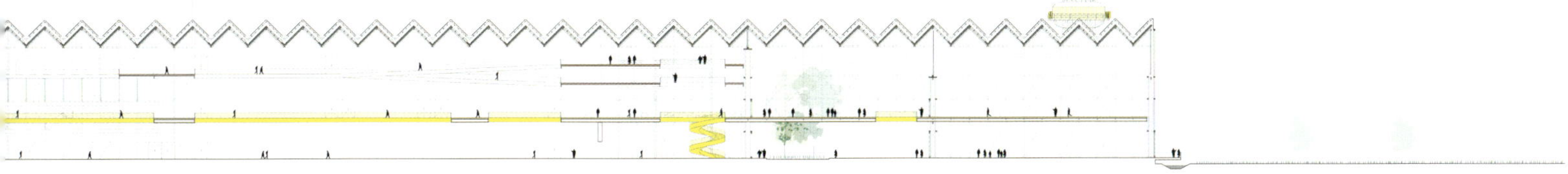

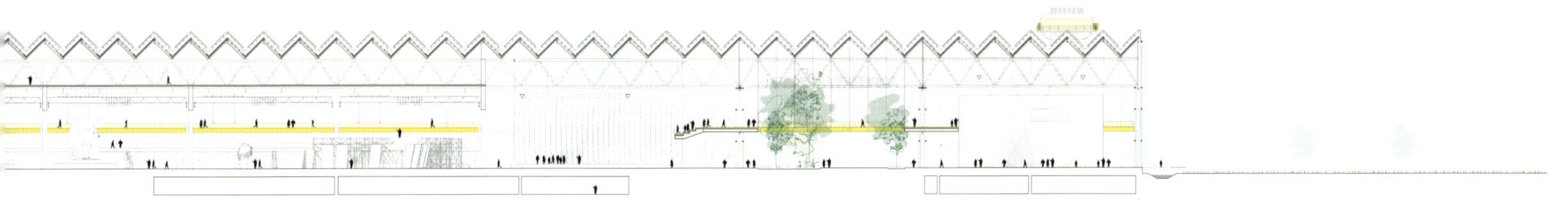

Magnifica Fabbrica >>

261

Winner _ 당선작

■ **Axonometric**
- General

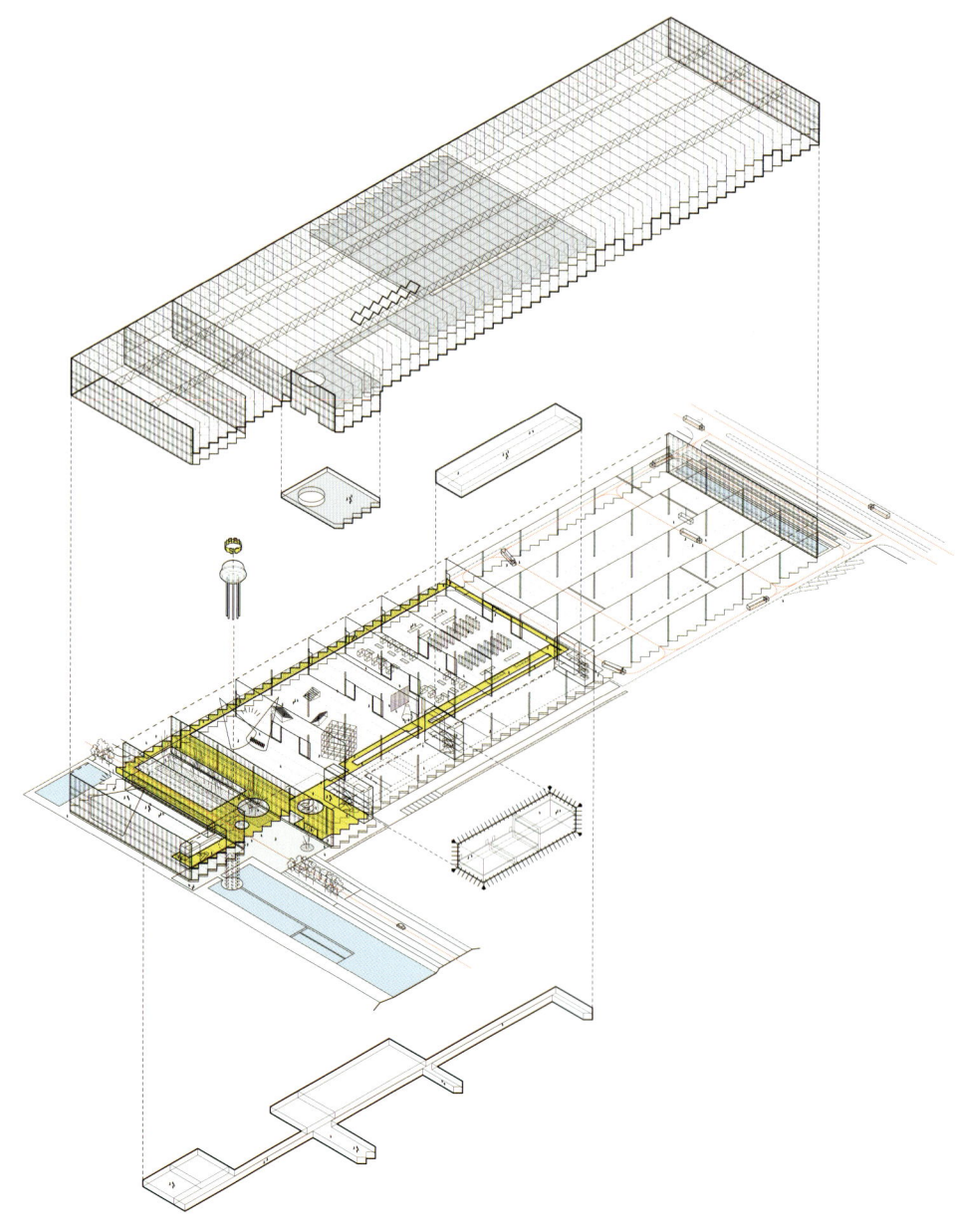

■ Axonometric

- Park

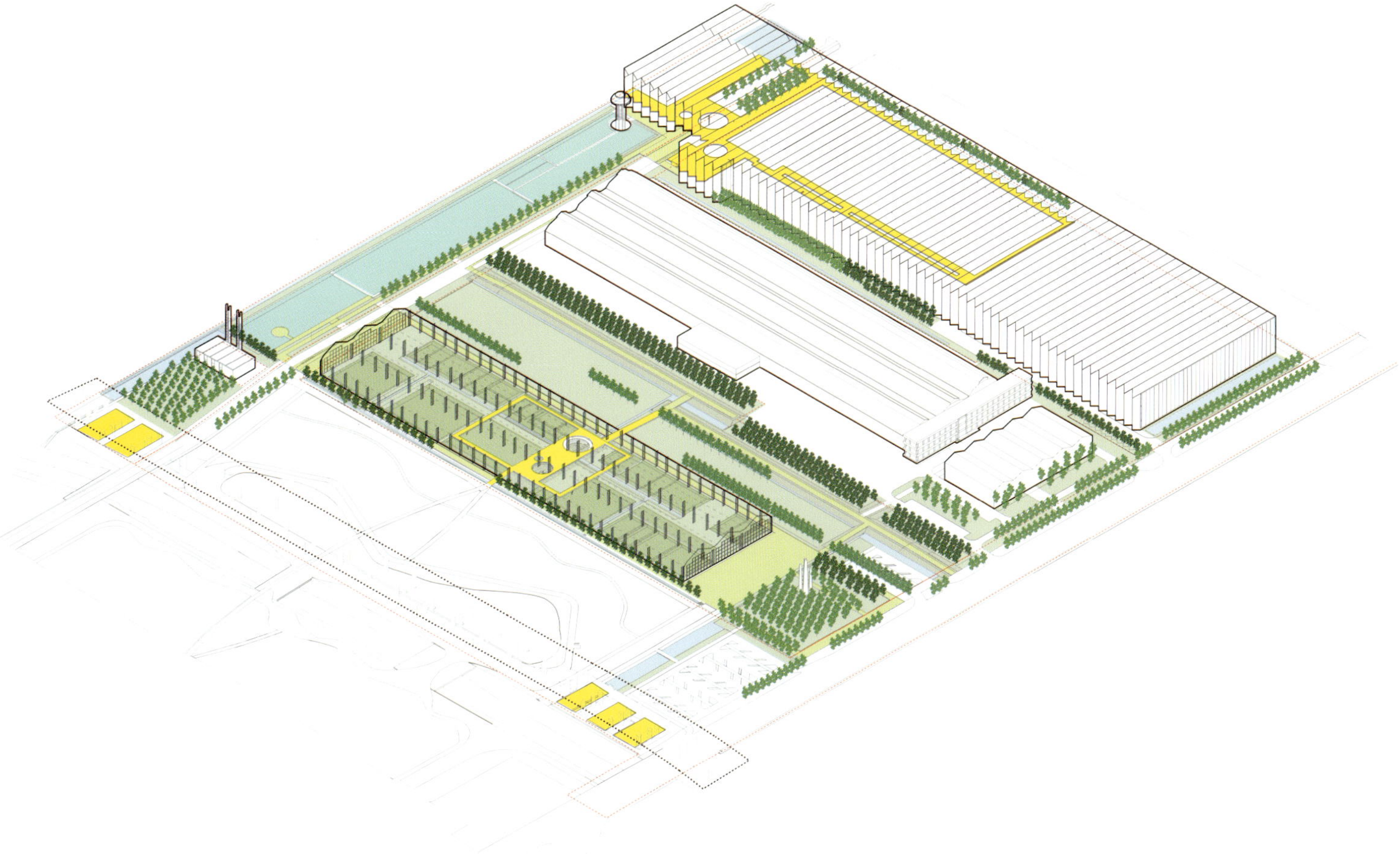

Winner _ 당선작

Magnifica Fabbrica >>

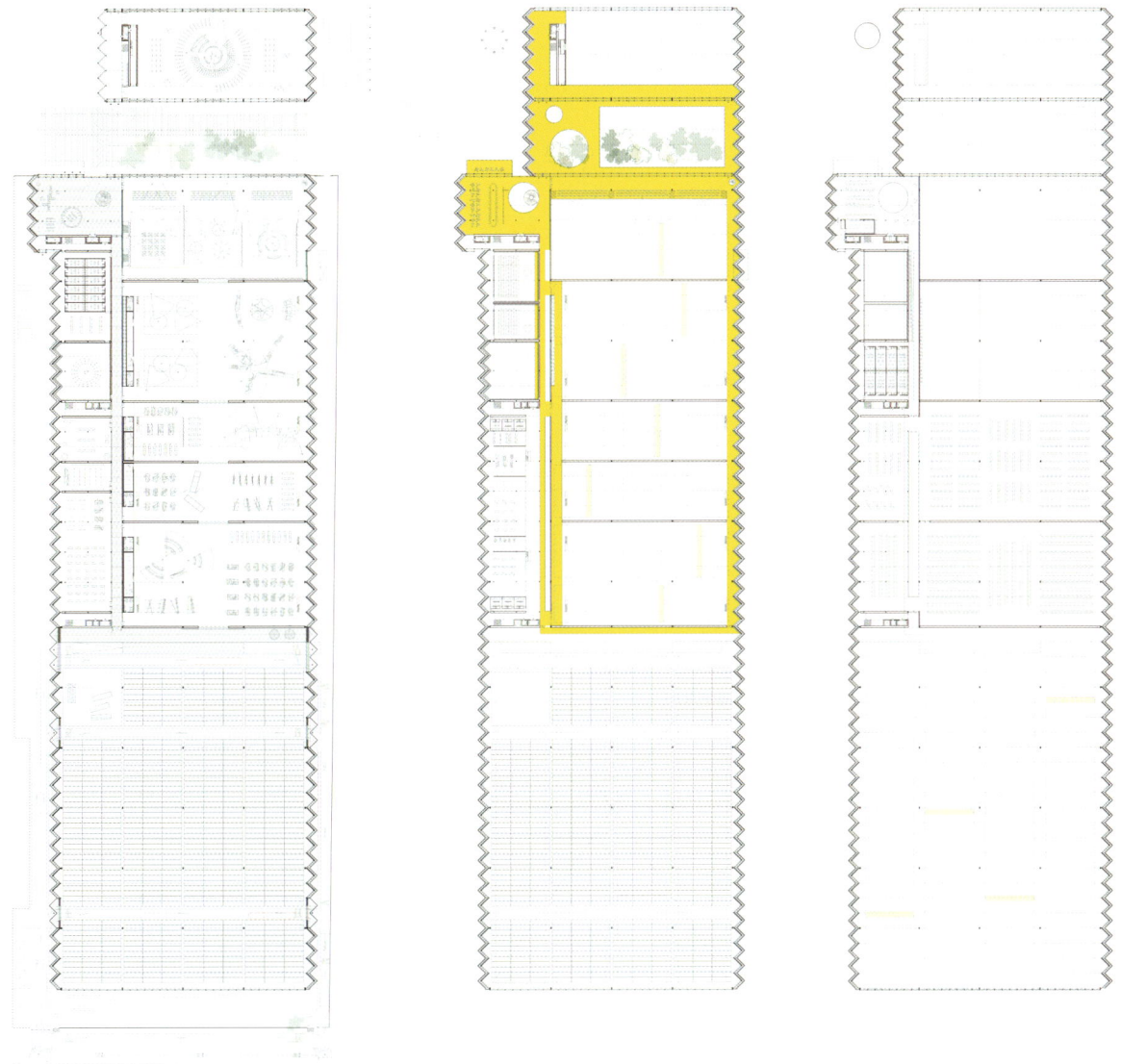

1st floor plan   2nd floor plan   3rd floor plan

Winner _ 당선작

Yantai Seafront Garden
옌타이 해안 정원

KCAP
KCAP

Yantai Seafront Garden comprises a planning strategy for the entire area, as well as a set of detailed spatial frameworks. KCAP created an overall scheme for a 95km long stretch of coastline, with 27,3km² consisting of coastal space. Within this larger territory, there is extra focus on an 18km key design area. KCAP drafted in-detail designs for a section of approximately 1km² within this 18km area.

The planning area includes four distinct 'habitats': the zone between the mountains and city, which will be deployed for water containment; the actual urban centre, set up as a sponge city; the coast, conceived as a resilient tidal area; and the sea itself.

Key to the proposal is restoring the coastal areas, by returning the artificial coastline and production facilities (i.e. fish ponds) to nature, and implementing measures that protect it from erosion while creating new ecological habitats.

Within the larger framework, there was specific focus on an 18km key design area because of its potential. Being the most open, green gateway it can act as a driving force for the overall development. Here, in Huangshi Talent Bay various public and social gatherings can take place, different lifestyles interact, and landscapes merge. Within this development, the team gave special attention to the design of one singular area: International Communication Bay. Within this 1km², detailed-landscape design area, 3 clusters were defined: the Talent Port, the Tidal Community and the Lagoon Community.

옌타이 해안 정원은 지역 전체에 대한 계획 전략과 일련의 세부 공간 프레임워크로 구성된다. KCAP는 95km의 해안선에 대한 전반적인 계획을 수립했으며, 27,3km²는 해안 구역으로 구성되어 있다. 이 큰 영역 내에서 18km의 주요 설계 면적에 특히 중점을 둔다. KCAP는 이 18km 면적 내에서 약 1km²의 단도면를 상세히 설계했다.

계획 영역에는 4개의 뚜렷한 '서식지'가 포함된다. 바다를 막기 위해 배치될 산과 도시 사이의 구역, 스펀지 도시로 설정된 실제 도심지, 탄력적인 조수 지역으로 인지되는 해안, 그리고 바다 그 자체이다.

제안의 핵심은 인공 해안선과 생산 시설(예: 양어장)을 자연으로 되돌려 해안 지역을 복원하고 침식으로부터 보호하는 동시에 새로운 생태 서식지를 만드는 방안을 구현하는 것이다.

큰 프레임워크 내에서 잠재력 때문에 18km의 주요 설계 면적에 특히 집중했다. 가장 개방적이고 친환경적인 관문으로서 전반적인 발전에 원동력으로 작용할 수 있다. 황스 탤런트 베이에서는 다양한 공동 및 사교 모임이 이루어질 수 있고 다양한 라이프스타일이 상호 작용하며 풍경이 어우러진다. 이 개발 과정에서 하나의 영역인 인터내셔널 커뮤니케이션 베이의 디자인에 특별히 신경을 썼다. 이 1km²의 상세한 경관 설계 영역 내에서 탤런트 포트, 타이달 커뮤니티, 라군 커뮤니티라는 3가지 클러스터가 정의되었다.

Location : Yantai, China I **Function** : Urban, Landscape design I **Client** : Yantai Economic and Technological Development Zone Construction and Transportation Bureau I **Collaborators** : Ecological consultant: LYQW Ecological Technology Co., Ltd.; Hydrological consultant: PADDI Environment Consulting

위치 : 중국 옌타이 | 용도 : 도시, 조경계획

Master plan

Site Analysis

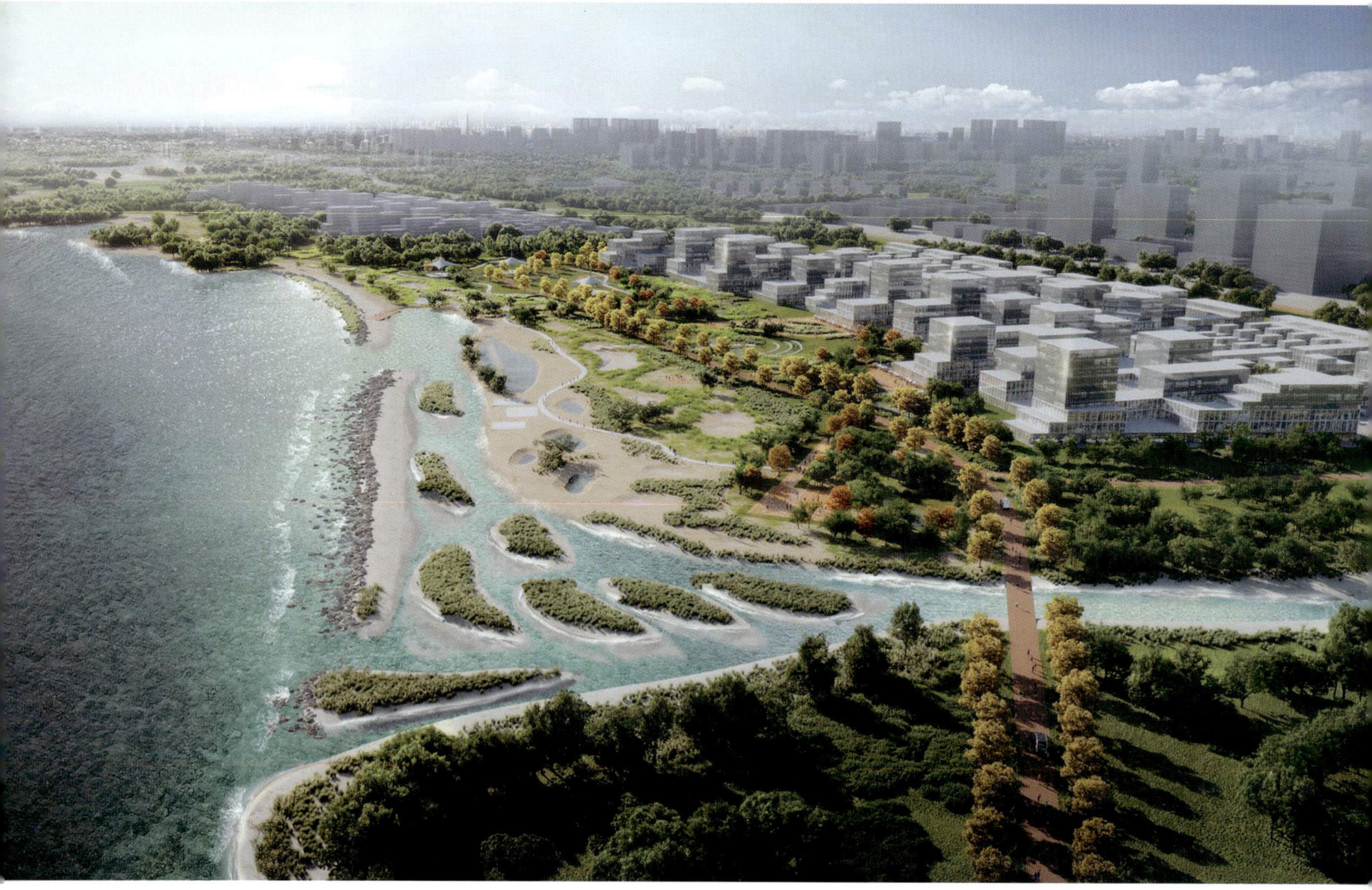

Coastal Development

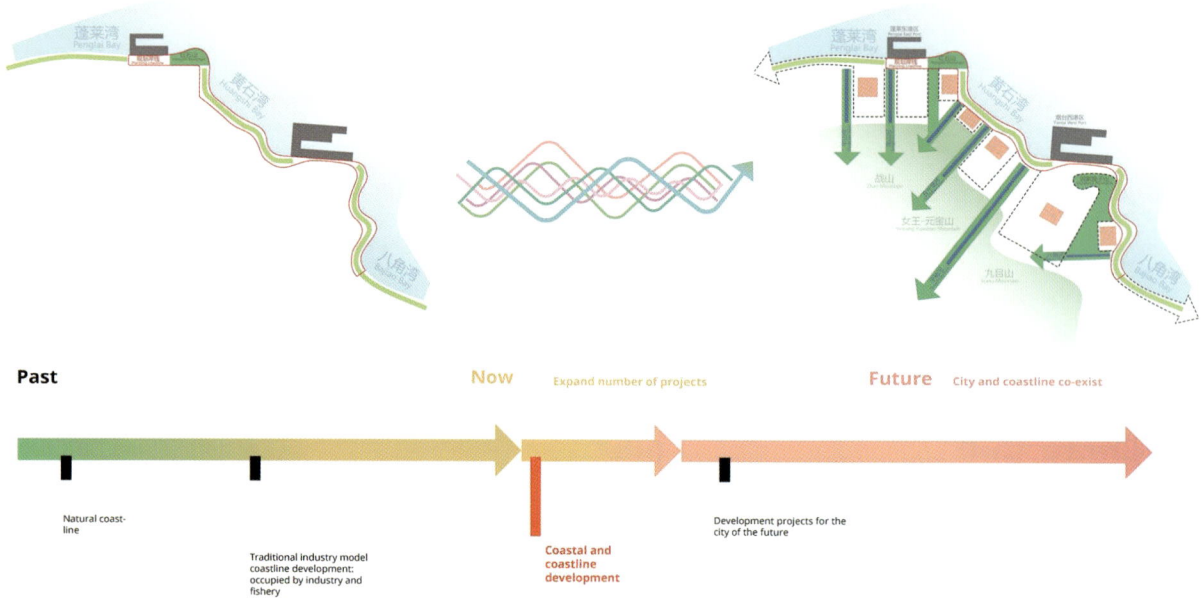

蓬莱湾
Penglai Bay
黄石湾
Huangshi Bay
八角湾
Bajiao Bay
Past
Now
Expand number of projects
Future
City and coastline co-exist
Natural coast-
line
Traditional industry model
coastline development:
occupied by industry and
fishery
Coastal and
coastline
development
Development projects for the
city of the future

Yantai Seafront Garden >>

■ Creeks & Landscape

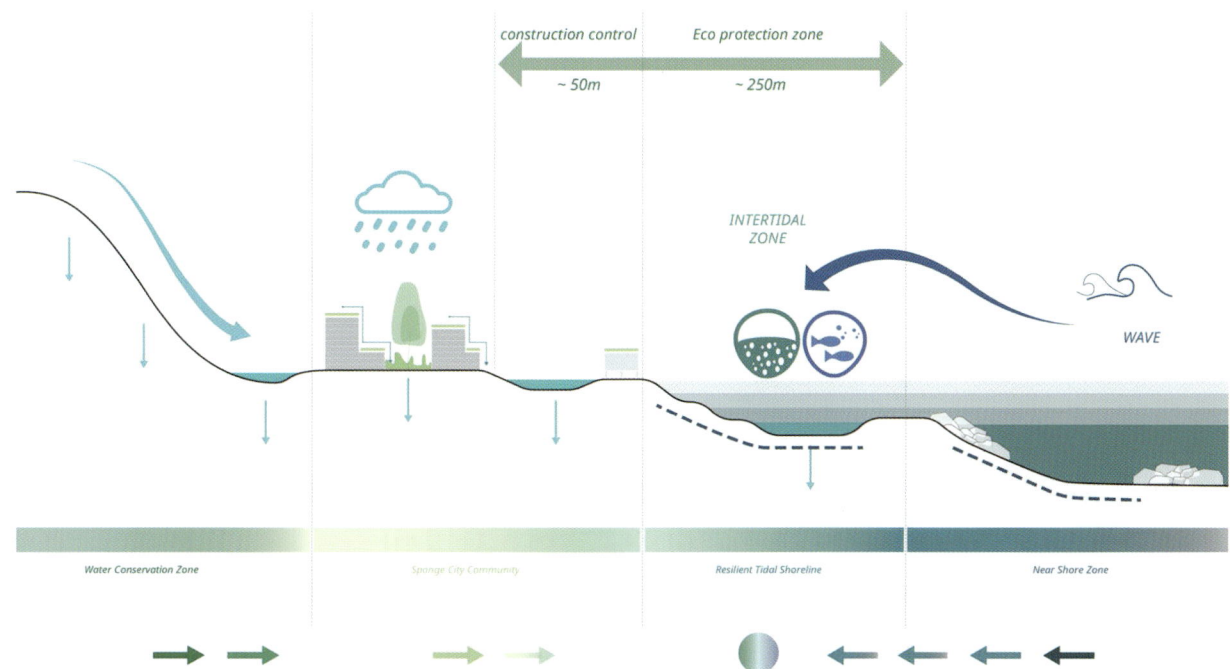

Gateway

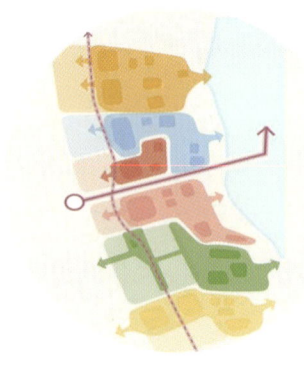 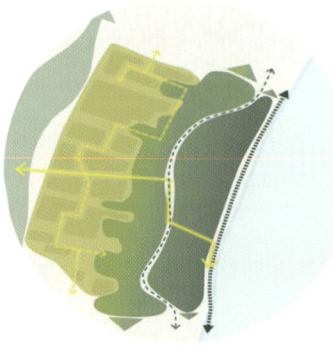 

Yantai Seafront Garden >>

■ Framework

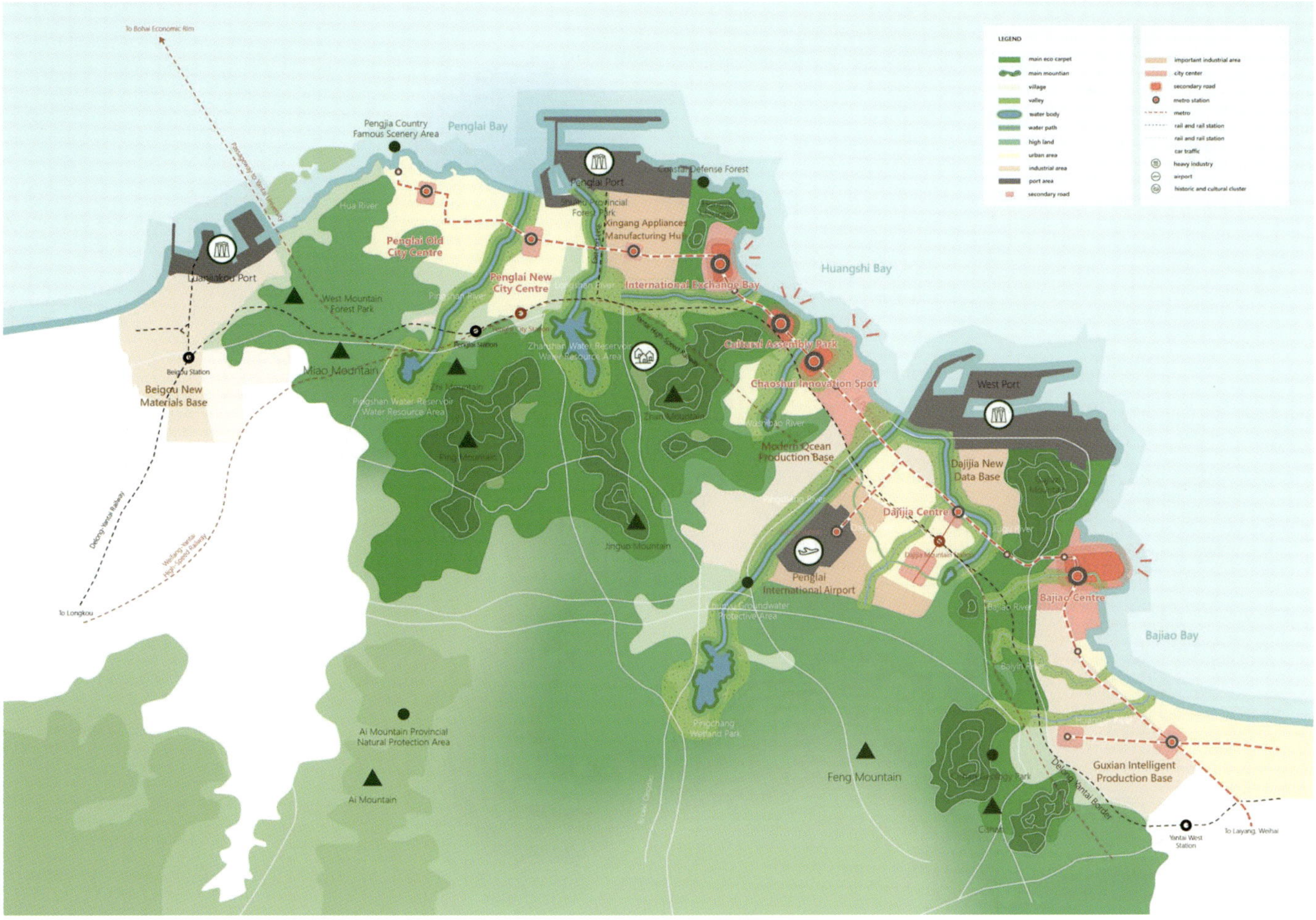

건축설계경기연감 XX

교육시설·도서관·어린이집 / 문화·전시시설 / 기타

Publisher	Jeong Kwang-young	정광영
Editor	Jeong Hye-young Lee Hye-kyung Noh Yoon-ju	정혜영 이혜경 노윤주
Designer	Cho Jae-hyon	조재현

Publishing ARCHIWORLD Co., Ltd.
건축세계(주)
5F World Bldg.
256 Neungdong-ro, Gwangjin-gu, Seoul, Korea
Tel 82-2-422-7392
Fax 82-2-422-7396
aid@archiworld1995.com
www.archiworld1995.com

자매지 건축세계저널(인터넷신문)

Price US $ 78 (Set US $ 156)
정 가 80,000원 (세트 160,000원)
ⓒArchiworld Co., Ltd.

Printed in Korea